Testing and Quality Assurance for Component-Based Software

For a listing of recent titles in the *Artech House Computing Library*,
turn to the back of this book.

Testing and Quality Assurance for Component-Based Software

Jerry Zeyu Gao
H.-S. Jacob Tsao
Ye Wu

Artech House
Boston • London
www.artechhouse.com

Library of Congress Cataloging-in-Publication Data
Gao, Jerry.
Testing and quality assurance for component-based software / Jerry Zeyu Gao, H. -S. Jacob Tsao, Ye Wu.
p. cm. — (Artech House computing library)
Includes bibliographical references and index.
ISBN 1-58053-480-5 (alk. paper)
1. Computer software—Quality control. 2. Computer software—Testing. I. Tsao, H.-S. J. II. Wu, Ye. III. Title. IV. Artech House computer library.

QA76.76.Q35G38 2033
005.1'4—dc22 2003057725

British Library Cataloguing in Publication Data
Gao, Jerry.
Testing and quality assurance for component-based software.
— (Artech House computing library)
1. Component software—Testing 2. Computer software industry—Quality control
I. Title II. Tsao, H.-S. J. (H.-S. Jacob) III. Wu, Ye
005.3'0287

ISBN 1-58053-480-5

Cover design by Igor Veldman

685 Canton Street
Norwood, MA 02062

International Standard Book Number: 1-58053-480-5
Library of Congress Catalog Card Number: 2003057725

10 9 8 7 6 5 4 3 2 1

To my wife Tracey and my lovely son Kevin
To my parents Ming Gao and YeFang Qin
—Jerry Zeyu Gao

To my mother Shu-Wen Pao, my brother Hsiao-Tzu Tsao,
my wife Hueylian, and my children Allison and Jason
—H.-S. Jacob Tsao

To my wife Wanjie and to my lovely daughter Ya Ya
—Ye Wu

Contents

Preface

What is this book about?

The widespread development and reuse of software components is regarded by many as one of the next biggest phenomena for software. Reusing high-quality software components in software development has the potential for drastically improving the quality and development productivity of component-based software. However, the widespread reuse of a software component with poor quality may literally lead to disasters. Component-based software engineering involves many unique characteristics, some of which have unintended consequences and side effects. The primary goal of this book is to elaborate on these characteristics, point out possible strengths and weaknesses, and develop methodologies to maximize the quality and productivity potentials associated with development and reuse of software components. Software testing is one of the most important phases in software engineering, and it plays a pivotal role in software quality assurance. This book focuses on testing and the larger context of quality assurance.

Part I provides a general discussion of software components, component-based software systems, the concomitant testing challenges, and the important issue of testability. Part II focuses on validation methods for software components. Part III covers the testing of component-based software systems, including integration testing, regression testing, and performance measurement. Part IV is devoted to quality assurance for software components and computer-based software systems.

Why has reuse of third-party software components become popular recently?

Building systems based on reusable components is not a new idea. It has been proven to be a very effective cost-reduction approach in the computer hardware industry. Today, that industry is able to build computer systems based on standardized high-quality hardware parts and devices reliably and quickly. Software engineers learned the value of this idea many years ago. They developed reusable software modules as internally shared building blocks for constructing different software systems, and they established repositories of such modules, but in an ad hoc manner. Many years of trial and error proved to many software developers that it was not easy to build software systems based on reusable software components. The major reasons include the lack of a well-defined discipline for component-based software engineering and the absence of a comprehensive collection of standardized high-quality software components within the company or on the market.

Recently, the increasing complexity of *information technology* (IT)-based software application systems and intensive competition in the marketplace forced software vendors and researchers to look for a cost-effective approach to constructing software systems based on third-party components. The major goal is to shorten the software development cycle and thereby reduce cost. Hence, component-based software engineering becomes a popular subdiscipline within software engineering because of its promising potential for cost and cycle-time reductions.

Why is testing software components and component-based software important?

As mentioned earlier, widespread reuse of a software component with poor quality may wreak havoc. Improper reuse of software components of good quality may also be disastrous. Testing and quality assurance is therefore critical for both software components and component-based software systems. A number of recent books address the overall process of component-based software engineering or specific methods for requirements engineering, design, and evaluation of software components or component-based software. We are not aware of any books focusing on testing, validation, certification, or quality assurance in general for reusable software components and component-based software systems.

Book organization

The book consists of four parts. We briefly describe their contents here.

Part I: Introduction provides background on software components, component-based software, as well as component-based software engineering. Moreover, it explains the importance of testing software components and component-based software. It consists of five chapters. Chapter 1 explains the basic concept of a software component, its main features, its classification, and the engineering process. It helps the reader understand the concept by comparing software components with traditional software modules. Chapter 2 discusses software component testing, including its problems and challenges, as well as its differences from traditional module testing. Chapter 3 provides an overview for component-based software, including its features, the engineering process, and different construction approaches, while Chapter 4 focuses on its testing. Chapter 5 is dedicated to the important issue of component testability.

Part II: Validation methods for software components consists of three chapters. They focus on validation techniques for software components. Chapters 6 and 7 review the applicable black-box and white-box testing methods for software components, respectively. Chapter 8 discusses test automation and automated tools for testing software components.

Part III: Validation methods for component-based software consists of four chapters, which focus on component-based software and discuss strategies and methods developed for integration, regression, and performance testing. Chapter 9 reviews the applicable software integration strategies and reports on current research results on component integration strategies and methods. Chapter 10 discusses regression testing of component-based software and related recent advances. Chapter 11 focuses on system performance testing and measurement metrics. Chapter 12 addresses system test frameworks for supporting the validation of reusable software components and component-based software.

Part IV: Quality assurance for software components and component-based systems consists of five chapters. Quality assurance for software components and component-based software systems is a big subject, and we can only provide an overview of the issues and their solutions in Chapters 13 and 14, which pave the way for the discussion on standards and certification of Chapter 15. Chapters 16 and 17 focus on quality verification and measurement and provide further details for some of the standards discussed in Chapter 15.

In our attempt to weave a coherent story about this big subject, we have tried to place a balanced emphasis on the published literature, particularly peer-reviewed journal articles as well as books written by recognized experts in the related fields, rather than relying merely on our personal experience or anecdotes. A common complaint about the current status of *software quality assurance* (SQA) is its predominant process orientation. We have attempted to achieve a balance between the process and product orientations.

In addition, we have attempted a balance between tailoring generic SQA methods and describing some recently proposed approaches motivated specifically by the development or reuse of software components. Moreover, in discussing software metrics, we have attempted to strike a balance between their theoretical justification and empirical validation.

Is this book for you?

This book is intended for professionals and students seeking a systematic treatment of the subjects of component-based software testing and quality assurance. We have attempted to provide a balanced report about the state of the practice and recent research findings regarding the new issues and their solutions. Therefore, researchers should also find this book useful.

This book can also be used for training software testing and quality assurance engineers because it provides them with a comprehensive understanding of software testing and quality assurance in general. In addition, this book can be used as a textbook for a software testing and quality assurance course at the graduate or undergraduate upper-division levels—the instructor may need to supplement the book with some class projects and/or homework exercises.

Acknowledgments

Many people have contributed to the completion of this book through technical discussions or encouragements. We gratefully acknowledge the contributions of

C. S. Ravi, P. Q. Cuong, and H. Q. Duong at Hewlett Packard; Richard Barlow, Pravin Varaiya, Adib Kanafani, and Mark Hansen of the University of California at Berkeley; Randolph Hall of the University of Southern California; Shu-Cherng Fang of North Carolina State University; Yasser Dessouky of San Jose State University; Stephen Chen and Jackie Akinpelu of AT&T Bell Laboratories; Mei-hwa Chen of the State University of New York at Albany; and Wen-li Wang of Penn State Erie.

We also thank Vickie Perrish and Lance Nevard for their editorial assistance on Parts I through III.

We thank Tiina Ruonamaa of Artech House for her superb management of this project every step of the way, which led to the on-time delivery of the manuscript. We also thank Rebecca Allendorf for her meticulous proofreading and editing. We are indebted to the anonymous reviewers of the original manuscript, who offered continuous support throughout the 16-month project and provided thorough reviews and insightful suggestions, which led to many improvements in not only the technical contents but also the structure of the book. Finally, all the support provided to the authors throughout the writing process by Tim Pitts, Commissioning Editor of Artech House, as well as his excellent selection of the anonymous reviewers made the writing process more and more enjoyable.

PART

I

Introduction

Today, the concept of software reuse has been widely accepted by the software industry. In the past two decades, many experts have exerted much effort to study software component reuse crossing projects, product lines, and organizations. Since software components are building parts for constructing software systems, creating highly reusable components is one of the important keys to the success of a software development project. With the increase of software component products in today's commercial market, many software vendors and workshops have begun to use a new approach, known as *component-based software engineering* (CBSE), to develop large, complicated software application systems based on available and reusable components. Its major objective is to reduce software development cost and time by reusing available components, including third-party and in-house grown components. This trend drives a strong demand for CBSE methodology, standards, and guidelines to help engineers and managers in software analysis, design, testing, and maintenance of component-based software and its components.

Recently, a number of books have been published to address component-based software development processes, analysis and design methods, and issues. However, there are no technical books that discuss new issues and challenges in validating software component quality, addressing solutions and techniques in component-based software testing, reporting software component testing standards and certification, and discussing quality assurance processes for component-based software. This book is written to cover these topics.

The first part of this book is structured to provide the fundamental concepts about software components and component-based programs. It discusses testing issues and challenges in component-based software engineering. The objective of this part is to provide readers with answers to the following questions:

- What are the major differences between conventional software modules and modern software components?
- What are the new issues and challenges in testing software components for component-based software engineering projects?
- What are the primary differences between conventional software and component-based software?
- What are the new issues and challenges in testing component-based software?

There are five chapters in Part I. Chapter 1 provides a necessary background and concepts about software components, including definitions, elements, and properties. In addition, the essential differences between software components and traditional modules are discussed.

Chapter 2 introduces software component testing concepts, and discusses its objectives, processes, issues, and challenges from the perspective of component-based software engineering.

Chapter 3 helps readers learn the essentials of component-based software, including development process and infrastructures. Moreover, the major differences between conventional software and component-based software are examined.

Chapter 4 focuses on the introduction of testing component-based software, including its basics, challenges, and issues. Finally, Chapter 5 discusses the detailed concepts of software testability for components and component-based programs.

CHAPTER

1

Contents

Introduction to software components

In past decades, we have used software components to refer to the building parts of software programs. What is a software component? When we ask people this question, we get different answers because they have different understandings about the software component concept. In this chapter, we introduce the concept of software components in component-based software engineering by comparing them with traditional software modules.

We first review the evolution of software components from subroutines to reusable software components. Next, we discuss software component definitions, basic elements, properties, and deliverables. We examine the differences between software components in *component-based software engineering* (CBSE) and conventional modules. Later, an overview of an engineering process for software components is discussed. Finally, we highlight questions and concerns in component validation and quality assurance in component-based software engineering.

1.1 The evolution of software components

In the early days of programming, programs were constructed by creating a main program to control (or invoke) a number of subroutines. Each subroutine was programmed as a specific part of the program based on the given requirements and function partitions. To reduce programming efforts, programmers in a project team reused subroutines during a project's implementation [1].

Hence, subroutine reuse is one of the earliest formats in software reuse. In 1968, M. D. Mcilory introduced the concept of *software components* at the NATO Software Engineering Conference [2]. Since then, the evolutionary history of software component technologies has gone through a number of stages.

From the late 1970s to the 1980s, engineers used the structure-oriented software development methodology to partition an application system into a number of modules based on its given functional requirements. After the modules were developed, they integrated them to form an application system. During that time period, many function libraries were developed in the Fortran programming language as reusable software packages to support the development of scientific applications. These software packages could be considered as the first generation of software components. Although many programmers had used these packages to develop various scientific application programs, they did not know how to practice cost-effective software reuse due to the lack of a disciplined component-based software engineering methodology.

In the 1980s, the introduction of object-oriented technology enabled software reuse in a broader scope, including the reuse of class analysis, design, and implementation. Many object-oriented C++ class libraries were developed as reusable software packages [3]. Thus, object-oriented technology steered the evolution of component technology from *reusable function libraries* to *object-oriented class libraries*. In the 1990s, many large corporations, such as GTE, IBM, Lucent Technologies, and Hewlett Packard, launched enterprise-oriented software reuse projects to develop domain-specific business components for product lines using object-oriented technology [4].

In 1989, component technology advanced to its new era—*reusable middleware*—when the *Object Management Group* (OMG) began to standardize an open middleware specification for distributed application systems and developed *Common Object Request Broker Architecture* (CORBA) [5]. Its major goal was to allow software middleware vendors (like IONA) to provide common reusable middle-ware to support distributed objects of application software to interact with each other over a network without concern about object locations, programming languages, operating systems, communication protocols, and hardware platforms. To provide high-level reusable components, the OMG also specified a set of CORBA Object Services that defined standard interfaces to access common distribution services, such as naming, transactions, and event notification.

Ten years later, with the increasing deployment of commercial middleware components in the software industry, the evolution of component technology reached a new stage: *reusable application framework and platform*. In 1999, Sun Microsystems published *Enterprise JavaBeans* (EJB) as a software component model for developing and deploying enterprise-oriented business applications [6]. The model provides component producers, customers, end-users, and component model producers with an open specification that can be used to develop application software systems employing software components. EJB technology is designed to provide application developers with a robust, distributed environment that is responsible for many complex features of distributed computing, such as transaction management and multithreading.

Meanwhile, as the latest incarnation of *Component Object Model* (COM) on the PC platform, Microsoft's COM+ [7] is the cornerstone of a framework of technologies designed to support the development of large-scale distributed applications on the Windows platform. Many experts believe that COM+ makes the development of scalable Windows 2000 applications easier because the COM+ run-time environment provides a foundation for building scalable distributed enterprise applications.

Recently, several well-defined component-based development methods have been published to support the development of component-based software systems. Cheesman and Daniel's UML Components [8], D'Souza and Wills' Catalysis [9], and Herzum and Sims' Business Component Factory [10] are typical examples. These component-based methodologies provide engineers with well-defined processes, analysis and specification models, and engineering guidelines.

1.2 Why is software component reuse important?

Although the concept of software component reuse was introduced in 1968, it did not receive much attention within the software industry and academia until the late 1980s.

After the mid-1980s, the fast increase of the software complexity in software application systems forced software vendors to look for cost-effective methods to construct complicated software systems in a short development life cycle to meet marketing needs. Software reuse based on high-quality components, therefore, has become popular in academia and the software industry because of the advantage of reducing development costs and time. Companies such as IBM, Lucent Technologies, and GTE [4] launched enterprise-oriented software reuse projects to develop in-house reusable component libraries. The major purpose was to speed up production lines of application systems based on in-house reusable components so that they could keep pace with market forces and increase their competitive edge in a business market and cope with dynamic business needs and fast updated technologies.

Since 1995, the information technology and service market has grown rapidly because of the explosive popularity of the Internet and fast technology advances [11]. For example, according to the *Japan Information Service industry Association* (JISA), the Japanese information service market exceeded ¥10 trillion ($100 billion) in 1999. This market change created a strong demand for new large-scale information service systems and e-commerce applications over the Internet. To fulfill this demand, the software market and industry had to change in order to deliver large Web-based information service systems and e-commerce applications on a very tight schedule based on *commercial-off-the-shelf* (COTS) components. For instance, the Japanese software market structure has been changed from custom software to shrink-wrapped package software and component-based software [11]. Meanwhile, the Japanese software industry structure has also been changed from a hierarchical structure to an open and competitive structure.

The tragic events of September 11, 2001, destroyed and damaged many business application systems and infrastructures. This has driven a strong demand for setting up new secured application systems and information services to replace out-of-date systems in businesses and government agencies. Most systems are global distributed information service systems with a complex structure and multiple platforms. They have strict system requirements for reliable and scalable functional services. It is very difficult or even impossible for software workshops to develop new IT-based application systems and service infrastructures without using COTS components. To reduce development cost and time, they must use a component-based software engineering approach to develop new systems or upgrade existing systems based on high-quality reusable components.

1.3 What is a software component?

What is a software component? You may get different definitions and answers from engineers, managers, testers, and researchers. One of the reasons is that a modern software component is different from a traditional module, although both capture and define the concept of "building parts" of a software system. Not long ago, we considered software modules as the default notion of software components. In recent years, however, as component-based software engineering advances, a number of experts have defined the term "software component" in different ways. This further contributed to different understandings of the concept of software components among engineers and managers.

Let us review some of the important definitions of software components given by the experts. One of the earliest definitions is given by Gready Booch [12]:

> A reusable software component is a logically cohesive, loosely coupled module that denotes a single abstraction.

This definition captures the idea that a reusable component is an encapsulated software module consisting of closely related component elements. Later, Clement Szyperski presented his well-known definition of a software component at the 1996 European Conference on Object-Oriented Programming [13]:

> A software component is a unit of composition with contractually specified interfaces and context dependencies only. A software component can be deployed independently and is subject to composition by third parties.

This definition is well accepted in the component-based software engineering community because it highlights the major properties of software components that are not addressed in traditional software modules, such as context independence, composition, deployment, and contracted interfaces. In 2000, a broader, more general notion of software components was defined by Alan W. Brown [14]:

> An independently deliverable piece of functionality providing access to its services through interfaces.

A similar idea is also expressed in the *unified modeling language* (UML) definition [15]:

> A component represents a modular, deployable, and replaceable part of a system that encapsulates implementation and exposes a set of interfaces.

Recently, Bill Councill and George T. Heineman gave a new definition to emphasize the importance of a consistent component model and its composition standard in building components and component-based software [16]:

> A software component is a software element that confirms to a component model and can be independently deployed and composed without modification according to a composition standard.

According to them, a *component model* defines the ways to construct components and regulates the ways to integrate and assemble components.

It supports component interactions, composition, and assembly. In addition, a component model also defines the mechanisms for component customization, packaging, and deployment.

To simplify the concept of software components, we classify the notion of software components into three types: (1) reusable modules, (2) reusable components, and (3) composite building blocks.

- A *reusable module* is an independent and deliverable software part that encapsulates a functional specification and implementation for reuse by a third party.
- A *reusable component* is an independent, deployable, and replaceable software unit that is reusable by a third party based on the unit's specification, implementation, and well-defined contracted interfaces.
- A *composite building block* is a reusable component that is developed as a building part to conform a well-defined component model and the accompanying composition standards.

Clearly, the concept of reusable modules given here comes from the traditional reuse idea, where functional module reuse is the major focus. It could be very useful for engineers and testers who deal with reusable software modules generated from a traditional software engineering process. The concept of reusable components given above emphasizes the quality production of reusable software parts with contracted interfaces. Since this component notion reflects the currently adopted status of component technology in industry, it is therefore very useful for engineers and managers involved in component-based software development projects. The given notion of composite building blocks not only focuses on component reuse but also emphasizes component composition. The idea of building blocks, therefore, is very important for component developers and testers in a component vendor because it represents the current and future trend of component technology.

1.4 Properties of software components in CBSE

Component properties refer to the essential characteristics of software components.

Today, people expect component-based software engineering to deliver ssoftware components that facilitate component-based software construction. According to recently published books on component-based software engineering [13, 14, 16], we summarize component properties into two groups: basic properties and advanced properties. Basic component properties refer to necessary features of a software component. Advanced component properties refer to the distinct features offered in modern software components.

A software component in CBSE must have the following basic properties.

- *Identity:* Each component must be uniquely identifiable in its development environment and targeted deployment environment. Without this feature, a large scale of component reuse is impossible. Current component technology ensures this by using a well-defined (or standard) naming scheme. CORBA, EJB, and Microsoft DCOM are typical examples.
- *Modularity and encapsulation:* Software components result from partitioning of a software system by focusing on system modularity. Each component encapsulates a set of closely related data elements and implements coherent functional logic to perform a specific task.
- *Independent delivery:* Software components must be delivered as independent parts that can be replaced under certain conditions. Each part must play an independent functional role in a system and support a specific task in a targeted operating environment. Third-party COTS components must be independently deliverable to component users as individual parts to form a component-based system.
- *Contract-based interfaces:* An interface between software components defines a contract between the client of an interface and a provider of an implementation for the interface [13]. Each contract specifies what the client must do to use the interface to access a component service function. It also defines what kind of services and implementations have to be provided to meet the service contract. This suggests that a component delivers its provided services only when its clients access the provided component interface in a right way.
- *Reusability:* Reusability of software components is the key to the success of component-based software engineering. A conventional software module usually has a very limited reuse scope because it is developed for a specific project. However, a modern software component provides multiple-level granularities for reuse on a large scope. The reusable elements of a software component include its analysis specification, component design and design patterns, source code, and executables. In addition, component deployment mechanisms and test support information (such as test cases and test scripts) are reusable items too. These

artifacts can be reused among different organizations, projects, and production lines.

Modern software components, such as reusable components and building blocks, not only demonstrate these basic properties, but also possess the following advanced properties.

- *Customizability and packaging:* This refers to the customization capability and packaging function that are built inside software components based on a well-defined approach. This feature ensures that software components can be customized and packaged to meet the given component requirements. This feature enhances component reusability. Components with this property are known as *customizable components* because they can be configured and packaged by selecting component metadata, functional features, and interfaces.
- *Deployable:* Software components are known as deployable if a well-defined deployment approach has been built inside a component to support its deployment after they are customized, packaged, and installed in an operating environment. The deployment approach usually supports the creation of an executable component instance in a targeted deployment environment.
- *Multiple instances and polymorphism:* A software component may have more than one deployed instance in an operational environment. Since all of the deployed instances implement the same set of functions, they must share some common features and interfaces. However, they should be viewed as different instances because they have different customizations, packaging selections, operational data, execution tasks (or threads), and dynamic status in an operational environment.
- *Interoperability:* Component interoperability refers to the component capability that supports communications and data exchanges between components. Components are interoperable if they are developed based on standard communication channels and data-exchange mechanisms defined in a component model. There are two types of component interoperability: local and remote. Local interoperability refers to component interoperability for components in a host-centered environment. Remote interoperability refers to component interoperability between components over a network. Remote interoperability is based on the concept of *remote method calls* (RMCs), which is an extension of *remote procedure calls* (RPCs) between different processes over a network.

- *Composition:* Components are composite if they are developed based on a standard composition method to support component composition. In other words, the composite aggregation relationship between components is supported in an operational environment so that a component instance is able to create and destroy its aggregated parts (which could be other components). Supporting this composition relationship requires a uniform identification mechanism for composite parts so that the transitive nature of the composition relationship between components can be easily managed. "Composition is transitive," means that if C is a part of A, and E is a part of C, then, E is also a part of A. Since components are the building parts for the development of component-based software, the composition property is very important for developing complex components.
- *Model conformity:* This means that component realization must follow a predefined component model and its related standards. In other words, components must be developed based on a well-defined component architecture model, interface styles, and mechanisms to support component interactions, composition, packaging, and deployment. To achieve component realization based on a model, engineers need a dedicated component quality control process to check component products based on a well-defined component model and standards.

Clearly, these two sets of component properties list the ideal and required features of reusable components to support component-based software construction. They raise many new needs, concerns, and challenges in component quality validation and control. They are summarized in Section 1.9.

1.5 Basic elements of software components

What are the basic elements of a software component? What are the deliverables of a software component development process? In this section, we try to answer these questions based on a comparative view of reusable modules, components, and composite building blocks.

A *reusable software module* usually includes five basic elements. The first is the module requirement specification, which specifies its function and performance requirements. The second is the module's interfaces, which specify how the module interacts with others. The third is the module design specification, which records the design decisions on its functional logic, data structures, and behaviors. The fourth refers to the module implementation, which includes its source code and executable code on an implicitly specified operating environment, such as a computer hardware platform and operating system. The last is the module test specification and report, which include its tests and unit test environment, as well as test reports. Figure 1.1(a) shows all deliverables of a software module. Among them, only the first two items are visible to its clients unless its source code is designed for reuse.

Similar to a reusable module, a *reusable component* also has its five elements: component requirement specification, design document, interfaces, source code and executable program, and validation documents and test reports. However, a reusable component differs from a software module in two areas. One of them has something to do with component interfaces. A conventional module's interfaces are designed to support external access to its internal functions and its interactions with other modules.

The interfaces of a reusable component are known as contract-based interfaces, which define a contract between a client of an interface and a provider of an implementation for the interface. A component's contract-based interfaces can be classified into two types: (1) *import interfaces,* through which the component obtains services from other components, and (2) *supply interfaces,* through which other components could obtain services from the component.

The other difference is related to component operating environments. In conventional software projects, a module's design and implementation frequently depends on an implicit operating environment and technology because it is created for a specific project. Component design and implementation, however, must always be performed based on explicit operating environments and technology.

As shown in Figure 1.1(b), a reusable component has two extra elements than in a reusable module. The first is a deployment facility (such as an installation and configuration package), which makes it an independent deployable

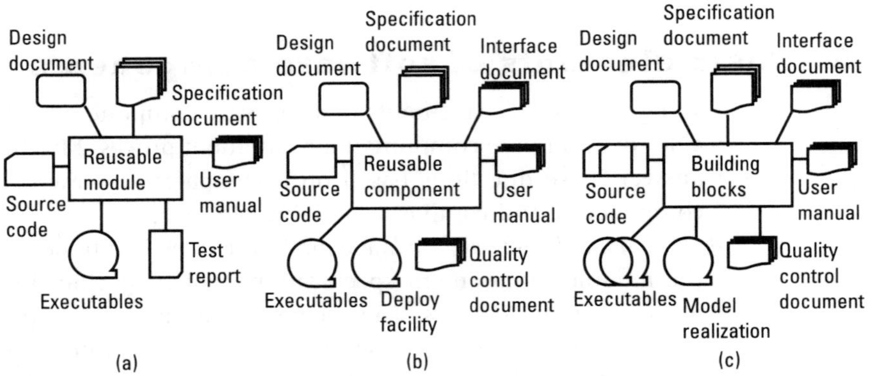

Figure 1.1 Component deliverables: (a) deliverables for reusable modules, (b) deliverables for reusable components, and (c) deliverables for composite building blocks.

product. The other refers to its quality control results, such as quality control documents, metrics, or a certification report. This suggests that a well-defined quality control process and standards are in place in a component vendor to control and monitor the quality of components as final products.

As shown in Figure 1.1(c), a *composite building block* contains a distinct element, known as *component model realization*. It refers to the selection and realization of component models to support component customization, packaging, and deployment. It conforms to the component development based on a well-defined component model and the concomitant standards [13] and defines the approaches and mechanisms for component customization, packaging, and deployment. Similar to a reusable component, a composite building block also consists of its specification, design, implementation, and interfaces. If a composite building block is customizable, then for the same component specification, it may contain more than one design and implementation for the same function features and interfaces. Each corresponds to a specific technology and a targeted operating environment.

Today, in real-world applications, test engineers and quality assurance groups frequently encounter the following types of reusable components in a component-based software development project:

- Commercial-off-the-shelf components with a clearly defined function specification, an interface document, and a user reference manual;
- In-house built reusable software modules with the access to their function specifications and interfaces;
- In-house built reusable software libraries with well-defined specifications and interfaces, as well as a user reference manual;
- Legacy modules with very limited function specifications and interface documents.

Since most of these modules and components do not provide any quality validation reports, metrics, and quality control data, it is very difficult for component users to understand, select, evaluate, and compare the quality of these reusable modules and components.

1.6 Software modules versus software components in CBSE

In the previous sections, we discussed the major differences between modules and reusable components, including concepts, elements, deliverables, and properties. Now, let us compare them from a development perspective. Table 1.1 summarizes the major differences.

In traditional software development, modules are developed by programmers (or engineers) to fulfill a specific functional part of a system. The major objective is to generate high-quality modules with good modularity and well-structured function logic to make sure that they are easy to use, integrate, and maintain as a part of the system. The development of software modules is conducted by focusing on its functions and internal structures instead of module interaction and composition. Hence, the resultant software modules usually lack interoperability and standard interfaces. Since most software modules are created in a system-oriented software development process, packaging and deployment approaches are not implemented as a part of the modules. Therefore, they usually are not developed to be final software products that can be delivered and deployed individually. Since most conventional software modules are developed for a specific targeted operating environment, their reusability on different context environments is limited.

In past decades, engineers have defined and used various rigorous validation methods, criteria, and quality control processes for safety-oriented and life-critical systems to check the quality of involved software modules. However, they are not good enough for quality validation and assurance of reusable components and building blocks in component-based software engineering because these validation methods, criteria, and quality control processes do not address the following issues:

- How does one validate component reusability?
- How does one validate component interoperability?
- How does one assure a developed component follows a component model?
- How does one check a component for its correct packaging and deployment?
- How does one check whether a component is customized properly and correctly?

In component-based software engineering, component engineers develop software components as highly reusable parts for third-party reuse across product lines and organizations. The major objective is to create reusable building blocks with contract-based interfaces to make sure that they are highly reusable, interoperable, easily packaged, and deployable. Component-based software development is a reuse-centered development process that focuses on component reuse and that concerns component interaction and composition.

Table 1.1 Software Modules Versus Reusable Components

Perspectives	*Software Components in CBSE*	*Traditional Software Modules*
Development purpose	Build high-quality software parts that can be reused by third parties for a long term	Construct quality functional parts for a specific system in a software project
Development objective	Develop high-quality software parts with standards and contract-based interfaces using a standard component model	Develop high-quality modules with good modularity and well-structured logic and behavior
Developers	Component development engineers	Programmers or software engineers
Users	Application system developers and integration engineers	Application system programmers
Development process	A dedicated component engineering process is in place to control and manage the construction of components	A system-oriented development process is used to control and manage module development
Design and analysis	Think in terms of reusable components Focus on black-box and interfaces Conduct domain analysis and modeling Follow a component model Apply a methodology (such as UML) for analysis and design Use design patterns Implement the design based on explicit context environments	Think in terms of functions, or subfunctions Focus on white-box design and coding Use an ad hoc domain model Apply structure-oriented or object-oriented analysis and design methodology Base on a specific technology under an implicit operating environment
Properties	Unique identification in an operational environment Contract-based interfaces specified in a standard format Modular and encapsulation of functional logic and data Reusability in different scopes Packaging and deployment Interoperability (for building blocks) Composition (for building blocks) Using a component model (for building blocks)	Unique identification in a particular application system Independent program unit and compiled unit Customer-built interfaces specified in an ad hoc format Function modularity Implicit dependency on a specific technology and a targeted operating environment
Standards	A component model and standards that stipulate consistent component interaction mechanisms, and standard packaging and deployment approach	There is no component model nor consistent approaches to support component interoperation, packaging, and deployment
Reuse granularity	Component abstracts, reusable function/class libraries, packages or modules, frameworks, platforms, application servers	Procedures/subroutines, classes, modules (or a group of classes or subroutines), and subsystems

Table 1.1 Continued

Perspectives	*Software Components in CBSE*	*Traditional Software Modules*
Reuse scopes	Large reuse scopes encompassing different projects, product lines, organizations, or even the programming endeavors of the general public	Limited reuse scope for a specific project, product line, or an organization
Deliverables	Component specification, design, source code, executables, interfaces, user reference manual, quality control metrics and report (Note: component design and implementation results usually are not delivered to clients)	Module specification, design, source code and executables, as well as a user reference manual
Quality control	A rigorous quality control process in place to control the quality of reusable software components	Only system quality control process used (in most cases, a quality control process is used to validate the quality of software modules by focusing their functions and behaviors)

To build highly reusable software building blocks, component engineers perform this process based on a well-defined component model with consistent component standards and approaches to support component interactions, customization, packaging, and deployment.

Before a component development process begins, component domain analysis and modeling is performed first to come out with a domain-specific business model to support the definition of component requirements. In acomponent development process, engineers use the component-oriented UML to define components by specifying component use cases, object-oriented structures, and dynamic behaviors.

The results of this process are various types of reusable parts, including reusable components, libraries, packages, frameworks, platforms, and application servers.

To assure component quality, a well-defined component quality validation and control process is needed. It checks component functions and behaviors and validates correct reuse in the targeted reuse contexts. For reusable building blocks, the quality validation and control process must confirm that the preselected component model and standards are correctly followed.

What are the differences between software classes and software components? A class in a program is a type of class objects. Each class is a program unit that encapsulates a set of functions and data attributes. Hence, it is a programming concept. In the component world, however, a class could be viewed as a reusable component without deployment support because it possesses the common properties of a software component, including identity, encapsulation, reuse, polymorphism, and composition. Classes are not composite building blocks because no standard models and mechanisms are involved to support class interactions, packaging, and deployment, although class templates allow programmers to customize classes in metadata types and functions.

1.7 An engineering process for software components

A well-defined software engineering process plays an important role in software project management. It enables project managers and engineers to control and manage software projects in a systematic manner. Similarly, building software components needs a well-defined component engineering process to control component development activities. A software component development process involves six groups of people:

- Component engineers, who perform component analysis, design, and implementation according to a given component model and its companion standards. They also perform component unit tests in a targeted operating environment.
- Technical managers, who are in charge of component development projects by managing and coordinating various project activities.
- Component test engineers, who perform functional validation and performance evaluation for generated software components in the specified operating contexts.
- Quality assurance personnel, who define component test plan, quality control process and standards, and quality evaluation metrics for software components. They also control the quality process to ensure the conformity of all constructed components to a preselected (or predefined) component model.
- Component technical supporters, who are assigned to perform software maintenance tasks for components after they are shipped to customers.
- Component technical writers, who are responsible for writing user-oriented documents.

Figure 1.2 shows a component engineering process for software components. It is an iterative process that consists of the following six phases.

- *Analysis:* In the analysis phase, all component requirements (including requirements about functions, performance, data and objects, use cases, interfaces, technologies and operating environments) are collected,

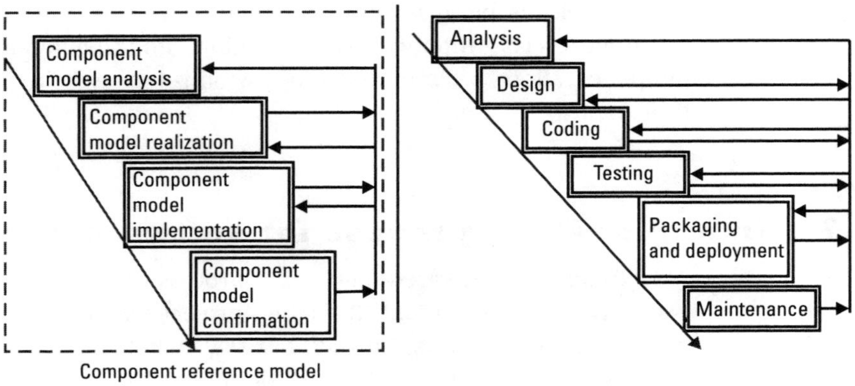

Figure 1.2 An engineering process for software components.

analyzed, and specified based on a well-defined methodology, such as UML [17]. The result of this phase is a component specification document.

- *Design:* In this phase, engineers conduct component design based on the component requirement specification from the previous phase. Component design includes three tasks. The first task is to conduct component design for functional logic and data objects and make trade-off decisions on technologies and operation environments. The next task is to follow a selected component model and work on component realization by providing data exchange mechanisms for component communications and interactions. The last task is to define consistent approaches to support component packaging and deployment. The outcome of this phase is a component design document.
- *Coding:* In this phase, component implementation is performed using a specific technology and programming language based on the given component design and targeted operating environments. It is possible for a component to be implemented for more than one operating environment. Each implementation depends on a specific technology set and a targeted operating environment.
- *Testing:* The major task of this phase is to validate a given software component based on its specification and design. During this phase, component testers perform software testing, such as white-box and black-box testing, to uncover various errors. Since software components are delivered as final products, component testing must include component usage testing,

performance testing, and deployment testing. The result of this phase is a component test plan, test design document, unit test bed and test suite, and test metrics and test reports.

- *Packaging and deployment:* This phase is required only when a component provides packaging and deployment capabilities. Testers and quality assurance groups must execute and validate different packaging facilities, customization features, and deployment approaches to check if they work fine under each configuration setting in every specific operating environment.
- *Maintenance:* This phase begins after shipping the first version of a software component to customers. In this phase, software components are updated and enhanced to meet customer requests, and to resolve discovered problems.

S. D. Lee et al. [17] present an UML-based component engineering methodology. They discuss the detailed tasks and modeling methods involved in component analysis and design. In modern component technology, software component construction must follow a well-defined component model and standards. To achieve this, a component reference model is needed to support the engineering activities in all phases.

1.8 Questions, concerns, and needs in component validation and quality control

In component-based software engineering, we construct application systems by reusing and integrating three types of components: (1) third-party components, (2) homegrown components or legacy modules, and (3) newly developed components. This suggests that the quality of reusable software components has a serious impact and ripple effect on software systems that adopt them. Therefore, testing and quality assurance of reusable components become extremely important to the success of component-based software development because a single defect of a component may affect many component-based systems in projects, product lines, and organizations. This raises some serious questions and concerns to software engineering researchers and industry practitioners.

Software engineering researchers and component testers must think about the following questions:

- What are the new issues and challenges in testing reusable software components? Do we have solutions to deal with them?

- Can we use the existing software testing models, strategies, methods, and criteria to validate reusable components? What are their limitations?
- How can we check special features of modern software components, such as reusability, interoperability, customization and packaging, and composition?
- How can we validate reused components in a new reuse context? What are the adequate testing criteria?
- How can we evaluate and measure component performance? What performance models and metrics can be used?
- How can we achieve component test automation? Can we use existing software tools to support component test automation?
- How can we provide a sufficient component quality report to component users for quality evaluation?

Component users encounter the following questions:

- How can we evaluate different software components with the same functionality in a cost-effective way?
- How can we compare the quality of software components with the same functionality in the commercial market?
- How much testing we need to perform for a third-party component in a component-based software project? What are the adequate test criteria? How can we do this in a cost-effective way?
- How to validate a customized reusable component? What are the suitable test criteria?

Quality assurance engineers encounter the following questions:.

- How can we perform quality assurance tasks in a component vendor?
- Can we use the existing software quality assurance process and control methods to deal with reusable software components? What are the limitations?
- How can we address the special features of reusable software components in a quality assurance process, including reusability, composition, interoperability, customization, and packaging?
- How can we define or set up a cost-effective quality assurance process for reusable components in a component vendor?
- How can we set up a quality assurance process for component-based software projects?

- How can we establish a component certification procedure and standards for software components?

The above questions have been encountered in component development projects. Recently, there have been a number of published technical papers addressing some of these issues, but there are no published books that do the same. In this book, we focus on these questions and provide discussions and answers based on the available research results and current advances in component-based software testing. In Chapter 2, we focus on the discussion of software component testing in CBSE, including issues, challenges, and needs. In Chapter 3 we introduce component-based software. In Chapter 4, we discuss the problems and challenges in component-based software testing. Part II of this book covers various validation methods of software components and test automation. In Part III of this book, we provide the issues and solutions in testing component-based software, including integration, regression, and performance validation.

It is clear that we need a well-defined quality assurance process to address special needs in component quality control, certification, and assessment. Part IV of this book is dedicated to the quality control processes, verification methods, certification standards, and quality measurement techniques of software components and component-based programs.

1.9 Summary

This chapter is written to help our readers understand software component concepts in component-based software engineering. It covers the evolutionary history of software components and the motivation of component-based software engineering. Moreover, it provides a basic introduction of software component concepts, including definitions, basic elements, properties, and deliverables. In this chapter, we classify and define reusable software components into three groups: reusable modules, reusable components, and reusable building blocks. The first group includes legacy reusable modules and most of homegrown modules that are reusable for specific projects. The second group includes all reusable components that are developed for reuse in a third-party environment. Most current third-party components developed for reuse belong to this group. The last group includes modern components that are developed as building blocks for third-party users. Components in this group usually are created based on a well-defined component model and standards.

In addition, we compare conventional software modules and reusable components from a development perspective. Furthermore, a component software development process is given to help readers to understand component development. Finally, we list many questions and concerns about component validation and quality control.

Although many people believe component-based software engineering could be a cost-effective way to construct large software application systems using reusable components, we must point out that more research work and experience reports are needed to prove the effectiveness of this methodology. Meanwhile, we need more research and test tools to cope with the listed issues and assist engineers in testing and controlling the quality of software components.

References

[1] Clements, P. C., "From Subroutines to Subsystems: Component-Based Software Development," *The American Programmer*, Vol. 8, No. 11, 1995.

[2] Mcilory, M. D., "Mass Produced Software Components," *Proc. of NATO Conference on Software Engineering*, Garmisch-Partenkirchen, NATO Science Committee, NATO, Brussels, 1969.

[3] Johnson, R. E., and B. Foote, "Design Reusable Classes," *Journal of Object-Oriented Programming*, June 1988, pp. 22–35.

[4] McClure, C., *Software Reuse Techniques: Adding Reuse to the System Development Process*, Englewood Cliffs, NJ: Prentice Hall, 1997.

[5] Wang, N., D. C. Schmidt, and C. O'Ryan, "Overview of the CORBA Component Model," in *Component-Based Software Engineering*, G. T. Heineman and W. T. Councill (eds.), Reading, MA: Addison-Wesley, 2001, pp. 557–572.

[6] Blevins, D., "Overview of the Enterprise JavaBeans Component Model," in *Component-Based Software Engineering*, G. T. Heineman and W. T. Councill (eds.), Reading, MA: Addison-Wesley, 2001, pp. 589–606.

[7] Ewald, T., "Overview of COM+," in *Component-Based Software Engineering*, G. T. Heineman and W. T. Councill (eds.), Reading, MA: Addison-Wesley, 2001, pp. 573–588.

[8] Cheesman, J., and J. Daniels, *UML Components: A Simple Process for Specifying Component-Based Software*, Reading, MA: Addison-Wesley, 2001.

[9] D'Souza, D. F., and A. C. Wills, *Objects, Components, and Frameworks with UML: The Catalysis Approach*, Reading, MA: Addison-Wesley, 1999.

[10] Herzum, P., and O. Sims, *Business Component Factory: A Comprehensive Overview of Component-Based Development for the Enterprise*, New York: John Wiley (OMG Press), 2000.

[11] Aoyama, M., "CBSE in Japan and Asia," in *Component-Based Software Engineering: Putting the Pieces Together*, G. T. Heineman and W. T. Councill, (eds.), Reading, MA: Addison-Wesley, 2001, pp. 213–226.

[12] Booch, G., *Software Components with Ada: Structures, Tools, and Subsystems*, 3rd ed., Reading, MA: Addison-Wesley, 1993.

[13] Szyperski, C., *Component Software—Beyond Object-Oriented Programming*, Reading, MA: Addison-Wesley, 1999.

[14] Brown, A. W., *Large-Scale, Component-Based Development*, Englewood Cliffs, NJ: Prentice Hall, 2000.

[15] OMG Unified Modeling Language Specification, Version 1.4, Object Management Group, 2000.

[16] Heineman, G. T., and W. T. Councill, (eds.), *Component-Based Software Engineering: Putting the Pieces Together*, Reading, MA: Addison-Wesley, 2001.

[17] Lee, S. D., et al., "COMO: A UML-Based Component Development Methodology," *Proc. of 6th Asia Pacific Software Engineering Conference*, Takamatsu, Japan, December 7–10, 1999, pp. 54–62.

CHAPTER

2

Contents

Software component testing

In the component-based software-engineering paradigm, component-based software is developed using a set of in-house and commercial-off-the-shelf components. These components are reused, adapted, and tailored to meet the specifications of a specific project in a given context, including system platform, technology, and running environment.

Since system quality depends on component quality, any defective component causes a ripple impact on all systems built on it. Hence, component validation and quality control is critical to both component vendors and users. To validate component quality, component vendors must perform a cost-effective test process and implement a rigorous quality process for all generated software components. Component users must go through well-defined processes to evaluate, validate, and accept third-party components before using them in a component-based software system.

Since 1990 many organizations have begun to develop large-scale information systems using reusable software components to reduce development cost and shorten the production cycle; and more people have begun to pay more attention to component testing and component-based software testing. In the past decade, there have been a lot of research efforts dedicated to component-based software engineering. Most of them have focused on component-based software development processes, and analysis and design methods. Only a limited number of published papers have addressed the problems and solutions in testing reusable software components and component-based software.

Current component-based software engineering and modern component technology not only introduce changes in the way to create components but also add new features and properties into software components. This shifts the software development paradigm from being construction-oriented to being reuse-oriented. These new features and properties have brought out new demands, issues, and challenges in testing components and component-based software [1–7]. In addition, reuse-oriented software development has led to a new focus of software component testing on component reuse. We need new methods to test and maintain software components to make sure they are highly reliable and reusable if we plan to deploy them in various software projects, products, and environments.

This chapter provides an introduction to software component testing from the perspective of component-based software engineering. It only focuses on the testing issues and challenges of software components. In Chapter 4, you will learn component-based software testing and related issues and challenges.

This chapter is structured as follows. Section 2.1 reviews the basic concept of component testing for modules in traditional software development. Section 2.2 discusses basics of component testing and test processes in component-based software engineering. Section 2.3 addresses the component testing issues and challenges encountered by component vendors and users. Furthermore, component testing myths and other needs are listed and specified in Section 2.4. Finally, a summary of this chapter is given in Section 2.5.

2.1 Component testing background

Generally speaking, software *component testing* refers to testing activities that uncover software errors and validate the quality of software components at the unit level [8]. In traditional software testing, component testing usually refers to unit testing activities that uncover the defects in software modules. Unit testing for modules is the first step in a software test process. Its major task is to check the quality of a module by performing unit tests to uncover its program errors. The major objective is to validate and confirm the quality of a software component by checking its functions, behaviors, and performance to make sure a module under test meets its functional requirements and specified design in a given operational environment.

In past decades, many test methods, strategies, and criteria have been developed to support testing of software modules at the unit level. These test methods can be classified into the following two categories:

- *Black-box test methods,* which focus on the validation of required functional features and behaviors of software modules from an external view. These test methods are developed to generate unit tests based on the given module specifications.
- *White-box test methods,* which focus on validation of program structures, behaviors, and logic of software modules from an internal view. These test methods are designed to drawn unit tests based on the given source code of the module.

The primary objectives in unit testing for a module are to:

- Design high-quality unit tests to uncover as many errors as possible;
- Conduct cost-effective unit tests to demonstrate its functions and behaviors matching the given requirements and design;
- Achieve adequate test coverage based on a predefined unit test model and criteria.

There are four major tasks in unit testing for a module. The first is to design and perform white-box testing[1] to validate its internal logic, data, and program structure based on the given design and source code. The second is to design and perform black-box testing[2] to validate its external accessible functions, behaviors, and interfaces based on the given requirements. The third is to validate and measure its performance (such as reliability and processing speed) based on the given requirements. The final task is to report testing results, program errors, and test coverage.

The major focuses in unit testing are:

- Checking incorrect internal module structure and functional logic;
- Checking incorrect internal data types, object classes, and data values;
- Checking incorrect functions, interfaces, and behaviors, as well as performance in a software module.

Engineers need the following means to conduct effective unit testing:

- Adequate test models to provide a fundamental basis for engineers to come out with test criteria and strategies;

1. White-box testing of modules refers to the unit testing that is performed using program-based tests to uncover the program errors in internal logic, data, and structures.
2. Black-box testing of modules refers to the unit testing that is performed using requirement-based tests to uncover the program errors in external functions, behaviors, and interfaces.

- Well-defined test criteria to support a quality control process;
- Cost-effective test strategies to support a test process;
- Efficient test methods to support test case design and test data generation;
- Systematic test automation tools to enhance test management and reduce test cost and efforts.

2.2 Component testing in CBSE

In component-based software engineering, both component vendors and users must test software components. Component testing, therefore, includes two types of testing efforts and activities:

- *Vendor-oriented component testing,* which occurs as one step of a component development process. It refers to a component test process and testing activities performed by a component vendor to validate a software component based on its specifications.
- *User-oriented component testing,* which occurs as a part of a component-based software development process for a specific application project. It refers to a component validation process and testing activities in a specific context to make sure every involved software component delivers the specified functions, interfaces, and performance. Moreover, component reuse is validated in the given context and operational environment.

These two types of component testing have different focuses. In vendor-oriented component testing, the primary purpose is to answer the following questions for component developers:

- Are we building a right component with high quality based on the given specifications?
- Are we building a component based on the specified standards and component model?

User-oriented component testing is performed to answer the following questions for component users:

- Are we selecting and deploying a reusable component that is right for a system?

- Are we reusing a component correctly in a system?
- Are we adapting or updating a component correctly for a project?

The following sections provide the basics and unit test processes of component testing.

2.2.1 Vendor-oriented component testing

Test engineers of a software component vendor implement a component test process based on well-defined component test models, methods, strategies, and criteria to validate the developed software components.

The vendor-oriented component testing has three major objectives:

- Uncover as many component errors as possible.
- Validate component interfaces, functions, behaviors, and performance to assure that they meet the given component specifications.
- Check component reuse, packaging, and deployment in the specified platforms and operation environments.

As shown in Figure 2.1, a component test process for a vendor consists of the following six steps.

1. *Component black box testing:* In the first step, component developers and test engineers use black-box test methods to check incorrect and

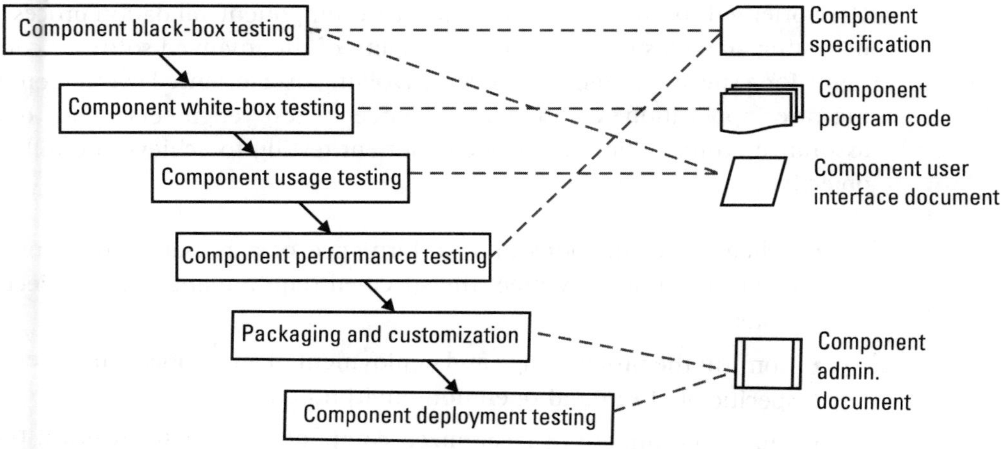

Figure 2.1 A test process in vendor-oriented component testing.

incomplete component functions and behaviors based on the given component specifications. Many traditional black-box test methods can be used here. The chapters in Part II of this book provide the details.

2. *Component white-box testing:* This is the second step of the component testing in which component developers use white-box test methods to uncover the internal errors in program logic and structure, data objects, and data structure. Many existing white-box test methods can be used here. The chapters in Part II of this book provide the details.
3. *Component usage testing:* Test engineers exercise various component usage patterns through component interfaces to confirm that the correct functions and behaviors are delivered through the given contract-based interfaces.
4. *Component performance testing:* Test engineers and quality assurance people validate and evaluate the performance of components.
5. *Packaging and customization:* This step is only useful for components that provide built-in customization features and packaging facility. Its testing focus is built-in customization features and an implemented packaging approach.
6. *Component deployment testing:* As the last step of a component test process, this validates the component deployment mechanism to make sure it is correctly designed and implemented according to a given component model.

2.2.2 User-oriented component testing

User-oriented component testing refers to a component validation process and its testing activities that confirm the quality of the involved software components for a specific system. Engineers involving user-oriented component testing are application component engineers, test engineers, and quality assurance groups. They perform component testing to achieve the following objectives:

- Validate the functions and performance of a reusable component to make sure that they meet the specified requirements for a project and system.
- Confirm the proper usage and deployment of a reusable component in a specific platform and operation environment.
- Check the quality of customized components developed using reused components.

- Test the quality of new components created for a specific project.

Testing reused components (such as COTS components) in a new context is necessary and critical to a component user even though component vendors have already tested them. The Ariane 5 disaster, reported in [9, 10], showed us that not testing a reusable component in a new reuse context and operation environment may cause serious consequences and failures. In the Ariane 5 project, developers had reused certain software components from the Ariane 4 system without substantial retesting because they assumed there were no significant differences in these parts of the two systems. As pointed out by E. J. Weyuker [6], reusing a component in a component-based software project needs to have considerable testing to ensure that it behaves properly in its new context environment.

User-oriented component testing checks component reuse. Engineers must devote their testing efforts to the validation of component reuse by answering the following questions:

- Is a reused component packaged and deployed properly in a targeted operational environment?
- Does a reused component provide the specified user interfaces accessible to a user?
- Does a reused component provide the correct functional features, proper behaviors, and acceptable performance when it is reused in a new context and environment?

Three types of components are used in a component-based software development project:

- *Completely reused components*, which are the reused components without altering or tailoring (COTS components are typical examples);
- *Adapted and updated components*, which are the components developed based on reusable components, such as in-house built components, and COTS components, to fit into a particular system context and to meet specific requirements;
- *Newly developed components*, which are the newly created components for a project to fulfill specific system requirements.

Validating newly developed components is similar to vendor-based component testing. The component test process described before can be used here.

However, validating reused and updated components has different focuses and limitations. For example, a component user usually has no access to the source code and artifacts of a completely reused component from a third party. The user has to validate the reused components using black-box testing without access to the personnel and expertise used to create it. This occurs as a part of a component evaluation process at the earlier phase of a component-based software development process.

Figure 2.2 shows a validation process for completely reused components from a third party. The process includes the following five steps.

1. *Component deployment:* This is the first step of the user-oriented component testing. It validates a component to see if it can be successfully deployed in its new context and operational environment for a project. It focuses on the built-in component deployment mechanism and supporting facility.
2. *Component customization and packaging:* This step checks to see if a component can be successfully packaged and properly customized using its built-in packaging and customization features in its new context environment. The major focus is on the built-in component packaging and customization mechanism, and supporting facility.
3. *Component usage testing:* In this step, a component user designs test cases to exercise different usage patterns of a component through its interfaces. The primary goal is to cover the important component usage

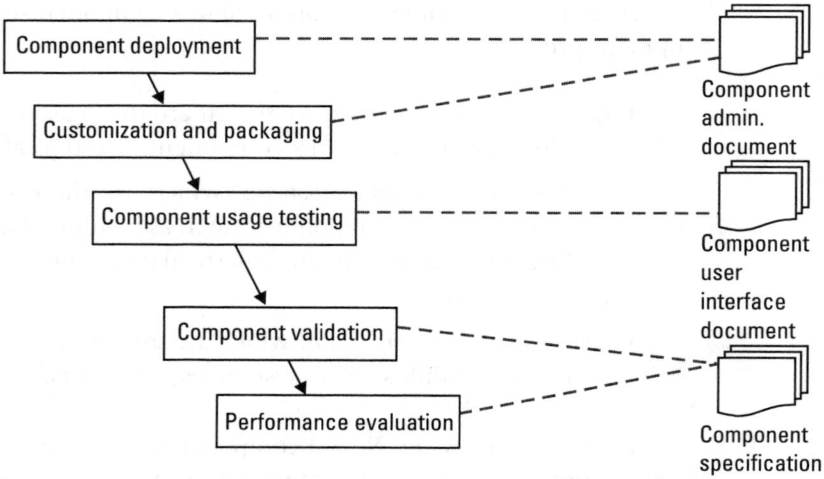

Figure 2.2 A validation process for completely reused components.

patterns in its new context and environment. Two typical examples are:

- Checking frequent function invocation sequences;
- Trying typical usage patterns on data parameters in component interfaces.

4. *Component validation:* The major task of this step is to perform component black-box tests to validate component functions and behaviors in the new reuse context and environment. Various existing black-box testing methods can be used here.
5. *Component performance evaluation:* This step evaluates and measures the performance of a component in a new context and operation environment to make sure that it satisfies the performance requirements. Performance testing focuses can be defined based on the major performance interests of a project. Typical examples are component operation (or function) speed, reliability, availability, load boundary, resource utilization, and throughput.

Validating adapted and updated components is another major task for component users. These components are known as customized components. They are developed based on reusable components by customizing and altering them. Since they contain reused components and newly added parts for a specific project and system, their validation process differs from the reused components. Figure 2.3 shows a validation process in five steps for customized components.

1. *Reused component validation:* A component user validates completely reused components, such as COTS components, based on the previous validation process.
2. *Black-box testing for customized parts:* Black-box tests are performed for those customized parts. The objective here is to uncover the function and behavior errors of the new and altered parts in a customized component based on the given specifications.
3. *White-box testing for customized parts:* White-box tests are performed based on the given source code to uncover program logic and structure errors of the customized parts, such as added functions, adapters, and tailored parts.
4. *Integration for customization:* Reused components are integrated with customized parts (such as an adapter, new function feature) to form

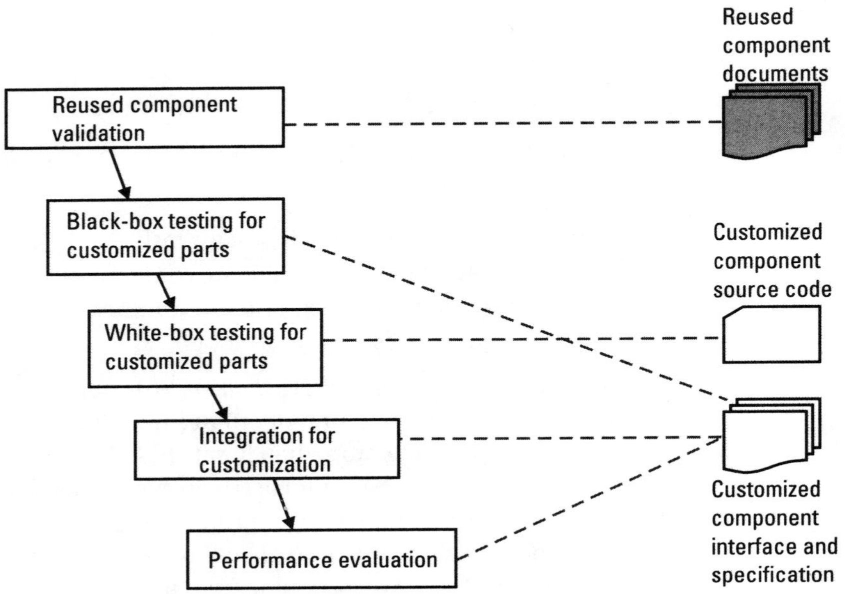

Figure 2.3 A validation process for adapted and customized components.

a specified customized component in a new reuse context and environment.

5. *Performance evaluation:* Concerning component performance, component users have to evaluate the performance of a customized component in a new context and environment based on its performance requirements.

2.3 Issues and challenges of component testing in CBSE

In the past decade, much research effort and many publications focused on how to analyze and design software components and component-based systems. Until now, there were only a few published research papers addressing the issues and solutions in testing software components and component-based programs. This section discusses the issues and challenges in testing software components. Part II of this book covers the existing component testing methods and techniques relating to the issues. Chapter 4 discusses the problems of testing component-based software, and Part III of this book reports the existing solutions to these problems.

In component-based software engineering, component testing involves two groups of people:

- Component developers, testers, and quality assurance people of component vendors;
- Component users, such as integration and application engineers, test engineers, and quality assurance people, who develop component-based systems.

Let us now examine component testing issues and challenges from the perspectives of these two groups.

2.3.1 Issues and challenges in user-oriented component validation

2.3.1.1 Issues in user-oriented component testing

In user-oriented component validation, component users, who are involved in component selection, validation, evaluation, alteration, and integration, encountered the following issues.

- *Issue #1: Difficult to perform component analysis and testing due to the lack of the access to component source code and internal artifacts.* This issue has been identified in [5, 6]. Since a third-party component user only has the user-level access to component specification, user interfaces, and reference manual, engineers encountered the following difficulties:
  - It is not easy to understand and comprehend the functions and behaviors of reusable components.
  - It is impossible to perform white-box program analysis to support debugging.
  - It is a challenge to conduct unit-level black-box testing without access to the component creators and related artifacts.
- *Issue #2: Testing reused components in a new reuse context and environment.* As we mentioned before, a reused component must be tested to make sure that it delivers the required functional features and demonstrates the correct behaviors in the new context environment for a project. The ARIANE 5 lesson [9] suggests that disastrous consequences might occur if a component (such as COTS component) is reused in a new context for

a project without sufficient testing. E. J. Weyuker in [6] reported her case study for component reuse in two different projects. She found that the two test suites for the same component in the two projects actually exercise the different parts of the source code with significantly different usage patterns. She argues, therefore, that we need cost-effective approaches to validate component reuse in a new context environment without access to the component source code.

- *Issue #3: Poor testability of reusable software components.* Due to the lack of standardized requirements for testability on reusable components, it is not unusual that component users may reuse or adopt components with poor testability. J. Gao in [1] discussed the four essential factors of component testability in terms of understandability, observability, controllability, and test support. He pointed out the fact that the current third-party components (such as COTS components) have poor testability for the following reasons:
    - *Poor traceability and observability [11]:* Components provide no or ad hoc traceability function for users to monitor the component behaviors and errors from a third-party view. This increases the difficulty of understanding component behaviors and reporting errors during component validation and integration.
    - *Poor test support and control capability:* Most components only provide an application interface without any support in component validation. They usually do not provide built-in test beds and test cases to allow users to perform black-box component testing. This is one of the contributing factors to the expensive cost of component validation and evaluation.
    - *Lack of quality and test coverage information [11]:* Commercial and in-house reusable components usually do not offer users any quality report or test coverage information. This increases the difficulty of understanding and evaluating the testing quality and coverage of a component.
- *Issue #4: Expensive cost of constructing component test bed, including test drivers and stubs.* According to J. Gao [11], third-party component users must spend much effort on constructing test beds to support component validation, because most COTS and reusable components do not provide any component test bed and facility supporting black-box testing. This cost includes the construction and maintenance overhead of component test drivers and stubs. There are two hurdles for engineers to set up a

component black-box test bed for a component. First, they have very limited understanding of the reused components due to the fact that they do not have access to the component creators and related artifacts. Second, current black-box commercial testing tools may not support the component's interfaces, the required technology, and the operational environment.

- *Issue #5: Adequate testing of component reuse [12].* Since a reused component must be validated in a new context environment, component users must deal with the issue of adequate testing for component reuse. In other words, they have to find out how much testing must be done to achieve enough test coverage in a new context. Clearly, this is related to two types of reused components: completely reused components and updated components. Although component users usually do not want to retest a reused component like a component vendor, they have to come out with an adequate test set to achieve some kind of test criteria to cover its reuse patterns in the new context.
- *Issue #6: Hard to reuse component tests for the third-party users [11].* Since current components, including COTS components, provide no test suite to users, it is impossible for them to reuse component tests in component validation and integration. This problem cannot be resolved until vendors deliver components with built-in tests and testing support facility.

2.3.1.2 Challenges in user-oriented component testing

There are three technical challenges for user-oriented component testing.

- *How does one perform adequate testing for a component in a new reuse context?* This is a new challenge for component users because we lack adequate test models, coverage criteria, and test generation techniques that address adequate testing of component reuse in a new context. The recent paper by D. Rosenblum [12] reported the first attempt to propose a formal adequate test model for component reuse based on the concept of application subdomains.
- *How does one test customized components using a systematic solution?* A software component can be customized in many different ways in a new context. Without the regulation of component customization, it is very difficult to find systematic solutions to support the validation of customized components due to the lack of new test methods and coverage criteria for customized components. H. Yoon and B. Choi in [13] discussed their approach using fault injection technique and mutation test criteria. Since

a customized component usually is an integration of reusable components and updated (or extended) parts, its systematic testing solution must support white-box testing, and black-box testing, as well as integration. Currently, we have a lack of effective test tools for customized components.

- *How does one automate the testing of completely reused software components in a cost-effective way?* Since component users must validate all reused components before adoption, they need a reusable unit test bed (or platform) to support component black-box testing to reduce the overhead of constructing test beds. This is a very difficult task because components have different application interfaces and depend on diverse technologies, programming languages, network protocols, and interaction mechanisms. This problem, therefore, cannot be resolved now unless vendors provide users with a component test bed and black-box test suites for each component. The alternative solution may be to define future component models by considering component testing and maintenance issues.

2.3.2 Issues and challenges in vendor-oriented component testing

2.3.2.1 Issues in vendor-oriented component testing

Unlike traditional modules, software components, generated from component vendors, are usually considered as products with a large-scale reuse scope. They are used as building blocks to form component-based software systems. Although component vendors can use the existing black-box and white-box testing methods, they will encounter a different set of issues and challenges than users encounter. Here we list typical ones.

- *Issue #1: Adequate testing of software components.* Adequate testing of modern software components is much more complicated than adequate testing of traditional modules because of their unique properties in reusability, interoperability, composition, packaging, and deployment. These new component properties extend the semantics and scope of adequate testing. For example, testing a traditional module always concerns one specific usage context and operation environment. A reusable software component, however, must have diverse usage contexts, and it may support more than one operating environment. This suggests that a vendor must test it for diverse reuse contexts and validate it under all specified operating environments to achieve adequate testing. The real difficulty here is how to understand and define an adequate test set for a component based on a well-defined test model that addresses the component properties of reusability, interoperability, and composition.

- *Issue #2: Ad hoc test suite management for software components [11].* Today test engineers and managers in the real world use diverse test management tools to create, manage, and maintain system-level test suites (including test cases, scripts, and data) for a software system in a test repository. Since software modules are not considered as final products in traditional software development projects, many engineers and managers do not pay enough attention to the management and maintenance of their unit test suites. However, components developed in a component development process are the final products. Each component must go through continuous updates in its evolution process to enhance its quality, features, operation environments, and reuse contexts to meet marketing and customer needs. This requires that engineers and managers consider the test suites as a part of artifacts for each component. They must carefully manage and maintain each component test suite using a systematic solution and tool.

- *Issue #3: Ad hoc construction and management of component test beds and test scripts.* In traditional software testing, engineers create and manage unit-level test beds (including test drivers and stubs) in an ad hoc manner. This kind of engineering practice is not cost-effective to component vendors. Since each component, as a product, must be updated and enhanced in a continuous evolutionary process, its unit-level test bed and test scripts must be updated according to its changes. Therefore, standardized solutions are needed to support the construction of unit test beds and test scripts to achieve the reduction of the involved cost and efforts. Meanwhile, a systematic configuration management solution is needed to effectively manage the component test beds and scripts.

- *Issue #4: Testing of component packaging and deployment.* With the advance of component engineering, component vendors have the tendency to create deliverable and executable components to support the plug-in-play concept. These modern components (such as building blocks defined in Chapter 1) usually provide a built-in packaging mechanism and deployment approach as a part of the component. Because of this, component developers, testers, and quality assurance groups must validate the provided packaging function and deployment features based on the given specification, design, and component model. Therefore, testing of built-in packaging and deployment becomes a necessary validation task. This task may be complicated and tedious when a component has complex built-in customization features and packaging functions.

2.3.2.2 Challenges in vendor-oriented component testing

Testing modern software components is more complicated and challenging than testing traditional modules due to the new component properties in reusability, interoperability, packaging, and deployment. There are a number of challenges in vendor-oriented component testing. They are summarized next.

- *How can we conduct adequate testing for software components?* Producing high-quality software components for third-party reuse needs cost-effective methods to perform adequate testing. Today, we lack research results and application experience on adequate testing of software components; developers and test engineers are looking for new effective approaches and guidelines to answer the following questions:
  - What is an adequate test model and coverage criterion for a modern software component?
  - How can we define and evaluate a test model to make sure that it addresses the new features of a modern software component in terms of its reusability, interoperability, packaging, and composition?
  - How can we design an adequate test set for a reusable software component?
  - How can we define a quality control standard and process to consider adequate testing issues of software components?

  M. J. Harrold et al. in [5] pointed out that component developers and test engineers need new test models, criteria, and test methods to address adequate testing issue for reusable software components.
- *How can we test customizable software components?* Customizable components refer to components with a built-in mechanism to allow users to customize and configure the functional features, metadata, and interfaces. Testing these types of components is a new challenge for a component vendor due to the lack of test models, coverage criteria, and systematic methods.
- *How do we construct testable components?* To reduce component test cost and effort for component vendors and users, we must pay attention to the

design for component testability to build *testable components* [14]. J. Gao et al. in [14] identified the basic requirements and discussed the architecture, interface, and methods in building a testable component. There are three basic questions related to design for component testability: How can we design and define its component model, architecture, and interfaces to enhance component testability; how can we control a component development process to monitor and verify component testability; and how can we control, monitor, and support the unit testing of a testable component?

- *How can we build a reusable component test bed?* In general, a program test execution environment consists of several supporting functions: test retrieval, test execution, test result checking, and test reporting. Clearly, a component test environment must include similar supporting functions. The primary challenge is how to come out with a new component test bed technology that is applicable to diverse components developed in different languages and technologies.
- *How can we construct reusable component test drivers and stubs?* Now, in real-world practice, engineers use ad hoc approaches to developing module-specific or product-specific test drivers and stubs based on the given requirements and design specifications. The major drawback of this approach is that the generated test drivers and stubs are not reusable because they are designed only for a specific project. It is clear that this approach leads to a high cost in the construction of component test drivers and stubs in a component engineering process. In addition, using this approach has the difficulty of dealing with customizable components due to the complex built-in customization features. Thus, we need new methods to construct reusable test drivers and stubs for components.

2.4 Component testing myths and other needs

There are a number of misunderstandings about component testing when we develop component-based software systems. Here we have collected several myths about component testing.

Myth #1: A software component can be tested once and reused everywhere.

Reality: A component must be tested whenever it is reused in a new context environment.

Discussion: A component user must validate a third-party software component before reusing it into a new context and environment. Testing a component by a user has two major purposes. The first is to validate the selected component to see if its functions and behaviors meet the given specifications in a new context environment. The other is to make sure that it is reused properly in the specific context. Reusing a component in a new context may cause a disaster. The ARIANE 5 project is a typical example.

Myth #2: If a vendor has adequately tested a component, then it can be used by a third party in a project without validation.

Reality: A component user must validate all COTS components to achieve adequate validation coverage for component reuse in a new context.

Discussion: In real-world practice, a component vendor cannot test a component exhaustively before releasing it to customers. What a vendor can do is to define and perform an adequate test set based on a predefined test model and criteria. This test set only can be executed in a limited number of context environments to exercise a few usage patterns. To achieve adequate testing for component reuse, therefore, a user must define an adequate test model and test set for the component in a new context to focus on its reuse patterns.

Myth #3: Only a component vendor needs a rigorous component test process.

Reality: A component user also needs a well-defined component validation process to validate the quality of all reused, adapted, and created components in a project.

Discussion: A component-based software system is developed based on three types of components: COTS, customized, and newly created components. A component user must establish an internal component validation process to assure the quality of the three types of components before they are integrated into a system. As we described before, a dedicated validation process is needed for reused components to focus on reuse functions and behaviors in a new context. Similarly, we also need a validation process to check the customized components to assure that the integration of adopted components and customized parts satisfies the given specifications.

Myth #4: Component testing in component-based software engineering is the same as the unit testing of software modules.

Reality: Component testing in component-based software engineering is different from traditional unit testing of modules although existing unit test methods are applicable.

Discussion: As we discussed before, component testing encompasses vendor-oriented component testing and user-oriented component testing. Similar to traditional unit testing, vendor-oriented component testing focuses on black-box and white-box testing of component functions, behaviors, and performance. Unlike traditional unit testing, in component testing by vendors, they must pay attention to the new features of modern components, such as reusability, interoperability, built-in packaging, customization, and deployment mechanisms. User-oriented component testing has different objectives and focuses. For example, component validation by users must be targeted at component reuse patterns in the next context. For customized components, component validation must check the adopted parts, updated parts, and their integration.

There are other needs relating to component testing. Here we highlight some of them.

- *Needs in component testability:* Component testability is an interesting topic relating to component testing. Current and future commercial markets need testable software components, so that component defects can be easily uncovered and observed, and component tests can be easily performed and controlled. To achieve this goal, component vendors need more practical research results to provide methods and guidelines to develop testable software components. Meanwhile, we need new standardized, cost-effective solutions to support component vendors to evaluate and measure the testability of software components. The measuring result of component testability is very useful for users to select and evaluate components. Chapter 5 provides more detailed discussion on this subject.
- *Needs in component quality assurance:* Component quality control is an important topic relating to component testing. To assure component quality, both component vendors and users need an effective component quality control process, well-defined standards, validation guidelines, and quality measurement methods. The chapters in Part IV of this book address these issues and report recent advances.
- *Needs in component certification:* Component certification is another subject relating to component quality validation. In many cases, the quality of components must be validated and evaluated by a third party. For example, building a COTS-based system for a government agency may need a third party to certify the quality of COTS components used for the project. Conducting effective component certification, we need more research studies and experiential reports about component certification processes, methods, standards, and guidelines. Chapter 15 provides a detailed discussion on this subject.

2.5 Summary

This chapter introduces the basic concept of component testing from the perspective of component-based software engineering. Similar to traditional unit testing of modules, component testing in CBSE is performed to validate the functions and behaviors of components based on the given specifications. Existing white-box testing methods can be used to uncover the internal errors in program structure, logic, data, and object types. Current black-box testing methods can be used to uncover the errors in external accessible functions, behaviors, and interfaces. Details about the applications of these testing methods can be found in Part II. Unlike unit testing of modules, component testing in CBSE is not only performed by component vendors, but also conducted by component users. This leads to two types of component testing: vendor-oriented and user-oriented component testing. Vendor-oriented component testing validates component features, functions, and behaviors based on the given specifications. User-oriented component testing has a new objective of validating component reuse in a new context and environment.

This chapter discusses three test processes. One of them is proposed for component vendors. It includes not only black-box and white-box testing, but also consists of testing steps for component usage, performance, packaging, customization, and deployment. The other two are defined for component users. One validation process is given here to validate COTS or completely reused in-house components. And the other is defined to support the validation of customized components.

This chapter provides detailed discussions about issues and challenges of component testing from the perspectives of component vendors and users. Finally, it presents several component myths in component testing, and other needs of component certification, testability, and quality assurance.

References

[1] Gao, J., "Challenges and Problems in Testing Software Components," *Proc. of ICSE2000's 3rd International Workshop on Component-Based Software Engineering: Reflects and Practice*, Limerick, Ireland, June 2000.

[2] Garlington, K., "Critique of 'Design by Contract: The Lessons of Ariane'," http://www.flash.net/~kennieg/ariane.html, March 1998.

[3] Ghosh, S., and A. P. Mathur, "Issues in Testing Distributed Component-Based Systems," *Proc. of 1st International ICSE Workshop on Testing Distributed Component-Based Systems*, Los Angeles, CA, May 17, 1999.

[4] Grossman, M., "Component Testing," *Proc. of 3rd International Workshop on Component-Oriented Programming*, 1998.

[5] Harrold, M. J., D. Liang, and S. Sinha, "An Approach to Analyzing and Testing Component-Based Systems," *Proc. of 1st International ICSE Workshop on Testing Distributed Component-Based Systems*, Los Angeles, CA, May 17, 1999.

[6] Weyuker, E. J., "Testing Component-Based Software: A Cautionary Tale," *IEEE Software*, Vol. 15, Issue 5, September/October 1998.

[7] Wu, Y., D. Pan, and M.-H. Chen, "Techniques for Testing Component-Based Software," *Proc. of 7th International Conference on Engineering of Complex Computer Systems (ICECCS2001)*, IEEE CS Press, 2001.

[8] Beizer, B., *Software Testing Techniques*, 2nd ed., New York: Van Nostrand Reinhold, 1990.

[9] Jezequel, J.-M., and B. Meyer, "Design by Contract: The Lessons of Ariane," *Computer*, Vol. 14, No. 1, January 1997, pp. 129–130.

[10] Lions, J. L., "Ariane 5, Flight 510 Failure, Report to the Inquiry Board," http://www.esrin.asa.it/htdocs/tidc/Press/Press96/ariane5rep.html, July 19, 1996.

[11] Gao, J., "Testing Component-Based Software," *Proc. of STARWEST'99*, San José, CA, November 1999.

[12] Rosenblum, D. S., *Adequate Testing of Component-Based Software*, Technical Report 97-34, Department of Information and Computer Science, University of California at Irvine, Irvine, CA, 1997.

[13] Yoon, H., and B. Choi, "Interclass Test Technique Between Black-Box-Class and White-Box-Class for Component Customization Failures,"*6th Asia-Pacific Software Engineering Conference*, Takamatsu, Japan, December 7–10, 1999.

[14] Gao, J., "On Buiding Testable Software Components," *Proc. of 1st International Conf. on Cost-Based Software System*, 2002, pp. 108–121.

CHAPTER

3

Contents

Introduction to component-based software

Component-based software comprises a collection of self-contained and loosely coupled components that allow plug-and-play. The components may have been written in different programming languages, executed on different operational platforms, and distributed across geographic distances. Some components may be developed in-house, while others may be third-party or *commercial-off-the-shelf* (COTS) components, with the source code unavailable. In this chapter, we first explore the basic concepts of component-based software, its characteristics, and its differences compared with traditional programs. Then, we discuss component-based software infrastructures, which are key elements of component-based software. Finally, a unique engineering process for developing component-based software is introduced.

3.1 Introduction

What is a component-based system? Generally speaking, a *component-based system* (CBS) is an assembly of software components, which follows an engineering process that conforms to component-based software engineering principles. A common misconception is that, once software components have been developed by following an engineering process, a component-based system can be developed by simply gluing available components together. Many facts [1, 2] have demonstrated that the development of component-based software is a complex engineering process, and component-based software of high quality must present the following three key elements [3]:

- *Software components:* Compared to the development of traditional software systems, the focus of developing component-based system has shifted from development to the integration of existing software components. These software components can be COTS components that are developed by other development teams, or reusable in-house developed software components.
- *Component model:* A *component model* is the "backbone" of a component-based system, which provides basic infrastructure for component development, composition, communication, deployment, and evolution [1, 4]. Currently, .NET/COM/COM+, CORBA, and EJB are three major commercial component models.
- *Component-based engineering process:* To develop and maintain highquality component-based software, a component-based engineering process needs to be adopted [5–7]. Component-based engineering shares many activities with traditional software engineering processes. It is an evolutionary process, and each cycle of the evolutionary process includes different phases such as requirement analysis, design, coding, and testing. Nevertheless, unique component-based software characteristics require different processing in each individual phase. For instance, during the requirement analysis and design phases, activities such as component identification, interface design, and component architecture design demonstrate significant differences, while the implementation phase includes many innovative activities, such as component selection, customization, and composition. Testing and maintaining software components and component-based software also involves new issues and challenges. Therefore, new methods and techniques are needed.

The detailed definition of a software component can be found in Chapter 1, and a detailed discussion of the characteristics of component-based software, component modeling, and component-based engineering processes is explored in the next three sections.

3.2 Component-based software versus traditional programs

The objective of software development is to ensure the high quality of a software product, such as high reliability, maintainability, reusability, and buildibility. To achieve these goals in component-based engineering, we first need to understand the characteristics of component-based software, its strengths, and weaknesses. Then we need to utilize these characteristics in component-based engineering practice. In this section, we first investigate properties of component-based software and then we briefly explore the differences between component-based software and traditional programs.

3.2.1 Properties of component-based software

Differing from traditional software systems, component-based software often demonstrates the following unique characteristics.

- *Heterogeneity:* Heterogeneity is one of the key features of component-based software. It allows component-based software to be built up from components that are implemented using different languages, executing on different platforms, even in multiple locations, over great distances. Some components may be developed in-house, while others may be third-party off-the-shelf components. On the one hand, this feature supports a high degree of expandability and flexibility; on the other hand it requires high interoperability of the components.
- *Source code availability:* When developing component-based software, developers prefer, rather than implementing the code, to adopt COTS components whenever suitable components are available. These COTS components are often delivered in binary form and the source code is not available. This feature may facilitate the development process, but it also requires additional efforts in properly designing, integrating (component selection, customization, composition), testing, and maintaining the system.
- *Evolvability:* The plug-and-play feature of component-based systems allows components to be dynamically added, removed, or upgraded without being recompiled or reconfigured. Component-based software includes components from diversified sources, running in various environments. More frequent upgrades are anticipated for component-based software than traditional programs. The evolvability characteristics of component-based software can ensure that the efforts to maintain

evolving systems will be held to the lowest level, while still guaranteeing its quality.

- *Distribution:* With the development of the Internet, more and more component-based software is now distributed across networks. Different components will be designed, developed, and integrated in a distributed environment. Therefore, component-based software often inherits both the strengths and weaknesses of distributed systems.
- *Reusability:* One of the main objectives of component-based software engineering is to promote software reusability, so that it can improve the quality of future products and, at the same time, reduce development costs. Reusable artifacts in component-based software include individual software components, such as user interface, business logic, and data management components, as well as component architecture.

When adequately utilized, these component-based system features of the software development process not only bring great benefits to the software development team but increase the quality of the software products as well. The benefits are:

- *Reduced time to market and reduced cost:* The development of component-based software not only adopts reusable components and COTS components, but can also include reusable component architecture patterns as well. Therefore, less time may be used in determining the architecture and components of a system in the requirement analysis and design phases, as well as developing software components in the coding phase. Less time will be used in the testing phase for individual components as well. Thus, reduced time to market and reduced cost are expected.
- *Higher-quality product:* Quality attributes include improved reliability, maintainability, evolvability, interoperability, and performance, as well as many others.
    - *Reliability:* Reliable components themselves do not guarantee the high reliability of an integrated component-based software system. A reliable component-based system has to be built by following a component-based engineering process. Therefore, a component-based system combines reusable component architecture patterns and adequate reusable or COTS components, which all have demonstrated high reliability. Thus, high reliability of the system can be expected. In addition, component models integrate many complex user tasks, such as resource management and thread control. Thus,

there is less implementation complexity for developers, contributing to the higher reliability as well.

- *Maintainability, interoperability, and evolvability:* Maintenance activities can be classified into corrective, perfective, adaptive, and preventive activities. First, in component-based software, interactions among components have to go through well-defined interfaces. Therefore, the effects of corrective changes will be controlled in a minimal scope. Second, when perfective and adaptive activities are performed, the new system can be used without recompilation and reconfiguration. On the other hand, the adopted component model manages many interoperability and evolvability issues. Hence, efforts of managing perfective and adaptive activities can be minimized.

- *Performance:* The performance of a component-based software can be managed at a different level in a component-based system. Most importantly, the component architecture should be carefully designed to ensure this performance. If we adopt a reusable component architecture with good performance, the performance of the new system can be guaranteed. If the performance of individual components needs to be improved, this can be done independently by the component providers. Moreover, during the component selection and customization phases, performance of individual components and the component-based system can be precisely evaluated. Therefore, early performance analysis can reduce the risk of poor performance from the final product. In addition, the component model may also potentially improve the performance. The component model is able to manage the life cycle of components and various resources within the system, such as database connections and thread control. Utilization of these services can improve the performance of the final system.

- *Complexity is managed better, thereby improving the quality of the solution*: The complexity of a component-based system can be managed at different levels: component level, component architecture level, and component model level. Moreover, different levels of management do not affect each other. At the component level, the focus is on deriving efficient, reliable, and reusable components; while the component architecture level focuses on overall interactions among components. Component models also play important roles in managing the complexities. Component models not only provide frameworks for the improvement of interoperability and evolvability, they also provide many user functionalities such as resource management, access control, and concurrent control, thus further reducing the complexity of the component-based software.

- *Better utilization of human resources:* When developing component-based software, many engineering activities can actually be done in parallel, which can potentially reduce development time and cost. Also, when developing component-based software, with proper management, different types of functionalities can be isolated in different components, so the talents of individual engineers can be better utilized. As a result, people who are good at user interface design, business logic, and data management, can work independently on their own area of specialty; as a result, an increase in productivity can be expected. And more importantly, the cost will be reduced, as people with multiple skills are not required.
- *High reuse for future projects:* When following a component-based engineering process, many different artifacts in the system can be reused—for example, different components and component architectures. Anything that is newly developed in a component-based system can potentially be reused in future systems. When more and more components are available for reuse, more and more systems can be built with greater efficiency and effectiveness.

Nothing comes without a cost. The above characteristics of component-based software certainly benefit component-based software development in many aspects. At the same time, they also introduce many new challenges. Here are some of the difficulties we face.

- *Interoperability issues:* Software components tend to be built via various technologies and under different platforms. When composing various components together, incompatibility issues may occur at different levels, and they may generate unexpected behavior.
  1. *System level interoperability issue:* In a component-based system different components may be built under different infrastructures—for instance, different operating systems or different sets of system libraries. The incompatibilities among those infrastructures may cause problems in component-based systems.
  2. *Programming level interoperability issue:* When components are written in different programming languages, incompatibility among programming languages may cause failure. For example, one of the incompatibility problems that is often encountered is due to the different float value processing of VC++ and VB.

3. *Specification level interoperability issues:* Different developers may misinterpret specifications, and there are many different ways that the specifications may be misunderstood:

   - *Data misunderstanding:* data that passes through the interfaces is misunderstood—for example, misunderstanding the types and values of input parameters or of return values.

   - *Control misunderstanding:* the patterns of interactions may be misinterpreted. This includes misunderstanding the execution sequences of interfaces. For example, many components have an initiating interface that has to be invoked prior to invocations of other interfaces, and failure to do so may cause failures.

- *Development time and effort for software components may increase significantly:* If new software components are needed, development efforts for these new components will be much greater than those for traditional modules. This is because component design, coding, and testing has to conform to the entire component architecture. In addition, it has to be adaptive to changes and demonstrate high reliability.

- *Reusable software components may be more complex:* When developing reusable components, developers have to consider all possible scenarios. Therefore, reusable components are prohibitively more complex than those components that deal with specific contexts, because additional constraints and assumptions will simplify the implementation. Also, the complexity will lead to extra operations, which may jeopardize the performance and reliability of software components, and consequently the whole system.

- *Lack of a standard:* Right now, there is no standard that specifies what is an adequate component-based software engineering process, what is the standard for component models, what is the standard of a component, and so forth. Therefore, we can only take the advantage of many of component-based software characteristics, such as evolvability, interoperability, and maintainability, in a limited scope. When applying component-based principles in a diversified context, unexpected behavior may occur.

3.2.2 Component-based software versus traditional programs

The previous section investigated characteristics of component-based software. In this section, we explore some major differences between component-based software and traditional software. Table 3.1 summarizes some of these major differences.

For traditional systems, the main objective is to deliver an efficient and reliable software system. Reusability, maintainability and evolvability are characteristics that are not among the top priorities. Nevertheless, as software systems get larger and larger, and more and more complex, these characteristics become more critical. Component-based systems must properly address these issues in order to achieve the following objectives: shorter time to market, lower cost, better reusability, higher reliability, and maintainability.

Another major difference between component-based software and traditional software is the roles played by the development teams. For traditional systems, very often, only one team is involved in the software engineering process, from requirement analysis to software maintenance. Therefore, that team has full control of everything. To the contrary, for component-based software, two different roles are involved in the engineering process, component providers and component users. Very often they work at different places, under different development environments. On the one hand, this can achieve parallel processing, and thus reduce time to market and cut the cost of software development. On the other hand, the component users do not have full control of the entire engineering process, which can introduce difficulties in component composition, selection, testing, and maintenance, as well as many other activities. For other differences, please refer to Table 3.1.

Table 3.1 Component-Based Software Versus Traditional Software

Perspectives	*Component-Based Software*	*Traditional Software*
Objective	To efficiently develop reliable, maintainable, reusable software systems	To develop efficient, reliable software systems
Infrastructure	Besides operating systems and programming languages, certain component models are needed. For instance .NET/COM+, EJB, or CORBA	Only need the support of operating systems and programming languages
Process	Component-based software engineering process—includes many unique activities such as component selection, customization, composition	Traditional software engineering process—includes requirement analysis, design, coding, and testing
Roles	Component providers Component users	Usually, only one development team participates in the entire process.
Portability	Relatively easy to achieve, because the adopted component model and component architectures will reconcile many incompatibility and interoperability issues	Difficult to achieve as the entire engineering process is often for one specific environment

Table 3.1 Continued

Perspectives	*Component-Based Software*	*Traditional Software*
Interoperability	1. Communications among components will go through interfaces and are managed by component model. Component model can manage some of the interoperability issues for you 2. When combining varied components, interoperability issue is more serious	Very often, traditional software systems are developed in a restricted context, and there are less interoperability issues. Once there are, it is hard to achieve as modules and subsystems are context dependent
Maintainability	1. When producer provides newer version, system can be easily upgraded 2. Less control over how and when components are maintained 3. May introduce more overhead when source code is not available 4. Versioning is handled by component providers independently, so potential conflicts may exist	1. When maintaining the software, any modification may require reconstruction of the entire system 2. Have full control of any maintenance activity.
Reusability	Reusability can occur at different level: component, component architecture	Difficult to reuse; for some OO systems, the reuse is limited to individual classes
Scalability	Easy to scale: the scalability can be automatically managed by component model	Hard to scale as any change in the code may require changes to all relevant modules or subsystems
Structure	Loosely coupled Often distributed	Usually tightly coupled and centralized system
Code availability	Some component source code may not be available	Source code is usually available
Customization	1. Many components are general- purpose components and need customization before composition 2. Introspection mechanism provided by the component systems to explore available services	When integrating modules into the final product, no customization is needed

3.3 Component-based software infrastructure: component model

A component model is the backbone of a component-based system. It provides critical support for component development, composition, communication, deployment, and evolvement. The component model is a key factor to achieving interoperability, evolvability, maintainability, as well as many other quality attributes. Currently, .NET/COM/COM+, CORBA, and EJB are three major commercial component models.

Enterprise Java Bean [8], developed by Sun Microsystems, is a component model for server-side component-based software construction. The Java bean is the basic reusable component in EJB. Java beans are classified into session beans and entity beans, where session beans model a business process and entity beans model data in underlying storage. The key mechanism that manages beans and their communications is the EJB container or EJB server. The EJB container not only provides services such as bean life cycle management, name service, and persistence, it can also provide transaction support security, load balancing, and many other services.

.NET [9] is the brand new architecture from Microsoft that supports component-based software development. .NET expands its services in many different ways as compared to COM/COM+, which is discussed later. The most significant improvement is the interoperability issue. The adoption of the .NET framework as a computing platform for different programming languages and the adoption of XML as the means for communication among different components greatly improve the interoperability issue that has hung over Microsoft for a long time. On the client side, .NET supports all kinds of devices, from Windows CE to Windows XP. Over the server side, all kinds of enterprise services have been integrated—for example, SQL server, BizTalk server, and commerce server.

The *Component Object Model* (COM) [10] is an older architecture for component-based software that was first proposed by Microsoft in 1995. The first COM model included only basic features for component integration, packing, and a binary standard to ensure the interoperability among different components. DCOM, an extension of COM, provides the ability to integrate components in a distributed environment. COM+ is an enterprise solution from Microsoft. Besides COM and DCOM, COM+ also includes additional features such as *Microsoft Transaction Services* (MTS), *Microsoft Message Queue* (MSMQ), Microsoft *Active Data Objects* (ADO), and Microsoft Cluster Server (wolfpack).

Common Object Request Broker Architecture (CORBA) [11], as defined by OMG, provides a common framework to integrate components regardless their different locations, programming languages and platforms. A critical part of CORBA is the *object request broker* (ORB), which provides the basic mechanism to achieve the transparency characteristic. In addition, another part of CORBA, object service, provides some advanced services such as persistency and transaction support.

We have summarized the basic characteristics of the three component models. To determine which model to adopt for component software, we depend on current system configurations. We also depend on many other services that are provided by component models, such as persistence, state management, security, and load balancing. Detailed information on the comparisons among COM+, EJB, and CORBA is shown in Table 3.2. In general, .NET and COM+ are component models for a more restricted

Table 3.2 Comparisons of .NET, COM+, EJB, and CORBA

Perspectives	*.NET*	*COM+*	*EJB/J2EE*	*CORBA*
Platforms	Windows CE – Windows XP	Win 2000 or XP	All	All
Programming language	Neutral	Neutral	Pure JAVA	Neutral
Basic component	.NET component	COM component	Java beans	CORBA objects
Distribution	.NET framework	DCOM	RMI	CORBA IIOP
Enterprise solution	.NET enterprise servers	COM+	EJB/J2EE	CORBA services
Transaction support	MTS, MSDTC	MTS, MSDTC	JTS,JTA	CORBA transaction support
Portability	Partial (only among Windows)	No	Yes	Yes
Interoperability	Yes	Yes	Yes	Yes
State management	Stateless model	Stateless model	Stateful and stateless bean	Five-level state management: method, transaction, request, session, global
Persistence	ADO .NET	ADO or OLE-DB (API)	Entity bean or session bean, automatic persistence support	CORBA components package
Database caching	ADO RecordSet has been split into smaller pieces: DataReader, DataSet and DataAdapter.	Only ResultSet caching	ResultSet caching Data object caching	—
Message model	MSMQ	MSMQ	JMS	CORBA IIOP
Security	SSL + SSPI PASSPORT	SSL + SSPI	SSL + JAAS	Common secure interoperability, CORBA security service, resource access decision facility
Scalability and load balancing	.NET framework provides services that support scalability	Component server maintains resource pool to support scalability, load balancing	EJB server/ EJB container provides pooling for database connections and EJB instances	CORBA provides more flexibility in manage component lifetime; it can mange it in many different levels
Performance	Excellent	Excellent	Very good	Very good
Cost to own the technology	Expensive	Expensive	Cheap	Cheap
Tool support	Excellent	Excellent	Very good	Very good

environment—in other words, the Microsoft development environment. The environment includes Microsoft-compatible hardware, operating systems, programming languages, and development tools. The EJB and CORBA component models provide the customer with more choices, more hardware, as well as operating system and programming language choices, making it much easier to implement and evolve the system. On the other hand, because COM+ and .NET are restricted to certain development environments, it has more potential to improve the performance and reliability, as it can be implemented in platform-dependent and programming language–dependent fashion. Platform and programming language–dependent features are then utilized to improve performance and reliability. At the same time, interoperability would be a less serious problem. For EJB/CORBA component models, in order to achieve better interoperability and evolvability, container or middleware technologies are adopted, and inevitably communications will have to go through these containers or middlewares, and thus performance degradation can be expected.

There is no simple rule to help you to make decisions about which component model to adopt. It is a process that involves balancing different functional and nonfunctional requirements, as well as the current system configurations. Nonetheless, no matter which model is chosen, improvement of quality can be achieved.

3.4 Engineering process for component-based software

Combining a group of reliable software components may not yield a highly reliable component-based software system. The key to success is the software process, "a framework for the tasks that are required to build high-quality software" [12]. Compared with traditional software engineering processes, engineering processes for component-based software include many unique activities. In this section, we first review some traditional process models, and then we discuss a process model for component-based software.

3.4.1 Process models for traditional software

For traditional software systems, different process models are adopted based on the nature of those applications. The *waterfall model* [12], the oldest process model, describes a sequential processing approach as shown in Figure 3.1. It includes analysis, design, coding, testing, and maintenance phases. Analysis is usually the first phase to be carried out for a large system; this specifies what the system will do. The design phase specifies how we are going to do it. After

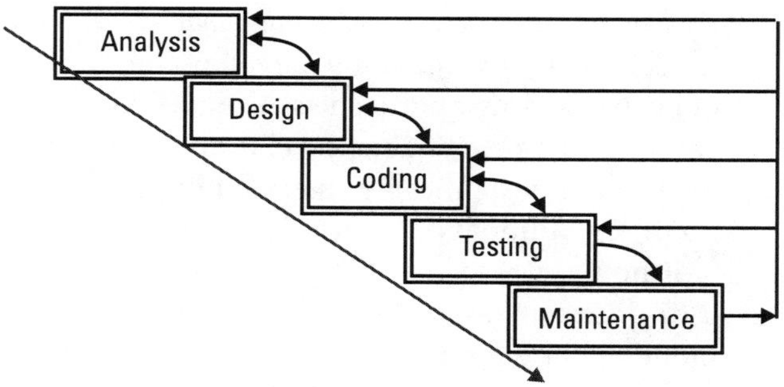

Figure 3.1 Waterfall model.

implementing the system in the coding phase, we need to go through a series of testing efforts, which include unit, integration, and system testing. Once the final system is delivered to the customer; maintenance activities are then inevitable. Maintenance activities include corrective, perfective, adaptive, and preventive activities. Each activity may require changes in requirement, design, or source code. Therefore, another iteration of similar processing might be necessary.

Even though the waterfall model has been successfully used for many software projects, limitations of the model have become more and more serious as software has become more and more complex:

- Software requirements are not properly tested until a working system is available.
- A working system will not be available until late in the process.
- The model requires the complete requirement during the analysis phase, which might be difficult to achieve. In addition, the waterfall model accommodates changes with great difficulty.

To overcome the difficulties of the waterfall model, different models are proposed, which address those issues.

The *throwaway prototyping model* [13] is adopted to bridge the gap in communications between customers and developers. Throwaway prototypes can be developed after the preliminary requirement is available. This type of prototype is used to clarify user requirements. Another type of prototype can be developed after the preliminary design is finished. This prototype can verify performance and other quality characteristics of the system. Generally, throwaway prototypes should not evolve into the final product.

The *evolutionary prototyping model* [13], a type of evolutionary process model, will incrementally develop the system into the final product. The objective of this approach is to have a subset of the work system available early in the process. Therefore, functional and nonfunctional requirements, as well as the design, can be verified early in the process. With proper management, evolutionary prototypes can evolve into the final product.

The *spiral model*, a risk-driven evolutionary process model, can be divided into different tasks, such as planning, risk analysis, and development. The model will iterate through these tasks until the final system is complete. During each iteration of the spiral model, risks will be analyzed and managed in order to accommodate the evolutionary nature of the software engineering process.

The *unified software development process* (USDP) is a "use case driven, architecture centric, iterative, and incremental process" [14]. USDP utilizes UML, an industry standard for the analysis and design of object-oriented software systems, to develop software systems. The life cycle phases of this process consist of inception, elaboration, construction, and transition. Each phase includes many iterations, and the core workflow for each iteration is requirements, analysis, design, implementation, and testing.

3.4.2 A process model for component-based software

Process models that are described in the previous section can be adopted for component-based software. Nevertheless, reusability, maintainability, and other major quality characteristics of component software have not been adequately addressed. To properly incorporate component-based software features into the processing model, we need to address this model in two aspects: a processing model for software components and a processing model for component-based software. In Chapter 1, a processing model for software components from the component providers' points of view was discussed. In this section, we focus on a processing model for component-based software. Component-based software will rapidly evolve over its lifetime, so a processing model for component-based systems has to be an evolutionary model. To properly address the evolvability issue, the evolutionary prototyping model, spiral model, or unified software development process model can be adopted. Here we will focus on the derivation of a proper workflow for each iteration in the process model.

Generally speaking, the overall engineering process for component-based software can be divided into the following phases, as shown in Figure 3.2 [5, 7]: requirements, analysis and design, and implementation. These phases include component selection, customization and composition, testing, deployment, and maintenance.

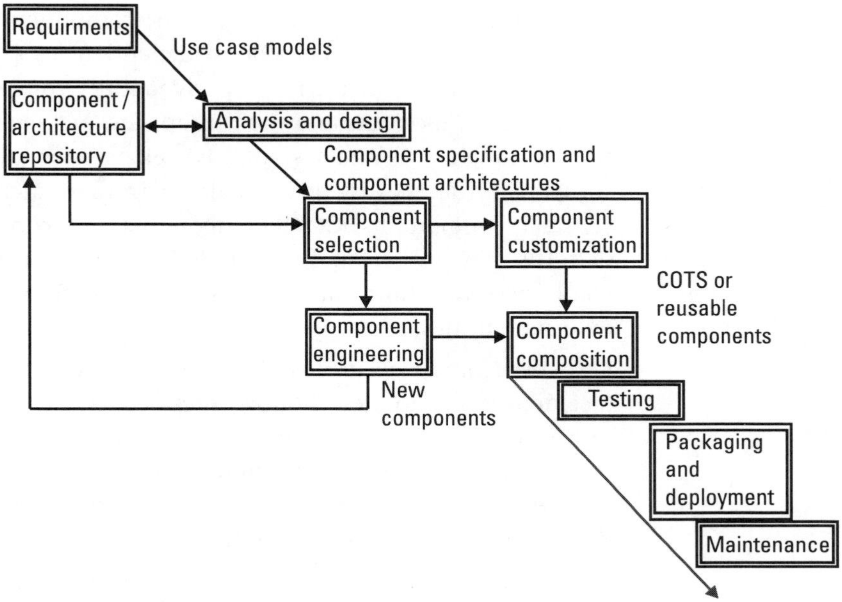

Figure 3.2 Engineering process for component-based software.

3.4.2.1 Requirements

Requirements analysis will derive specifications for the system under development. During this phase, we follow conventionally required engineering approaches; and the UML use-case model is a typical document of this phase.

3.4.2.2 Analysis and design

This is one of the most important parts of our process model. In this phase, software components and their specifications will be determined. The interactions, dependencies, and structural relationships among these software components will be decided as well. Differing from conventional analysis and design models, component-based analysis and design are significantly affected by the repository of software components and component architecture patterns. To adequately address these new changes, we further divide the process into the following steps. Note, that these steps are not totally isolated from each other. They are iterative and interactive.

- *Component and component architecture identification.* During this phase, the component architecture and an initial set of interfaces and component specifications will be determined. *Component architecture* is a set of application-level software components, component structural relationships, and component behavioral dependencies [1]. The emphasis at this stage is on discovery. To properly identify component architectures, candidate components and their interfaces, we need to analyze interactions among actors and systems, as well as interactions among internal software artifacts. Instead of starting from scratch, the repository of reusable components, COTS components, and component architecture patterns can provide helpful guidelines to determine what can be a component and how components interact with other components and external entities.

- *Component interaction identification.* This is a process to refine existing interface definitions, to identify how interfaces will be used, and to discover new interfaces and operations. After we derive an overview of the component architecture and all candidate components, in this step, we need to identify a complete architecture model and a comprehensive interaction model among different components.
- *Component specification.* During this phase, we will precisely define the component interface and as well as the specifications for the dependencies among different component interfaces. Specifications of interfaces usually include operations that will be performed in the interfaces. For each operation, input parameters, return values, potential effect on components (postconditions), and some constraints that may apply (preconditions) are specified. Besides specifications for each individual interface, the specification for a component should also specify constraints and dependencies among different interfaces. For example, in an ATM system, a sequence constraint needs to be specified to assure initial interfaces will be invoked before accessing any other interfaces. Data dependence relationships can be specified to reflect data dependence relationships among different interfaces. For instance, a withdrawal interface depends on the deposit interface.

3.4.2.3 Component selection, customization, and development

Given the component architecture and component specifications, component selection is the process of finding matching components from the repository of reusable and COTS components. It is actually a much more complex process than what you might intuitively think. To determine whether two components match with each other, functional requirements, nonfunctional requirements, and deployment constraints all need to be considered. Therefore, the selection process is usually an iterative process, which includes the following subtasks: check the candidate list of components, access the potential components, and determine appropriate components [5]. The inquiry-the-candidate-list-of-components task obtains a list of potential components by walking through all the reusable and COTS components. Based on the list from the inquiry step, the access-the-potential-components task will thoroughly evaluate each candidate component. It examines the interface specifications, run-time requirement analysis, security features, and exception handling. If the result of the second step reveals more than one candidate component, determine-appropriate-components task will be activated to select the best-fit component.

Even though the component selection process might be able to find a group of candidate components, a direct adoption often requires a perfect match of the candidate components with component architecture, component specification, and external environment. Any mismatch in the above areas requires customization of components. When source code for a component is available, a direct modification of the source code will resolve conflicts. When source code is not available, a component wrapper is usually developed to bridge the gap.

If, after the process of component selection, no qualified candidate is identified, a new component will be developed according to the component specifications. Once the newly developed component is proven to be reliable, proper changes to the repository will be performed.

3.4.2.4 Component composition

Component composition is the process of integrating all available components together, which includes all newly developed components, customized reusable and COTS components. To properly compose these components, component architecture and a specific component model needs to be taken into consideration.

3.4.2.5 Testing

The testing of component-based software should be viewed in two aspects. The testing of individual components, which has been addressed in Chapter 2, and the testing of component-based software, which is discussed in the next and subsequent chapters. In general, once each individual component has been adequately tested, the testing of component-based software will mainly focus on interactions among different components. The major difficulty lies in the fact that many of the components are third-party components, where the source code is not available. In addition, different components may be developed under different platforms: using different programming languages and running at different locations. Interoperability, incompatibility, distributed features, and other characteristics also throw tremendous difficulties on the development of test model and framework. Different strategies to address these difficulties are explored later in the book.

3.4.2.6 Deployment

A component-based software can be used as a component for another system. Therefore, it is necessary to provide packaging and deployment capacity. In this phase, different configurations under different contexts need be verified.

3.4.2.7 Maintenance

A component-based system contains components from various sources, as well as various architectural styles and different component models. After the system is delivered, all of these artifacts may be changed when bugs are identified, technology evolves, or business logic changes. Therefore, maintainability is critical to the success of component-based software. However, maintenance for component-based software may not be as easy as it looks, just plug-and-play. Many new challenges may occur. For instance, when a component provider changes the component, changes are transparent to component users, but some of the changes may affect the integrated system. Sometimes when new versions are added into the system, multiversions of the same service may exist in the software, which will not happen for traditional software systems. So the maintenance for component-based software focuses on how to adequately model the various modification activities and how to determine how much effort is adequate.

A well-defined quality assurance process is required to control and monitor the quality of the component-based software that is derived from a component-based engineering process. Information about component quality control for the component-based engineering process can be found in Part IV of this book.

References

[1] Heineman, G. T., and W. T. Councill, (eds.) *Component-Based Software Engineering: Putting the Pieces Together*, Reading, MA: Addison-Wesley, 2001.

[2] Weyuker, E. J., "Testing Component-Based Software: A Cautionary Tale," *IEEE Software*, Vol. 15, No. 5, September/October 1998, pp. 54–59.

[3] Hopkins, J., "Component Primer," *Communications of the ACM*, Vol. 43, No. 10, October 2000, pp. 27–30.

[4] Szyperski, C., *Component Software—Beyond Object-Oriented Programming*, Reading, MA: Addison-Wesley, 1999.

[5] Cheesman, J., and J. Daniels, *UML Components: A Simple Process for Specifying Component-Based Software*, Reading, MA: Addison-Wesley, 2001.

[6] Wallnau, K. C., S. A. Hissam, and R. C. Seacord, *Building Systems from Commercial Components*, Reading, MA: Addison-Wesley, 2001.

[7] Sparling, M., "Lessons Learned Through Six Years of Component-Based Development," *Communications of the ACM*, Vol. 43, No. 10, October 2000, pp. 47–53.

[8] Sun Micro, Enterprise JavaBeans Technology, http://java.sun.com/products/ejb/, 2002.

[9] http://www.microsoft.com/net/default.asp.

[10] Microsoft, COM+ Component Model, http://www.microsoft.com/com, 2002.

[11] Object Management Group, CORBA Component Model Joint Revised Submission, 1999.

[12] Pressman, R., *Software Engineering: A Practitioner's Approach*, New York: McGraw-Hill, 2001.

[13] Gomaa, H., *Designing Concurrent, Distributed, and Real-Time Applications with UML*, Reading, MA: Addison-Wesley, 2001.

[14] Jacobson, I., G. Booch, and J. Rumbaugh, *The Unified Software Development Process*, Reading, MA: Addison-Wesley, 1999.

CHAPTER

4

Contents

Testing component-based software

The objective of component-based software engineering is to efficiently and effectively develop high-quality software products. Under ideal situations, with minimal testing efforts, integration of reliable software components should produce high-quality software. In reality, however, as described in Chapter 3, many component-based software characteristics, such as heterogeneity, implementation transparency (source code unavailability), and evolvability, impose great difficulties on achieving this goal.

To thoroughly understand the difficulties and key issues in testing and maintaining component-based software, and thereby apply adequate strategies, we focus in this chapter on the answers to the following questions:

- What component-based software characteristics prevent us from achieving the objective of efficient and effective development of high-quality software products and why?
- How can we overcome these difficulties?
- Are we able to adopt traditional approaches?
- If we can, are those approaches sufficient?
- If not, what else should we do?

We first provide an in-depth discussion about why adequate testing and maintenance of component-based software are necessary. Then we further investigate the issues and difficulties that arise due to component-based software characteristics. To adequately overcome these issues and difficulties, in Section 4.3 we analyze the key factors in combining software components into an integrated software system. Then a test model is provided to serve as the basis for subsequent testing and maintenance activities. In Sections 4.4 and 4.5, brief discussions about how to utilize the test model to test and maintain a component-based software system are presented (more detailed discussion about this issue can be found in Chapters 9 and 10).

4.1 Introduction

A component-based system is an assembly of software components, which follows an engineering process that conforms to component-based software engineering principles. A CBS often consists of a set of self-contained and loosely coupled components that can easily be disconnected, reconnected, and exchanged. The components may be written in different programming languages, executed in various operational platforms, and when on the Web, distributed across vast geographic distances. Some of the components are developed in-house, while others may be third-party or COTS components, whose source code is usually not available to developers. These component-based software characteristics facilitate the development of complex systems by allowing integration of reusable components from different vendors and by providing flexibility for dynamic software evolution. At the same time, they introduce new problems in testing and maintaining component-based software systems as well [1–4].

To ensure the delivery of quality software, effective and efficient testing is a key part of software development. Test methodologies are generally categorized into two types: black box and white box. Black-box approaches, such as functional testing and random testing, do not require knowledge of the implementation details, but when applied to component-based software systems they may encounter problems similar to those found in the testing of traditional programs. Random testing can be performed only when the operational profile of the whole system is predictable and measurable, which excludes its applicability to many application domains. Functional testing has commonly been applied in system testing; however, complete functional testing is often infeasible because of the complexity of the actual combination of functions present in a system. Thus, white-box approaches are often used to complement functional testing. However, component-based software components often do not have the source available (*implementation transparency property*), and the architectural heterogeneity raises difficulties when traditional white-box techniques are used to test component-based software. These difficulties may be encountered for the following reasons: First, with implementation transparency the source code of some components is often not available, and in such circumstances white-box testing, which is code-based, may not be possible. Second, even when the source code of every component is available, components may be written in different programming languages. Performing white-box approaches such as statement, branch, data-flow, and mutation testing often relies on adding *instrumentation* to the source code in order to trace and record the parts of the program covered by the tests. Finally, components can easily be plug-and-play, and some third-party components may be frequently upgraded. White-box approaches that rely on coverage information obtained in regression testing from the re-executions of previous tests are not appropriate for evolving systems [4], since such information does not reflect the current version of the software.

So far, many studies have attempted to adopt traditional techniques such as data flow testing [5, 6], mutation testing [7], and graph-based testing for testing component-based software. Before we begin exploring these technologies, we need to alert readers that adequate testing and maintenance of component-based software are not as easy as they appear. To understand why they are difficult and how to overcome the difficulties, we start with a more detailed discussion covering the issues and challenges.

4.2 Issues and challenges of testing and maintaining component-based software

It is very common for people to have the following question in mind: the software components have been individually tested, many of these components have been deployed and have demonstrated high quality, so why are further testing efforts required to ensure the quality of the integrated system? In this section, we first discuss why adequate testing is necessary, and then we discuss what difficulties we face in trying to adequately test and maintain evolving component-based software systems.

4.2.1 Why is adequate testing for component-based software necessary?

Why does an integrated system of reliable software components need to be further tested? Some blame can be attributed to the lack of adequate components, which require a lot of reconfiguration or even code modifications of the original components; therefore, appropriate efforts to reassure the reliability of the modified components are necessary. But as time moves on, more and more software components will be available on the market. Nevertheless, even when "perfectly matched" components are integrated, the following issues may still be encountered during the testing and maintenance of component-based software, which again require further testing efforts.

4.2.1.1 Inconsistent infrastructure and environment

Heterogeneity is one key feature of component-based software. Heterogeneity may introduce incompatibilities among different programming languages, operating systems, databases, and external operational environments that are used to develop software components or support the execution of software components. For instance, the failure of the European Ariane 5 launch vehicle was due to a reused module in Ariane 4 failing to convert a 64-bit floating-point value to a 16-bit signed integer. Even though the module had been proved to be reliable in the old Ariane 4 environment, when applied to a new European Ariane 5 environment, absence of adequate testing failed to reveal the incompatibility failure, and ended in a catastrophe. Even though much effort is going in this direction, in order to solve the problem by defining a lot of different standards in an attempt to resolve different types of incompatibility issues, in the real world, backward compatibility is still needed and will still cause inconsistency issues in component-based systems for years to come. In another example, Microsoft has made its efforts in this direction by releasing the .NET framework. In this framework, they define an intermediate language: *Microsoft Intermediate Language* (MSIL). Other high-level programming languages such as Visual C++, Visual Basic, Visual J++, Java, and COBOL can be easily translated into MSIL, and then the .NET framework will interpret and execute the intermediate language code. With the .NET framework, components that are developed using different languages and under different platforms can be easily combined without worrying about the incompatibility issues. Nevertheless, so far, this will work only for components that are developed using the latest versions of these programming languages. For components that were developed using previous versions or some old version of a Microsoft operating system, such as Windows 95, the .NET conversion and transformation may not be able to be performed automatically, or even carried out at all.

4.2.1.2 Inconsistent interaction model

For many people, the integration of software components is just a process that is similar to adding procedure calls. For many simple cases, it is true that software components can be viewed as a point of service. In other words, a software component will receive requests from clients and produce corresponding replies. For more complex software components, control protocols and data models need to be taken into account. Control protocols specify how different components interact. Their interactions can go through a procedure call-like style, or their communications can go through brokers or other mechanisms. Data models define the contents and format of the interactions in the control protocol. Usually, software components are developed by different groups of people, who may make different assumptions of how components interact and what details are involved in their interactions. Moreover, a component may expose multiple interfaces, which may have varied constraints and different types of relationships with each other. For instance, an ATM server's components may expose verification, deposit, withdrawal, query, and logout interfaces. Among these interfaces, instead of allowing arbitrary invocation sequences, a verification interface has to be invoked at the very beginning of each transaction, and logout has to be performed at the end of each transaction. Different invocation sequences may generate different results.

4.2.1.3 Other factors

As component-based systems are always built under a distributed environment, which will then inherit all the issues of distributed systems, such as race conditions and deadlocks. These distribution related issues can only be detected during the integration phase. Moreover, component-basing may even introduce new issues such as the multiversion issue, which is caused by the coexistence of two different versions of a component in the system.

In short, adequate testing and maintenance of component-based software are definitely required.

4.2.2 Difficulties in adequate testing and maintenance for component-based software

The previous section discussed why adequate testing and maintenance are important for component-based software. We know it is important, but what prevents us from performing adequate testing and maintenance? What are the challenges? In the following sections we list some major factors that affect our testing and maintenance activities.

4.2.2.1 Code availability

With more and more COTS components becoming available, more and more systems will adopt third-party components. For COTS components, source code is very often not available, and the interaction among components will have to go through predefined component interfaces. The lack of source code causes a lot of problems. First of all, all white-box based testing and maintenance techniques are unavailable. Only black-box based techniques can be adopted. Even when black-box based approaches can be adopted, very often they are not able to provide enough confidence to the user. Moreover, many times black-box techniques are difficult to perform due to the lack of precise specifications for the software system, or other knowledge that is required to use the technique. For example, without an operational profile, adequate random testing cannot be performed. Second, without source code and corresponding specifications, other important quality assurance activities such as performance analysis and reliability analysis may not be able to accurately proceed. This is because the analysis of these quality features may rely on the results from the individual component. These results have to be adjusted according to the usage of this component in the integrated environment. Without the source code, this type of adjustment cannot be performed. Finally, without source, many other nontechnical issues may arise to affect the operation of your software product. For instance, if COTS components are maintained by another company, and that company goes out of business. It could affect the operation of your product when problems are encountered in that component. And even worse, there is nothing you can do to prevent it from happening. In the following, there are more discussions about the effects of source code unavailability.

4.2.2.2 Performance and reliability analysis

When testing a component-based software, correctness issues often draw most of the attention. Actually, the assurance of many other quality issues is also challenged in component-based software engineering. Performance is one of the quality features that are heavily affected by component-based software features. Heterogeneity allows components that are developed by different programming languages, under different platforms, to be integrated together. To achieve this goal, one of the very popular approaches is to standardize the communications among components. Standardization does help to reduce the incompatibility issues. Nevertheless, additional overheads are introduced in converting to and from standardized messages during the communication and end up with a degradation of performance. For instance, XML is quickly becoming the industry standard for modeling interactions among different software components. Compared with traditional systems, which use binary data and specific interaction models to communicate, XML not only introduces a tremendous amount of redundant data due to the various user defined tags, it also introduces more overhead in constructing and parsing

XML messages through the XML parser. Another example is J2EE and .NET, which are the two most popular component models and both of which advocate enhanced interoperability. But behind the scenes, the interoperability is achieved by sacrificing performance in the Java Virtual Machine and .NET framework. Besides the performance overhead caused by standardization, generalization can also introduce overhead in performance. In the case of COTS components, they are usually developed to be used in a different environment; therefore, additional code, which is developed for configuration, error checking, and auditing, may be redundant for your environment. Sometimes, in order to utilize a component's functionality, you have to include all the other functionalities in that component. This can potentially jeopardize your performance, resource management, and maintenance. Besides the performance quality features, other features such as reliability and usability also need to be reexamined. To analyze these quality features of component-based software, a key issue is how to reuse the results provided by the software components. For instance, to evaluate the reliability of a component-based system, for each component, there could be some reliability results from the component developer. But, that result is obtained based on the component developer's environment and assumptions. Therefore, the results might not match those for the component user's environment. In general, how we pass the quality analysis results from component developer to component user is a major challenge in testing and maintaining component-based software.

4.2.2.3 Adequacy

Test adequacy is one of the toughest issues in testing component-based software. On one hand, not all traditional test adequacy criteria can be used, especially for those source code–dependent criteria. For instance, coupling-based criteria [8] such as all-coupling-def, all-coupling-uses, and all-coupling-paths, all require access to the source code of both caller and callee's programs. Without the source code, these types of criteria cannot be used. On the other hand, even though some black-box based criteria can be adopted, the issue is: Do we still want to apply those criteria? If we do, the overall cost of the development may not be effective, since many components have been used in other systems, and proved to be reliable. Using traditional strategies may duplicate much of the original work. More importantly, new features introduced by component-based software may require additional testing efforts, which will further increase the overhead. For example, interfaces are the only means of interactions, so how do we adequately test interfaces that are not addressed for traditional programs? The interoperability issue, which is a great concern in component-based systems, is not adequately addressed in the traditional domain. Some current solutions to these issues will be discussed in Chapter 9. If we decide not to apply the entire strategy and only apply some basic criteria, since many components have proven reliability, then how much testing is adequate and necessary to ensure the adopted components fit in the new environment?

Another challenge to the adequacy is that the adequacy should be applicable to the overall component-based system. If a software system includes components without source code, obtaining unified adequacy criteria is a challenge. In a component-based system, many different types of components are integrated, including newly developed components, reusable in-house developed components, and COTS components. Do we apply the same criteria or different criteria for all these different components? How do we measure different quality attributes for this system?

4.2.2.4 Maintenance

Component-based software engineering has been increasingly adopted for software development. Such an approach using reusable components as the building blocks for constructing software, on one hand, embellishes the likelihood of improving software quality and productivity; on the other hand, it consequently involves frequent maintenance activities, such as upgrading third-party components or adding new features. The cost of maintenance for conventional software can account for as much as two-thirds of the total cost, and it can be even more for maintaining component-based software. Thus, an effective maintenance technique for component-based software is very desirable.

When a component in a component-based software is modified or upgraded, a maintenance activity occurs [9, 10]. Due to many of the characteristics of component-based software, difficulties can be encountered when traditional maintenance approaches are applied. For example, in a component-based system, many third-party packages are adopted and the source code of these components is often unavailable. A traditional code-based approach that relies on the analysis of program code may not be applicable. In addition, adoption of a third-party component also includes the risk of depending on other components that are beyond your control. If the company that supports the component goes out of business or stops supporting their legacy systems, then your system is in jeopardy. Moreover, since component-based systems may be heterogeneous, some faults that are related to interoperability may be encountered, and most traditional approaches do not take interoperability into consideration. Beside these difficulties, there are some other issues that are quite unique in the component-based domain. For instance, when upgrading a component, unlike the traditional software system in which the old piece will stop its service when the new one is in place, in a component-based system the old component can still provide its service; as is the case if you upgrade Microsoft COM objects, old and new components can coexist in the system. This actually happens quite often in the real world as many people get used to their old system, and are reluctant to transfer to a new system. The issues that occur when two versions of the same component coexist at the same time in the same system are novel and quite a challenge for the maintenance of component software. The detailed strategies of how to tackle these issues will be discussed in detail in Chapter 10.

4.3 Testing model for component-based software

To date, there are many different component models, and new models and technologies for enhancing component-based software engineering are emerging and changing everyday. To efficiently, effectively, and adequately test and maintain component-based software, we require the following two abstract models to represent characteristics of component-based software, regardless of the component models and technologies:

- *The interaction model,* which depicts how software components interact with each other within a component-based software;
- *The behavior model,* which depicts component internal structure and behavior. The behavior model can be used to carry component information, such as internal control structure, reliability information, and performance information. The information then can be used in the testing and maintenance of integrated component-based software systems.

4.3.1 Interaction model

The interaction model emphasizes the interactions among components in component-based systems and tries to detect as many interoperability faults, incompatibility faults, and other faults as possible. In a component-based system, a component is only able to interact with other components through interfaces that are exposed by those components. Nevertheless, the invocation can occur in different ways: direct invocation of the interfaces (procedure-call–like invocation), invocation by exception, or from a user action that triggers an event that invokes an interface. The first type happens in a predictable manner, while the rest happen in an unpredictable manner. Besides invocation, a component may also interact with other components indirectly through a sequence of events. Before we move forward, we reemphasize the definitions of "interface" and "event" in our context.

Interface: interfaces are the basic access means via which components are activated. Therefore, in integration and system testing it is necessary that each interface be tested in the integrated environment at least once.

Events: every interface that can be invoked after deployment must be tested at least once. This goal is similar to the traditional test criterion that requires every function/procedure to be tested at least once. However, when the same interface is invoked by different components within different contexts, there may be different outcomes. Thus, to observe all the possible behaviors of each interface during runtime, every unique invocation of the interfaces, the *event,* needs to be tested at least once. Normally, events can be classified into three kinds: normal interface invocation events, exception events, and user action events.

The issue of how to specify interfaces, events, and how to conduct detailed testing and maintenance will be discussed in Chapters 9 and 10.

Besides the direct interactions between two components, the overall interaction architecture of all components needs to be taken into account as well. The overall interaction architecture can be examined in two different ways: control structure and data interaction relationships.

4.3.1.1 Control structure

Interface and event testing ensure that every interaction between two components, client and server, is exercised. However, when execution of a component-based software system involves interactions among a group of components, not only the direct interaction between two components counts, the relationships among interactions need to be taken into account as well. To capture the interrelationships among various interactions, we define a control structure for a component-based system. A control structure can also be viewed as a control dependence relationship that is similar to the control flow dependence relationship in traditional software. Apparently, two different elements are involved in the control structure, interfaces and events; as a result, there will be four different types of control structures:

- *Interface–interface structure:* This is similar to a traditional procedure call, which within the execution of one interface, invokes another interface.
- *Interface–event structure:* This occurs when executing an interface. Exceptions are generated, which can be unexpected system behaviors or user defined exceptions.

- *Event–interface structure:* This simply indicates that when events are generated, as when user actions occur, certain interfaces are called.
- *Event–event structure:* This indicates that when one event is generated, it may trigger another event. An example of this type of control relationship is the closing of a console window, where the close-window event may trigger another close-window event of its subwindow.

The control structure will help users to clarify all the different interactions among the different components. Therefore, testing strategies based on this type of structure can help us in identifying interoperability faults, incompatibility faults, and many other types found in component-based systems.

4.3.1.2 Data interaction relationships

The control structure examines how interfaces are invoked, but it does not take the data relationships among invocations into consideration. For instance, in our previous example about the ATM server component, the query interface apparently depends on the withdrawal and deposit interface. The invocation of the query interface before and after the execution of the withdrawal or deposit interface will demonstrate different values. Data interaction relationships, on the other hand, will examine this type of relationship. The data dependence relationship is similar to that of data flow analysis in traditional programs, but we can investigate this type of relationship in different granularity. For instance, variable level, function level, and interface level.

To properly utilize these key elements in our subsequent analysis, an abstract notation of these elements is needed. Currently, many testing and maintenance techniques, as well as many other analytical approaches, such as performance and reliability analysis, are graph-based approaches. We can construct a component interaction graph to depict all these elements and utilize them for our subsequent analysis. In the graph, nodes are interfaces and events, while edges can reflect control and data relationships among different components. A detailed example can be found in [4].

4.3.2 Behavior model

In the interaction model, various key elements are discussed. But how are those elements derived? For COTS components where source code may not be available, how do we specify these elements? Without source code, how do we specify the behavior of a component?

To answer these questions, we need a behavior model to specify the interfaces, events, and control and data structures of a component when source code is not available. Even with source code, the behavior model can present characteristics of a component in a more abstract manner, which therefore can be processed more efficiently. More importantly, this model can provide a unified way to evaluate test adequacy and other quality attributes.

Based on our previous discussion, to derive a behavior model, we need a tool that has the following capabilities:

- The capacity to effectively and precisely specify the internal behavior of a component;
- Widely accepted, which means, when combining components from different vendors, minimal efforts of integration are required and incompatibility issues can be managed within a minimal scope;
- Easily obtainable, which means, when preparing the behavior model by the component providers, overhead can be managed at a very low level;
- Easily evolvable, which means, as components can be easily added, deleted, or upgraded, component models may evolve frequently, the tool should be able to be easily upgraded to reflect those changes.

Fortunately, UML can serve exact the role as we specified earlier. UML [11, 12] is a language for specifying, constructing, visualizing, and documenting artifacts of software-intensive systems that can be used. There are several advantages to adopting UML. First, UML provides high-level information that characterizes the internal behavior of components, which can be processed efficiently and used effectively when testing. Second, UML has emerged as the industry standard for software modeling notations, and various diagrams are available from many component providers. Third, UML includes a set of models that can provide different levels of capacity and accuracy for component modeling, and thus, can be used to satisfy various needs in the real world. For instance, in UML, collaboration diagrams and sequence diagrams are used to represent interactions among different objects in a component. Statechart diagrams, on the other hand, are used to characterize internal behaviors of objects in a component. Based on the statechart diagram, we further refine the dependence relationships among interfaces and operations that are derived from collaboration diagrams. Finally, UML is extensible, which means that if additional artifacts are found to be important to component-based software systems, they can be added into UML.

4.4 Testing and maintenance methodologies

The test model has provided us with various "ingredients" for testing and maintenance, the detailed discussion about how we utilize this information and what are the criteria to determine test adequacy will be found in Chapter 9. Here we just provide an overview of the different strategies we are able to adopt.

- Interfaces and events are the fundamental elements that need to be tested. To test interfaces, the first issue is how to specify interfaces. Interfaces can be specified in many different ways. Some people specify them through the statechart, and some through interaction protocols, preconditions, postconditions, and input and output parameters. Based on the specifications, various testing strategies then can be applied. These approaches include mutation testing [7], formal methods such as model-checking, and many other approaches [13, 14].
- The testing should not only focus on interactions between two components. More importantly, architectural level interactions among different components need to be examined [1, 13, 15]. According to our interaction test model, two types of elements need to be tested: control structures and data dependence relationships. To verify these two types of elements, some modified traditional approach such as mutation testing, data flow testing, and graph-based testing can be adopted; at the same time, many other novel approaches, such as architectural-based testing and scenario-based testing [16, 17] can be used as well. Detailed discussion about these approaches can be found in Chapter 9.
- When we test component-based software with COTS components and we do not have enough confidence in those components, the behavior model becomes necessary to specify the internal behavior of a component. It is then used to test whether those components fit in the new environment. As we have mentioned, the behavior model can include information about the component at different levels. Normally, collaboration diagrams are available to specify the internal control structure. Statecharts are sometimes available to specify the internal behavior, and class diagrams may be used to derive detailed interface specifications and their dependence relationships. To further rigorously characterize the constraints of a class, the formal language that is provided in UML, the *object constraint language* (OCL), can be adopted. For more critical scenarios, the OCL can be adopted to formally specify the behavior and constraints of a component as well.

Software maintenance activities can be classified into corrective, perfective, and adaptive maintenance. Normally, perfective and adaptive activities are the major maintenance activities. For component-based software, these two types of activities may occupy even larger shares. In the maintenance phase, despite different types of maintenance activities, two issues are always involved: (1) test selection from an existing test suite for reconforming program behaviors to those from before the modification, and (2) test generation for validating newly introduced features. To evaluate different techniques, different measurement models are proposed. Among different criteria, precision, efficiency, and safeness are the three major considerations. Precision is the measurement of the ability of a method to eliminate the test cases that do not need to be retested. Efficiency assesses the time and space requirements of the regression testing strategy. Safeness is used to indicate the ability to retest all test cases that have to be retested, and it is a very important criterion for life-critical systems. To maintain a component based software, according to our interaction and behavior model, different types of strategies can be performed. For instance, an efficient and safe strategy is to retest all test cases that execute at least one modified interface. But this strategy may include too many redundant test cases, which will decrease the precision and require longer reexecution time. On the contrary, an efficient but unsafe strategy is to select one test case to cover one modified interface. It is efficient, but may omit some test cases that could potentially do harm to the system. In the real world, which strategy to choose really depends on many factors such as delivery schedule and budget. Besides these traditional maintenance issues, version control and other novel aspects of component-based software require novel maintenance techniques as well [2, 6, 18–20]. Detailed discussion about this issue can be found in Chapter 10.

4.5 Enterprise-based test process for component-based software

For traditional software system, the general process includes unit testing, integration testing, system testing, acceptance testing, and regression testing. Actually, there are some unstated assumptions behind the process, which include, first, the assumption that source code is accessible throughout the process. Second, the entire process is maintained by one group of people, and therefore, interactions among different phases can be unified in management, and the development environment can be managed in a relatively stable and simple manner. On the contrary, for component-based software, many of the previous assumptions do not hold. This not only requires novel testing technologies, but requires adequate management processes as well.

For component-based software, two different roles are involved in the development process: component developers and component users. The process for testing and maintaining software components has been discussed in Chapter 2. Once software components are tested and delivered to component users for assembly, the first step for the component user is to make sure that the component has been configured for the new environment and demonstrates the required quality. The next step is integration testing, which makes sure the integration of these components will not adversely affect the component system. During the integration testing, an important issue is the order of integration. In traditional integration testing, either top-down, bottom-up, or both will be adopted, but the order actually relies on the function decomposition. This structure-dependent integration might not fit in the component-based context, as component-based software can demonstrate dynamic interaction among different components. Actually, our interaction and behavior model serves as a means to determine the incremental testing order. During the integration testing, it mainly focuses on functional testing. For system testing, other nonfunctional requirements such as performance requirements and reliability requirements will then be tested.

Software maintenance, traditionally, is managed solely by the development team, and it has always been conducted after the product is delivered. In the component-based context, this no longer holds. First, third-party components or even some reusable components may be maintained by another group of people, which consequently causes unexpected maintenance activities. It can occur at any time, not necessarily after the delivery of the product. Second, the development team might not be able to access the detailed information of the modification, and thus might not be able to follow the traditional routine for performing software maintenance.

4.6 Summary

This chapter looks over different issues and challenges in the testing and maintaining of component-based software. This will not only clarify many misconceptions about testing and maintaining component-based software, but also provide clues as to how we should perform testing and maintenance activities. Based on these observations, a testing model is then discussed. The model serves as a foundation for all subsequent testing and maintenance activities. In this chapter, we only briefly discuss basic testing and maintenance strategies; detailed information can be located in Part III of this book.

Despite of all these discussions, the reader should keep in mind that testing and maintaining evolving component-based software is a very difficult task. So far, for many issues, such as the interoperability issue, we are only able to provide approximate solutions. We still have a long way to go to achieve our initial goal.

References

[1] Garlan, C., R. Allen, and J. Ockerbloom, "Architectural Mismatch or Why It's Hard to Build Systems Out of Existing Parts," *Proc. of 17th International Conference on Software Engineering*, 1995, pp. 179–185.

[2] Voas, J., "Maintaining Component-Based Systems," *IEEE Software*, Vol. 15, No. 4, July/August, 1998, pp. 22–27.

[3] Weyuker, E. J., "Testing Component-Based Software: A Cautionary Tale," *IEEE Software*, Vol. 15, No. 5, September/October 1998, pp. 54–59.

[4] Wu, Y., D. Pan, and M. H. Chen, "Techniques of Maintaining Evolving Component-Based Software," *Proc. of 2000 International Conference on Software Maintenance*, October 2000, pp. 236–246.

[5] Harrold, M. J., D. Liang, and D. Sinha, "An Approach to Analyzing and Testing Component-Based Systems," *1st International ICSE Workshop on Testing Distributed Component-Based Systems*, Los Angeles, CA, 1999.

[6] Orso, A., M. J. Harrold, and D. Rosenblum, "Component Metadata for Software Engineering Tasks," *Proc. of 2nd International Workshop on Engineering Distributed Objects (EDO 2000)*, November 2000, pp. 126–140.

[7] Delamaro, M. E., "Interface Mutation: An Approach for Integration Testing," *IEEE Trans. on Software Engineering*, Vol. 27, No. 3, March 2001, pp. 228–247.

[8] Jin, Z., and J. Offutt, "Coupling-Based Criteria for Integration Testing," *Journal of Software Testing, Verification, and Reliablity*, Vol. 8, No. 3, September 1998, pp. 133–154.

[9] Cook, J. E., and J. A. Dage, "Highly Reliable Upgrading of Components," *International Conference on Software Engineering*, Los Angeles, CA, 1999, pp. 203–212.

[10] Orso, A., et al., "Using Component Metacontents to Support the Regression Testing of Component-Based Software," *Proc. of IEEE International Conference on Software Maintenance (ICSM2001)*, 2001, pp. 716–725.

[11] Booch, G., J. Rumabugh, and I. Jacobson, *The Unified Modeling Language Guide*, Reading, MA: Addison-Wesley, 2000.

[12] Fowler, M., and K. Scott, *UML Distilled*, Reading, MA: Addison-Wesley, 2000.

[13] Bertolino, A., et al., "An Approach to Integration Testing Based on Architectural Descriptions," *Proc. of 3rd International Conference on Engineering of Complex Computer Systems (ICECCS'97)*, IEEE CS Press, 1997, pp. 77–84.

[14] Liu, W., and P. Dasiewicz, "Formal Test Requirements for Component Interactions," *Proc. of the Canadian Conference on Electrical and Computer Engineering*, Edmonton, Canada, May 10–12, 1999, pp. 295–299.

[15] Kozaczynski, W., et al., "Architecture Specification Support for Component Integration," *Proc. of 7th International Workshop on Computer-Aided Software Engineering (CASE'95)*, IEEE CS Press, 1995, pp. 30–39.

[16] Kim, Y., and C. Robert, "Scenario Based Integration Testing for Object-Oriented Software Development," *Proc. of 8th Asian Test Symposium (ATS '99)*, 1999, pp. 283–288.

[17] Yoon, H., B. Choi, and J. O. Jeon, "A UML Based Test Model for Component Integration Test," Workshop on Software Architecture and Components, *Proceedings on WSAC'99*, Japan, December 7, 1999, pp. 63–70.

[18] Mezini, M., and K. Lieberherr, "Adaptive Plug-and-Play Components for Evolutionary Software Development," *Proc. of Conference on Object-Oriented Programming, Systems, Languages, and Applications*, 1998, pp. 97–116.

[19] Vidger, M. R., W. M. Gentleman, and J. Deans, "COTS Software Integration: State-of-the-Art," available at http://wwwsel.iit.nrc.ca/abstracts/ NRC39198.abs, 1996.

[20] Wu, Y., D. Pan, and M. H. Chen, "Techniques for Testing Component-Based Software," *Proc. of 7th International Conference on Engineering of Complex Computer Systems (ICECCS2001)*, IEEE CS Press, 2001, pp. 222–232.

CHAPTER

5

Testability of software components

Contents

Over the past decades, people have learned that it is impossible to conduct an exhaustive testing process for software products. To increase the quality of software products, we need:

- Cost-effective test generation methods;
- Efficient test automation tools;
- Software components and systems with high testability;
- Effective ways to measure the quality of software and components.

This chapter discusses the software testability of components in the component-based software engineering paradigm. The chapters of Parts II and III of this book cover test methods, coverage criteria, and tools for testing software components and component-based programs.

The structure of this chapter is as follows. Section 5.1 reviews the basic concepts of software testability. Section 5.2 discusses the different factors and perspectives of component testability in component-based software engineering. Section 5.3 focuses on the issues and solutions in designing highly testable software components. In Section 5.4, component testability evaluation and verification is discussed, and the existing approaches are reviewed. Furthermore, component testability measurement and analysis methods are summarized in Section 5.5. Finally, a summary of this chapter is given in Section 5.6.

5.1 Basic concepts of software testability

Testability is a very important quality indicator of software and its components since its measurement leads to the prospect of facilitating and improving a software test process. What is *software testability*? According to IEEE Standard [1], the term "testability" refers to:

> the degree to which a system or component facilitates the establishment of test criteria and the performance of tests to determine whether those criteria have been met; the degree to which a requirement is stated in terms that permit the establishment of test criteria and performance of tests to determine whether those criteria have been met.

This definition indicates that software testability is a measurable quality indicator that can be used to measure how hard it is to satisfy a particular testing goal, such as a coverage percentage or complete fault eradication. In fact, it depends on our answers to the following questions:

- Do we construct a system and its components in a way that facilitates the establishment of test criteria and performance of tests based on the criteria?
- Do we provide component and system requirements that are clear enough to allow testers to define clear and reachable test criteria and perform tests to see whether they have been met?

After conducting a literature survey on software testability, we found a number of articles addressing this concept from different perspectives [2–9]. For example, R. S. Freedmen [2] defines his *domain testability* for software components as a combination of component observability and controllability. In his definition, "*observability* is the ease of determining if specific inputs affect the outputs - related to the undeclared variables that must be avoided," and "*controllability* is the ease of producing specific output from specific inputs - related to the effective coverage of the declared output domain from the input domain." In addition, B. Beizer in [10] provides some useful practical advice on improving testability from the perspective of the various testing techniques.

In the past, much research on software testing has concentrated on methods for selecting effective sets of test data based on program specifications, program structures, or hypotheses about likely faults. Testing tries to reveal software faults by executing the program and comparing the expected output with the one produced so that it can guarantee program correctness. Much effort has been put into answering the question: What is the probability that the program will fail? Software testing, however, cannot show the absolute absence of failure unless it is exhaustive. Therefore, if testing is to be effective at all, then nonexhaustive testing must be performed in a manner that offers to answer to some degree the question: What is the probability that the program will fail if it is faulty?

J. M. Voas and K. W. Miller [3] view *software testability* as one of three pieces of the software reliability puzzle. They pointed out that software testability analysis is useful to examine and estimate the quality of software testing using an empirical analysis approach. In their view, software testability is "prediction of the probability of software failure occurring if the particular software were to contain a fault, given that software execution is with respect to a particular input distribution during random black box testing." This not only enhances software testing by suggesting the testing intensity, but also estimates how difficult it will be to detect a fault at a particular location. Software testability here examines a different behavioral characteristic: the likelihood that the code can fail if something in the code is incorrect—the probability that the code will fail if it is faulty. It is not only concerned with finding sets of inputs that satisfy coverage goals; it is trying to quantify the probability that a particular type of testing will cause existing faults to fail during testing. The focus is on the semantics of the software, how it will behave when it contains a fault. Testability is hence the sensitivity to faults. A. Bertolino [4] provided a similar definition of software testability:

> The probability that a test of the program on an input drawn from a specified probability distribution of the inputs is rejected, given a specified oracle (which decides about the test outcome by analyzing the behavior of the program against its specification and thus raise the alarm when the program has entered an erroneous state) and given that the program is faulty.

It is clear that analyzing and measuring software testability during a software test process provides test managers and quality assurance people a useful reference indicator about the reliability and quality of a system and its components to help them make a product release decision. However, it is too late to enhance the testability of constructed software components and systems and to identify the problems of defined test criteria and test sets. In a practice-oriented view, therefore, we need to focus more on the following areas:

- *How to increase software testability by constructing highly testable components and systems.* In component-based software engineering, a component-based program is made of three types of software components. They are COTS components, in-house-built components, and newly generated components. Clearly, software testability depends on component testability. Therefore, increasing component testability is a primary key to improving and enhancing the testability of component-based software. As pointed out in [6, 7], increasing component testability is very important and a challenge to component developers. To achieve this goal, we must find effective software development methods to address component testability issues by providing and enforcing various artifacts in a component development process to facilitate component testing and operations in the testing phase. The major objective is to make sure that the constructed components can be easily tested, operated, and measured in both vendor-oriented and user-oriented component testing processes. The primary benefit is in reducing the software testing cost and increasing the testability of component-based software.

- *How to analyze, control, and measure the testability of software and components in all phases of a software development process.* As pointed out by S. Jungmayr [9], testability is the degree to which a software artifact facilitates testing in a given test context. In our view, software testability is not only a measure of the effectiveness of a software test process, but also a measurable indicator of the success of a software development process. It is directly related to testing effort reduction and software quality.

Poor testability of components and programs not only suggests the poor quality of software and components, but also indicates that the software test process is ineffective. Like other requirements and design faults, poor testability is expensive to repair when detected late in the software development process. Therefore, we should pay attention to software testability by addressing and verifying all software development artifacts in all development phases, including requirements analysis, design, implementation, and testing. Table 5.1 lists all types of tasks and activities relating to component testability in different phases of a software development process. These activities and tasks must be carried out based on a well-defined set of the standards and methods that provide clear requirements, guidelines, and criteria on how to increase, verify, and measure software testability in each phase.

Components and software with good testability not only increase the quality of software, but also reduce the cost of software development and testing. According to [9], software testing usually costs more than 40% of the

Table 5.1 Different Tasks and Activities for Component and Software Testability

Software Development Phase	*Tasks and Activities Relating to Component and Software Testability*
Requirements analysis	Define clear functional requirements that allow testers to define clear and achievable test criteria, and design and perform tests to see whether they have been met
	Specify measurable nonfunctional requirements to make sure that test criteria can be easily established and tests can be designed and performed to measure and validate these requirements
	Clearly address the requirements of software artifacts at the component and system levels to facilitate software testing and operations
	Review and evaluate the generated component and software requirements to make sure that they are testable and measurable
Software design	Conduct design for testability of software and components
	Design required built-in facilities and tools to support integration and software testing
	Establish design artifacts addressing software testability, such as design patterns, guidelines, principles, and standards for testable components
	Review and evaluate software design concerning testability based on the established design patterns, guidelines, and standards
Software implementation	Implement testable components and programs based on the design for testability
	Generate required software testing facilities and tools
	Review and verify program code based on the established coding standards addressing software testability
Software testing	Define clear and achievable test criteria based on requirements
	Design and perform tests based on well-defined test criteria
	Set up and develop required test-support facilities and environment to support software testing, operations, and automation
	Evaluate, verify, and measure the testability of components and software
Software maintenance	Update test criteria and tests according to software/component changes
	Update the built-in facilities and required test tools based on all updates
	Evaluate, verify, and measure the testability of updated components and software

software project budget during the life cycle of a software product. Thus, designing highly testable software components becomes a very important and challenging task for component developers. Similarly, verifying and measuring the testability and quality of software components is also very important and challenging for application engineers and component users.

5.2 Understanding component testability

What is software component testability? In the past, only a few articles addressed the concept of component testability. For example, R. S. Freedman [2] defines his component testability from a function domain. In his paper, he considers two important component factors (component *observability* and component *controllability*) when verifying and measuring component *domain testability*. His basic idea is to view a software component as a functional black box with a set of inputs and corresponding outputs. Measuring component domain testability becomes a task to measure component observability and component controllability. Observing and checking the specific inputs of tests and their effect on the corresponding outputs carry out the testability measurement. The focus of his paper is on the measurement of component domain testability.

R. V. Binder [5] discusses the concept of testability of object-oriented programs from six primary testability factors: "representation, implementation, built-in test, test suite, test support environment, and software process capability." Each of the six factors is further refined into its own quality factors and structure to address the special features in object-oriented software, such as encapsulation, inheritance, and polymorphism. Clearly, his concept of object-oriented software testability provides insights from a more structure-oriented view, instead of a black-box view.

Similar to software testability, *component testability* is two-fold. First, it refers to the degree to which a component is constructed to facilitate the establishment of component test criteria and the performance of tests to determine whether those criteria have been met. Second, it refers to the degree to which testable and measurable component requirements are clearly given to allow the establishment of test criteria and performance of tests. Therefore, studying component testability focuses on two aspects:

- Studying component development methods, guidelines, principles, and standards that construct testable and measurable software components;
- Studying verification and measurement methods that facilitate the verification and measurement of component testability based on established component test criteria and tests in a component development process.

As shown in Figure 5.1, J. Gao in [6] discussed component testability in terms of five factors: component understandability, observability, traceability, controllability, and testing support capability. We believe component testability can be verified and measured based on these five factors. In this section, we provide more detailed discussions about them.

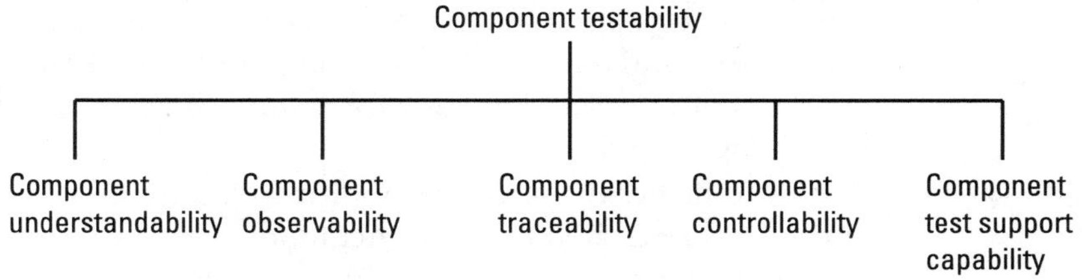

Figure 5.1 Dependent factors of component testability.

5.2.1 Component understandability

Component understandability depends on two factors. The first is the availability of component information. The other is the understandability of the provided component information. In other words, it refers to how well the component information is presented. A software component usually provides two sets of component documents. The first set is written for users. It include component user manual, user reference manual, component function and interface specifications. Component engineers write the others, which include component analysis and design specifications, component source code, testing documents, and quality report. Component testing documents include component test plan, test criteria, designed test cases, and testing environment information (i.e., component test scripts for component test drivers and/or stubs). Component quality information consists of component testing and problem reports, test coverage information, and quality metrics. Most current third-party components do provide users with a user manual and application interface specifications. Only some of them, however, provide a user reference manual, and most of them do not provide any information on component quality and test coverage. Although it seems reasonable for component vendors to withhold detailed test information and problem information, in the near future, customers will expect vendors to provide quality information, an acceptance test plan, and even test suites for components.

Component understandability is an essential factor for component testability because highly understandable components facilitate component test engineers and users in obtaining a better understanding of function requirements of components so that component test criteria can be easily defined and quality tests can be generated.

5.2.2 Component observability

According to R. S. Freedman [2], *software observability* indicates how easy it is to observe a program based on its operational behaviors, input parameter values, and actual outputs for a test case. The same concept is also applicable to software components to check component observability based on component interfaces to see the mapping relationships between the inputs and corresponding outputs.

According to Freedman [2], a software component is observable "if distinct outputs are generated from distinct inputs" so that, if a test input is repeated, the output is the same. If the outputs are not the same, then the component is dependent on hidden states not identified by the tester. Freedman calls this an *input inconsistency*. Clearly, his concern about component observability only focuses on the observation of a component's external function behaviors during testing in a black-box view.

In our view, component observability refers to the degree to which components are designed to facilitate the monitoring and observation of component functions and behaviors of component tests. Component observability is twofold:

- The first is to check the mapping relationships between inputs and corresponding outputs for each test. This could be done in two different stages. A test design review is the first stage, in which all component tests are verified to find the hidden inputs and missing outputs, and incorrect mapping relationships between inputs and outputs. Test result verification during component testing is the second stage, where an effective test tool could be an ideal solution to check this in a systematic manner.
- The other is to track and monitor all component tests and test results. There are two ways to achieve this goal. An effective component test bed is a natural solution. To reduce the testing cost in setting up a specific component test bed for each component, we need to find effective methods to construct testable components with good observability.

5.2.3 Component traceability

In the practice of component engineering, we found that component traceability is a very important factor that affects component testability. *Traceability* of a software component refers to the extent of its built-in capability that tracks functional operations, component attributes, and behaviors. Component traceability can be verified and evaluated in two different aspects:

- *Black-box traceability:* This refers to the extent of the component's built-in capability that facilitates the tracking of component behaviors in a black-box view, including external component interfaces, accessible function behaviors, external visible object states, and events. This is not only important to component developers during component testing, but also necessary to component users in component evaluation, component integration, and system validation.

- *White-box traceability:* This refers to the extent of the component's built-in capability that facilitates the tracking of component behaviors in a white-box view, including component internal function behaviors, business logic, object states, and program structure. This is very helpful for component developers in component testing and bug fixing during a component evolution cycle.

As discussed in [8], we must add a built-in component tracking code into components to increase component traceability. The objective is to generate the following types of program traces (see Figure 5.2):

- *Operational trace:* it records the interactions of component operations, such as function invocations. It can be further classified into two groups: (1) internal operation trace, which tracks the internal function calls in a component, and (2) external operation trace, which records the interactions between components. External operation trace records the activities of a component on its interface, including incoming and outgoing function calls.
- *Performance trace:* it records the performance data and benchmarks for each function of a component in a given platform and environment. Performance trace is very useful for developers and testers to identify the performance bottlenecks and issues in performance tuning and testing. Using performance traces engineers can generate a performance metric

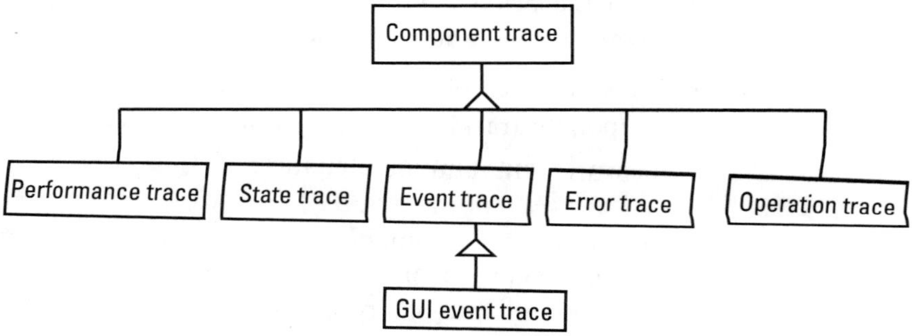

Figure 5.2 Different types of component traces.

for each function in a component, including its average speed, maximum, and minimum speed.

- *State trace:* it tracks the object states or data states in a component. In component black-box testing, it is very useful for testers to track the public visible objects (or data) of components.
- *Event trace:* it records the events and sequences that have occurred in a component. The event trace provides a systematic way for GUI components to track *graphic user interface* (GUI) events and sequences. This is very useful for recording and replay of GUI operational scenarios.
- *Error trace:* it records the error messages generated by a component. The error trace supports all error messages, exceptions, and related processing information generated by a component.

In the past two decades, software developers have learned that building programs with good traceability not only assist program testing and debugging, but also reduce a lot of software testing and maintenance cost. Today, however, component engineers and users have encountered a number of challenges relating to component traceability in their practice of component-based software engineering:

- Poor black-box traceability is due to the fact those current reusable components (including COTS components) do not provide enough component tracking capability for component users. This not only increases the difficulty of fault detection and problem isolation in component integration and system function validation, but also increases the cost of component integration and system development.
- Component developers use ad hoc component tracking mechanisms and inconsistent tracking message formats.
- There is a lack of well-defined standards in component tracking requirements, format, and solutions.
- There is a lack of a systematic approach and environment to support component tracking in testing integration, system testing, performance measurement, and maintenance.

Since a systematic component tracking capability is necessary for component testing, integration, and system testing to support fault detection and isolation, component traceability is one of the factors relating to component testability.

5.2.4 Component controllability

Component controllability is one of the important factors of component testability [2, 6]. According to R. V. Freedman [2], component controllability is the ease of producing a specific output from a specific input—related to the effective coverage of the declared output domain from the input domain. In [6], J. Gao extended this concept by looking at the *controllability* of a software component from three aspects: (1) component behavior control, (2) component feature customization, and (3) component installation and deployment. The first has something to do with the controllability of its behaviors and output data responding to its functional operations and input data. The next refers to the built-in capability of supporting customization and configuration of its internal functional features. The last refers to the control capability of component installation and deployment.

Here, we define *component controllability* as the extent to which a component is constructed to provide component control capability to facilitate the control of components. Component control capability includes the following five types (see Figure 5.3):

- *Component environment control:* This refers to the built-in component capability that supports component environment installation, setup, and deployment.
- *Component execution control:* This refers to the built-in component execution control capability that enables executions of a component in different modes, such as test mode, normal function mode, and control mode. With this capability, users and testers can start, restart, stop, pause, and abort a program execution as they wish.
- *Component state-based behavior control:* This refers to the built-in capability that facilitates the control of component state-based behavior, such as resetting component states, triggering a state transaction from one state to another.

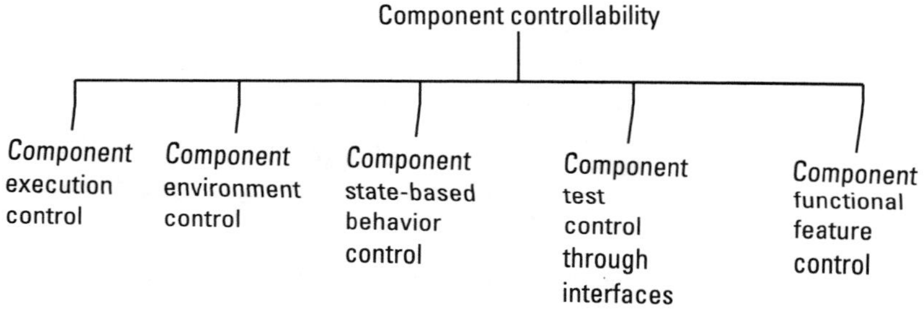

Figure 5.3 Different factors in component controllability.

- *Component test control:* This refers to the built-in capability that facilitates the control of component tests through component interfaces. Setting up different input parameter data for component tests is a typical example.
- *Component function feature control:* This refers to the built-in capability that facilitates the selection and configuration of component functional features. It is only applicable to software components with customizable functional features that allow component users to select and configure its features based on their needs.

Since these control capabilities are essential for supporting component testing and operations, component controllability is an important contributing factor for component testability. To increase component controllability, component developers need to learn how to construct software components with high controllability. The major challenge for them is that they are used to applying ad hoc approaches to supporting program controllability at a certain level to meet the specific requirements for program control and test control. They need well-defined software controllability concepts and supporting methodology to help them address component controllability in all phases of a component-based software engineering process. More research work is needed to assist component testers and customers to verify and measure component controllability, so that component testability can be easily evaluated and measured.

5.2.5 Test support capability

As shown in Figure 5.4, the test support capability for software components can be examined in the following aspects:

- *Component test management capability:* This refers to how well a systematic solution is provided to support the management of various types of test information. As we discuss in Chapter 12, today there are a number of testing management tools available to test engineers and quality assurance people to allow them to create, update, and manage all types of test information in a test process. However, the major problem they encounter is the lack of seamless integration between test management tools and current test suite technology. This creates an obstacle in component test automation. A detailed discussion is given in Chapters 8 and 12.
- *Component test generation capability:* This refers to the extent to which component tests and test scripts can be generated using systematic test

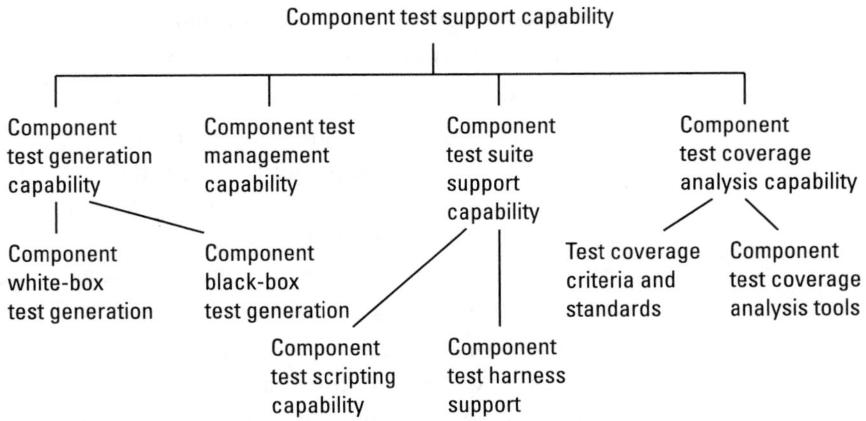

Figure 5.4 Different factors of component test support capability.

generation methods and tools. Although today a number of software test tools (discussed in Chapter 8) can be used to generate white-box and black-box tests for software components and systems, a large percentage of software test generation is performed manually by engineers. The major causes behind this are as follows:

- Current software systems are much more complex than traditional systems because many of them are distributed software systems supporting distributed user access, function services, networking, and database repositories. This increases the difficulty of software test generation due to the fact that the existing test generation methods are designed to focus on host-centered software.

- Existing white-box test generation methods and tools only address control-oriented program structures. For component-based software, engineers need new white-box and/or gray-box test generation methods and tools to deal with function-based scenario structures, and component-based interaction structures in a distributed environment.

- Existing black-box test generation methods and tools only focus on data-oriented equivalence partitions and related value boundaries. This is not enough to deal with distributed functional features, component interacting sequences, and different types of input and output data, and multimedia messages. For software components, there is a lack of domain-specific test generation methods and tools to cope with complicated domain-specific functions and features inside components.

- *Test suite support capability:* This refers to how easy it is for test engineers and test developers to create, maintain, and execute test scripts, test drivers, and stubs. This capability is essential to component testing and evaluation for users. As discussed in Chapter 8, a number of software test tools are available for engineers to create, edit, update, and run various test scripts for software and components. They do not, however, provide enough support for generating test drivers and stubs.
- *Test coverage analysis capability:* This refers to the extent to which component test coverage could be easily measured, monitored, and reported. To increase this capability, engineers need two things. The first is a set of well-defined component coverage criteria and standards, and the other is test coverage analysis function in test tools. As we discussed in the chapters of Parts III and IV, a number of existing test coverage criteria and standards are available to software components. However, component engineers and testers need new test coverage criteria, standards, and analysis tools to address special features of modern software components and COTS, such as component reuse, customization, composition, and deployment.

5.3 Design for testability of software components

In the component-engineering paradigm, one of the primary goals is to generate reusable software components as software products. Third-party engineers use them as building blocks to construct specific component-based software systems according to the requirements given by customers. To generate high-quality component-based software, we must increase component testability in component construction. In the real world, engineers are seeking effective ways to increase the testability of software components and component-based programs. In this section, we discuss the concept of testable components, the different approaches involved, and the ongoing research efforts on increasing component testability.

5.3.1 Understanding of testable software components

In [7], J. Gao et al. introduced a new concept of testable components, called *testable beans*. In this paper, a testable bean is defined as follows:

> A testable bean is a testable software component that is not only deployable and executable, but is also testable with the support of standardized component test facilities.

As pointed out by J. Gao, testable beans have the following unique requirements and features:

1. A testable bean should be deployable and executable. A JavaBean is a typical example.
2. A testable bean must be traceable by supporting basic component tracking capability that enables a user to monitor and track its behaviors. As defined in [8], traceable components are ones constructed with a built-in consistent tracking mechanism for monitoring various component behaviors in a systematic manner.
3. A testable bean must provide a consistent, well-defined, and built-in interface, called component test interfaces, to support external interactions for software testing. Although different components have diverse functional interfaces, we need two types of test interfaces for component testing. The first one is a *test information interface* between components and a test management system. The other is a *test operation interface* that supports the interactions between the component and a component test bed. With this interface, testers are able to set up and execute tests, and validate test results. These two interfaces are very important to the test automation of software components.
4. A testable bean must include built-in program code to facilitate component testing in two areas. The first is to provide the interaction layer between the defined component test interfaces and a component's functional interfaces. The other is to support the component test interfaces to interact with external testing environment and facilities for test setup, execution, and result collection.

5.3.1.1 Why do we need testable beans?

The major goal of introducing the concept of testable beans is to find a new way to develop software components so that they can be easily observed, traced, tested, deployed, and executed. Testable beans have a number of advantages in component test automation:

- Increasing component testability by enhancing component understandability, observability, controllability, and test support capability;

- Standardizing component test interfaces and interaction protocols between components and test management systems and test suite environments;
- Reducing the effort of setting up component test beds by providing a generic plug-in-and-test environment to support component testing and evaluation;
- Providing the basic support for a systematic approach to automating the derivation of component test drivers and stubs.

5.3.2 Methods to increase component testability

How can we construct testable components? In [7], J. Gao et al. discuss three different approaches. Table 5.2 lists the three basic approaches to increase software component testability.

5.3.2.1 Method 1: Framework-based testing facility

In this approach, a well-defined framework (such as a class library) is developed to allow engineers to add program test-support code into components according to the provided application interface of a component test framework. It is usually a test-support facility program or a class library. Component engineers are able to use this framework to add built-in program code into

Table 5.2 Comparisons of Three Different Approaches

Different Perspectives	*Framework-Based Testing Facility*	*Built-In Tests*	*Systematic Component Wrapping for Testing*
Programming overhead	Low	High	Very low
Testing code separated from source code	No	No	Yes
Software tests inside components	No	Yes	No
Test change impact on components	No	High	No
Software change impact on component test	No	Yes	No
Component complexity	Low	Very high	High
Usage flexibility	High	Low	Low
Applicable components	In-house components and newly developed components	In-house components and newly developed component	In-house components and COTS as well as newly constructed components

components by accessing component test interface to interact with external component test tools, such as a component test repository, and a component test bed. This approach is simple and flexible to use. However, there are several drawbacks. First, it has a higher programming overhead. Second, it relies on engineers' willingness to add test-support code. Moreover, this approach assumes that component source code is available, which makes it difficult to deal with commercial components (COTS) because they do not provide any source code to component users.

5.3.2.2 Method 2: Built-in tests

This approach requires component developers to add built-in tests inside a software component to support self-checking and self-testing. Usually, this approach needs a well-defined built-in mechanism and coding format. Developers must follow them to add tests and built-in test functions inside components. This causes a high programming overhead in component development. In this approach, component tests are built in inside components; hence, engineers can perform component tests without any external support from a component testing bed and test repository. Clearly, it is a good idea to include component acceptance tests inside components, like COTS. However, this approach has three major problems. The first problem has to do with high development overheads and component complexity because the built-in tests usually need an extra programming overhead and increase component complexity. The other problem has to do with the high change impact on software components due to component updates and test changes. Furthermore, only limited types and numbers of tests can be included inside components. Hence, this approach may not be suitable for components with complex functions and features.

5.3.2.3 Method 3: Systematic component wrapping for testing

This approach uses a systematic way to convert a software component into a testable component by wrapping it with the program code that facilitates software testing. Like the first method, this approach usually uses a well-defined test framework to interact with test tools or detailed testing functions. However, instead of adding the built-in code manually, this approach uses a systematic way to insert the wrapping code. Unlike the *built-in test* (BIT) components, in this approach, no component tests are included as a part of components. Compared with the first two methods, this approach has several advantages. First, its programming overhead is low because test-support code is inserted into components in a systematic way. Next, since the test-support code (which facilitates testing) is separated from the normal functional code inside components, engineers can set up two different program-execution modes (test mode and normal mode) to reduce the system overhead because the inserted test-support code could be ignored as comments under a normal execution mode. Third, there are no built-in component tests in a component; thus, there is no test change impact on software components. Clearly, this method can be used for developing in-house reusable components, newly developed components, and commercial components (COTS).

In real practice, we need to use these approaches together. To design and construct testable components, engineers need well-defined component models and consistent test interfaces concerning component testability. Here we list the basic principles of building testable components.

- It is essential to minimize the development effort and program overhead while we increase component testability by providing systematic mechanisms and reusable facilities.
- It is important to standardize component test interfaces of testable components so that they can be tested in a reusable test bed using a plug-in-and-play approach.
- It is always a good idea to separate the component functional code from the added test-support code in a component to facilitate component testing and maintenance. This implies that we should separate the normal component functional interfaces from its testing and maintenance interfaces.
- It is essential to control the component overhead on facilitating component testing and maintenance by providing the built-in mechanism to enable and disable the execution of the inserted test-support code inside components.

5.3.3 Current research efforts on building testable components

With the advance of component-based software engineering, more researchers are beginning to pay attention to component-based software testing. Recently, a number of published technical papers and reports have discussed different ways to increase component testability. We summarize them next.

5.3.3.1 Built-in test software components

Y. Wang, C. King, and H. Wickburg in [11] proposed an approach to constructing BIT components for maintainable software. In this approach, built-in tests are explicitly inserted into a software component at the source code level as extra function members to support component testing. Based on the built-in tests, a software program operates in two modes: the normal mode and the maintenance mode. In the normal mode, a software program with built-in tests has the same behavior as its original program. In the maintenance mode, the built-in tests can be activated in a corresponding component as its member functions. Figure 5.5(b) shows the basic BIT component structure. As we mentioned before, this approach is useful in providing a self-testing feature and acceptance tests inside components. However, it requires many system resources and needs a lot of programming overhead. In addition, it causes a high component change impact and test change impact on components during a component evolution cycle.

S. Edwards in [12] proposed a framework to provide BIT wrappers for component testing based on a model-based specification language called RESOLVE. In his approach, BIT tests are built into components using the specification language. His method and architecture focus on how to use a systematic way to detect component interface violations in component-based software. Each component provides a simple "hook" interface (with no run-

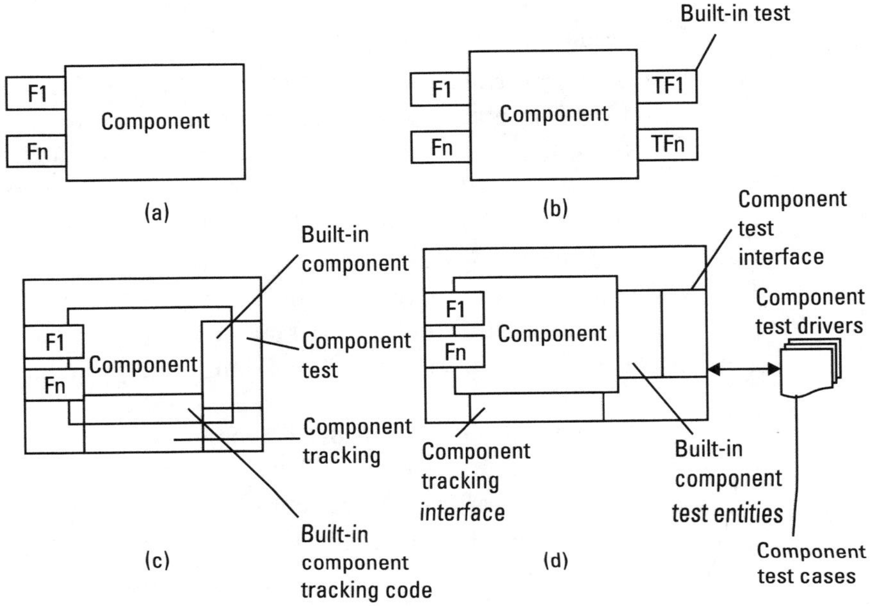

Figure 5.5 Building different types of testable components: (a) a software component, (b) a built-in test component, (c) a testable bean, and (d) a fully testable component.

time overhead) that can be used to provide components with self-checking and self-testing functions.

5.3.3.2 Framework-based testable beans

In [7], J. Gao et al. proposed a new concept, called *testable beans*, to enhance component testability. As mentioned before, intuitively, a testable bean is a software component, which is designed to facilitate component testing. Figure 5.5(c) shows the architecture model of a testable bean. A testable bean consists of two parts: part I includes its functional code that supports the required functional features and data, and part II consists of the parts supporting component testing.

- *Test interface:* This refers to a well-defined common test operation interface for all testable beans. It allows testers, test scripts, or test tools to interact with a testable bean to control test operations, including test setup, test execution, and test result validation. The details are described in [13, 14].
- *Built-in test code:* This refers to the built-in program code that accesses a component test interface to invoke basic component test functions, such as setup tests, run tests, and comparing test results based on the expected results. The built-in test code can be enabled and disabled as needed.
- *Tracking operation interface:* This refers to a well-defined common interface for all testable beans. It allows a component tester (or a supporter) to interact with a testable bean to control tracking functions, such as setting-up and selecting various tracking levels, and monitoring component trace and test results. The details are reported in [8].
- *Built-in tracking code:* This refers to the built-in program code that interacts with the component-tracking interface to monitor component behaviors. The built-in tracking code can be enabled and disabled as needed during program. The details are reported in [8].

Testable beans differ from BIT components in the following ways:

- Since there are no component tests inside testable beans, the programming overhead for built-in tests is reduced.
- Although software components have different application functions and external interfaces, a testable bean always provides a consistent component test interface and tracking interface to regulate its interactions between components and external test support facilities, such as

component test management systems, testing tools, and test scripts. This increases component interoperationability and compatibility with various test tools, and reduces the cost and efforts in setting up a specific component bed.

To support component testing at the unit level, engineers need a reusable component test bed to support component test operations and test suite accesses. J. Gao and his students [14] have worked on a research project for several years to set up a generic component test environment for various components by providing a common component test bed and generating component test drivers/scripts in a systematic way. As shown in Figure 5.5(d) a fully testable component is constructed using well-defined mechanisms by converting a normal software component into an executable, testable, and traceable test bed. It can be easily operated with systematic generated component test scripts. These test scripts (or test drivers) are generated using two different methods. The first is the *static approach,* in which developers and component test engineers use a well-defined component test-support framework to built testable components with well-defined architecture model and standard component test interfaces. The second is the *dynamic approach,* in which a component test wrapper and test scripts (or test drivers) are generated using an automatic method based on a provided component test framework, component profiles and test case information. The details can be found in [14].

5.4 Verification and evaluation of component testability

As pointed out in [9], it is important to verify component testability of components generated from a component development process. Before verifying component testability, it is always a good idea to evaluate the maturity of a component development process concerning component testability. In our view, the maturity level of component testability in a component development process can be evaluated based on the following levels.

- *Level 1—Initial:* At this level, component developers and testers use an ad hoc approach to enhance component testability in a component development process. The major problem at this level is no control of component testability for all generated components. In most cases, these components have poor testability.
- *Level 2—Standardized:* At this level, component testability requirements, design methods, implementation mechanisms, and verification criteria

are defined as a part of software component standard. In addition, a consistent standard is defined for components to cover test case formats, tracking message formats and mechanisms, built-in test techniques, and interaction interfaces with software test tools. When developers follow the defined standard, they can generate testable components to reduce the testing cost and increase component testability. However, in real-world practice, engineers may not follow the predefined standard and test criteria during component development and implementation; this suggests that the establishment of component testability standards and criteria is not enough.

- *Level 3—Systematic:* At this level, a well-defined component development and test process is used to facilitate component testability at all phases. In this process, component developers are provided with:
    - Well-defined component requirements that can be used to easily establish achievable test criteria and test cases;
    - Clearly predefined testability requirements, design methods, and standard component test interfaces to assist engineers to construct highly testable components;
    - Systematic testing facility and tools to facilitate component testing and test automation.
- *Level 4—Measurable:* At this level, component testability can be evaluated and measured using systematic solutions and tools in all phases of a software component development process, including component testing.

Currently, there are two basic approaches to evaluating component testability.

- *Static verification approach:* Here, component testability is verified in all phases of a component development process, including requirement analysis and specification, construction, and testing.
- *Statistic verification approach:* – In this approach, statistical methods are used to analyze and estimate component testability by examining how a given component will behave when it contains faults.

5.4.1 Static verification approach

As discussed in [9], component testability can be verified in all phases of a component development and test process by reviewing all component artifacts. Here are our observations.

- During the requirement analysis phase, engineers must make sure that component requirements are clearly specified so that tests and test criteria can be easily specified based on the provided component interface, function features, and logic. To achieve this, a well-defined component specification standard should be established to enhance the quality of component requirements so that they are understandable, testable, and measurable. Predefined guidelines also need to be included to address component testability requirements and facilitates. These standards and guidelines will be used as a base to conduct a component requirement review.
- During the design phase, component engineers must be provided with a component architecture model concerning component testability, well-defined design patterns, and a consistent test interface to facilitate them to design testable components in a systematic way. Design for component testability should be checked in the aspects of architecture model, interfaces, and built-in mechanisms for testing support.
- During the implementation phase, a component implementation review should be conducted at the source code level to check if the component testability artifacts are properly incorporated inside the component source code; for example, the component tracking code, and built-in tests.
- During the testing phase, test engineers must verify component testability in two ways. The first is to review the targeted component requirements for testing to make sure that they are testable and measurable. The next is to pay attention to the verification and measurement of component tests based on the established component test criteria.

5.4.2 Statistic verification approach

In the past, software testability has been used informally to discuss the ease with which some input selection criteria can be satisfied during testing. For example, if a tester wanted full branch coverage during testing and found it difficult to select inputs that cover more than half the branches, the software would be classified as having poor testability. In [3], J. Voas proposed a verification approach to examine how a given program will behave when it contains a fault. The major objective is to predict the probability of software failure occurring if the particular software were to contain a fault, given that software execution is performed based on a particular input distribution during random black-box testing. He believes this verification approach enhances software testing in the following ways:

- It suggests the testing intensity or estimates how difficult it will be to detect a fault at a particular location.
- It suggests the number of tests necessary to be confident that no faults exist.

J. Voas [3] uses the *sensitivity analysis* method to quantify behavioral information about the likelihood that faults are hiding. It repeatedly executes the original program and mutations of its source code and data states using two assumptions: a single fault in a single location and a simple fault, which is not distributed throughout the program. It involves three estimations at each location.

1. *Execution probability:* This is estimated for some given input distribution by determining the proportion of cases that execute the location of interest.
2. *Infection probability:* This is estimated by introducing mutations of the location and determining the proportion of input cases that give a different state from the nonmutated location.
3. *Propagation probability:* This is estimated by determining the proportion of cases for which perturbed data states at the location propagate and give rise to faulty results.

The result of sensitivity analysis is the estimated *probability of failure that would result* if the particular location has a fault. This estimate is obtained by multiplying the means of the three estimates from the analysis phase. To sharpen an estimate of the *true probability of failure,* testability analysis can be combined with the testing results. This essentially is a cleanup operation—a method to assess if software has achieved a desired level of reliability. In this case, the probability of failure is treated as a random variable and probability density function for it is estimated conditioned on an input distribution.

Sensitivity analysis allows testers to determine when testing can stop with an acceptable confidence that no faults are hiding. This is important information to have if it is known that a particular component is likely to be reused in differing environments. Software reuse can be more readily enabled if the testing of the code in the previous environment is still applicable to the new environment. The reusability of previous verification efforts is an important parameter in assessing the immediate reusability of the code. The ability to reuse verification is a factor to consider in determining the reusability of code. Sensitivity analysis can be used for the quantitative assessment of the reusability of previous verification.

5.5 Software testability measurement

As we mentioned in Section 5.1, software testability measurement is a part of software testability study. Here, we use a very limited space to discuss software testability measurement and related methods.

5.5.1 What is software testability measurement?

Generally speaking, *software testability measurement* refers to the activities and methods that study, analyze, and measure software testability during a software product life cycle. In the past, there were a number of research efforts addressing software testability measurement [2, 3, 15–19]. Their focus was on how to measure software testability at the beginning of a software test phase. Once software is implemented, it is necessary to make an assessment to decide which software components are likely to be more difficult and time-consuming in testing due to their poor component testability. If such a measure could be applied at the beginning of a software testing phase, much more effective testing resources allocation and prioritizing could be possible. For example, the measurement result could be considered as a feedback to testers so they would be able to know which component has poor testability. This would help them to add and/or improve test cases to those components. Hence, the measurement result of component testability could help testers and component users decide if a testing process is an effective process, and when it could be finished.

As we understand, the objective of software testing is to confirm that the given software product meets the specified requirements by validating the function and nonfunction requirements to uncover as many program problems and errors as possible during a software test process. Unlike software testing, the major objective of software testability measurement is to find out which software components are poor in quality, and where faults can hide from software testing. In other words, testability measurement complements formal verification by providing empirical evidence of software testability evaluation, which formal verification cannot do. It is clear that testability information cannot replace testing and formal verification, but neither should developers rely exclusively on testing and formal verification. To produce highly reliable software, we must construct software with high testability, check it using effective verification techniques, and validate it using an adequate test set.

5.5.2 How to measure software testability

In the past few years, a number of methods have been proposed to measure and analyze the testability of software [4, 16–19]. They can be classified into the following groups:

- Program-based measurement methods for software testability [16];
- Model-based measurement methods for software testability [17–19];
- Dependability assessment methods for software testability [4].

5.5.2.1 Program-based testability measurement

Since a fault can lie anywhere in a program, all places in the source code are taken into consideration while estimating the program testability. J.-C. Lin et al. [16] proposed a program-based method to measure software testability by considering the single faults in a program. The faults are limited to single faults and are limited to faults of arithmetic expressions and predicates.

- *Arithmetic expressions:* limited to single changes to a location. It is similar to mutations in mutation testing;
- *Assignment predicates:* an incorrect variable/constant substitution, for example, a variable substituted incorrectly for a constant, a constant substituted incorrectly for variable, or a wrong operator;
- *Boolean predicates:* a wrong variable/constant substitution, wrong equality/inequality operator substitution, or exchanging operator *and* with operator *or*.

The basic idea of this approach is similar to software mutation testing. To check software testability at a location, a single fault is instrumented into the program at this location. The newly instrumented program is compiled and executed with an assumed input distribution. Then, three basic techniques (execution, infection, and propagation estimation methods) are used to compute the probability of failure that would occur when that location has a fault.

The computation result is the testability of that location. Finally, this approach takes the location with the lowest nonzero testability to be the testability of an overall program. Authors in [16] reported a commercial software testability tool, known as PISCES. It is written in C++ and operates on programs written in C. PISCES produces testability estimates by creating an *instrumented* copy of the program. Then, this instrumented copy (which is about 10 times as large as the original source code) is compiled and executed with inputs that are either supplied in a file or generated based on a random input value distribution. PISCES implements *propagation, infection, and execution* (PIE) techniques. The detailed descriptions of these three techniques can be found in [16].

Since the program source code structure and semantics are used as a base for program executions, two programs with the same function may have different measurement results for software testability. The major problem of this approach is its high measurement cost and complexity considering the fact that all locations of the program source code could have faults. With the increasing complexity of today's components and programs, this approach does not seem practical enough to deal with real complicated components and programs.

5.5.2.2 Model-based testability measurement

Another measurement approach of software testability is proposed based on a well-defined model: such as a data flow model [17]. This approach consists of three steps:

- *Step #1: Normalizing a program before the testability measurement using a systematic tool.* Normalizing a program can make the measurements of testability more precise and reasonable. A program, after being normalized, must have the same semantics as the original one. This is done mechanically. Two types of normalization are performed here. They are *structure normalization* and *block normalization*. In the structure normalization, the program's control flow structure is reconstructed to make it regular to facilitate analyzing and property measuring. In the block normalization, an optimized compiler technique is used to simplify the problem and eliminate local variable definitions and uses in a source code block. In conventional structural tests, successful executions by test inputs should exercise the flow graph to achieve predefined test coverage (such as the node coverage) to gain the confidence in program quality. It is possible to measure the testability from program syntax and assess the amount of efforts required during testing.

- *Step #2: Identifying the testable elements of the targeted program based on its normalized data flow model.* The elements include the number of noncomment lines, nodes, edges, p-uses, defs, uses, d-u paths (pairs), and dominating paths.
- *Step #3: Measuring the program testability based on data flow testing criteria.* These data-flow testing criteria include: ALL-NODES, ALL-EDGES, ALL-P-USES, ALL-DEFS, ALL-USES, ALL-DU-PAIRS, and ALL-DOMINATING PATH.

Though there is no correlation between the measurements and the number of faults, this approach can be used to check how easily software modules can be tested. The measurement result can be used as a valuable quality indicator for programmers. This method can be easily practiced because only a static analysis of a program source code is required. Its primary limitation is that the measurement result only suggests software testability in a white-box view. Black-box tests and test criteria are not considered here.

C. Robach and Y. Le Traon [19] also used the data-flow model to measure program testability. Unlike the previous approach, their method is developed for co-designed systems. The data flow modeling is adapted to functional analysis where hardware testability measurements are performed. Hence, hardware design testability is applied to executable co-designed data-flow specifications. Several tools are available to predict hardware testability. One typical example is SATAN (an acronym for System Automatic Testability Analysis), which presents data information transfers using a data-flow model, which is obtained automatically from a behavioral/structural description of a given circuit. Each node in a model corresponds to a function of the circuit, and each link in the model presents the possibility to transfer information from one module to another. SATAN tool first generates a graph, known as an information transfer graph. Then, it performs testability analysis based on the program functional specification and the generated information transfer graph.

5.5.2.3 Dependency-based testability measurement

Clearly, the two previous approaches need program source code and/or a program-based model to support software testability measurement. A. Bertolino and L. Strigni [4] proposed a black-box approach, where the software testability measurement is performed based on the dependency relationships between program inputs and outputs. The basic idea is to perform an oracle in a manual (or systematic) mode to decide whether a given program behave correctly on a given test. The oracle decides the test outcome by analyzing the behavior of the program against its specification. In particular, an input/

output (I/O) oracle only observes the input and the output of each test, and looks for failures. A program is correct with respect to its specification if it is correct on every input setting; otherwise the program is faulty. If the program generates an incorrect output, then the test has failed. If the oracle output is approved, then the test is successful.

Based on this concept, the program testability is measured by estimating the probability that a test of the program on an input setting drawn from a specified probability distribution of the inputs is rejected, given a specified oracle and given the program is faulty. This approach can only be used to measure software testability in a black-box view when the given program specification unambiguously decides whether the program's output is correct for a given program input.

5.6 Summary

This chapter first introduces the basic concept of software component testability from the perspective of component-based software engineering. We revisit a number of definitions of software testability and discuss its contributing factors and artifacts. In our view, component testability depends on component understandability, observability, traceability, controllability, and test support capability.

In our opinion, the poor software testability issue should be addressed in all of the phases of a software development process. According to practitioners in the real world, managers and engineers would like to know software testability issues earlier, even before software testing operations, so that more of the software enhancement costs and testing efforts can be reduced. We believe increasing component testability and building testable components are essential in the component-based software engineering. In response to this need, we introduce the concept of testable software components and discuss the related objectives and requirements. Three methods have been discussed to increase component testability. Detailed ideas and comparisons are provided.

In Section 5.4, we discuss the concept of testability verification for software and components. Two different approaches are examined. They are the static approach, and the statistical approach. In the static approach, software artifacts are verified in various reviews to check the testability issues during a component development process. In the statistical approach, some empirical measurements are collected and studied. The verification method, proposed by J. Voas and K. W. Miller, is highlighted here.

Finally, in Section 5.5, we discuss software testability measurement and review three existing methods. They are program-based, model-based, and dependency-based measurement methods for software testability. Although these proposed methods could be useful to assess the testability of software components and component-based programs, we need more practical, cost-effective, and systematic methods to measure and analyze the testability of software components and component-based programs.

References

[1] "IEEE Standard Glossary of Software Engineering Terminology," ANSI/IEEE Standard 610-12 –1990, IEEE Press, New York, 1990.

[2] Freedman, R. S., "Testability of Software Components," *IEEE Trans. on Software Engineering*, Vol.17, No. 6, June 1991, pp. 533–564.

[3] Voas, J. M., and K. W. Miller, "Software Testability: The New Verification," *IEEE Software*, Vol. 12, No. 3, May 1995, pp. 17–28.

[4] Bertolino, A., and L. Strigini, "On the Use of Testability Measurement for Dependability Assessment," *IEEE Trans. on Software Engineering*, Vol. 22, No. 2, February 1996, pp. 97–108.

[5] Binder, R. V., "Design for Testability in Object-Oriented Systems," *Communications of the ACM*, September 1994, pp. 87–101.

[6] Gao, J., "Challenges and Problems in Testing Software Components," *Proc. of ICSE2000's 3rd International Workshop on Component-Based Software Engineering: Reflects and Practice*, Limerick, Ireland, June 2000.

[7] Gao, J., et al., "On Building Testable Software Components," *Proc. of 1st International Conference on Cost-Based Software System*, 2002, pp. 108–121.

[8] Gao, J., E. Zhu, and S. Shim, "Tracking Software Components," *Journal of Object-Oriented Programming*, October/November, 2001.

[9] Jungmayr, S., "Reviewing Software Artifacts for Testability," *Proc. of EuroSTAR '99*, Barcelona, Spain, November 10–12, 1999.

[10] Beizer, B., *Software Testing Techniques*, 2nd ed., New York: Van Nostrand Reinhold, 1990.

[11] Wang, Y., G. King, and H. Wickburg, "A Method for Built-In Tests in Component-Based Software Maintenance," *Proc. of 3rd European Conference on Software Maintenance and Reengineering*, IEEE Computer Society Press, 1999.

[12] Edwards, S. H., "A Framework for Practical, Automated Black-Box Testing of Component-Based Software," *Software Testing, Verification, and Reliability*, Vol. 11, No. 2, June 2001.

[13] Gao, J., et al., "Design for Testability of Component-Based Software," *Proc. of International Conference on Quality/Internet Week*, San Francisco, CA, 2000.

[14] Pham, P., *A Component Testing Framework for Component-Based Software*, Master Project Report, Department of Computer Engineering, San José State University, San José, CA, 2002.

[15] Bache, R., and M. Mullerburg, "Measures of Testability as a Basis for Quality Assurance," *Software Engineering Journal*, Vol. 5, March 1990, pp. 86–92.

[16] Lin, J.-C., I. Ho, and S.-W. Lin, "An Estimated Method for Software Testability Measurement," *Proc. of 8th International Workshop on Software Technology and Engineering Practice (STEP '97)*, 1997.

[17] Lin, J.-C., and P.-L. Yeh, "Software Testability Measurements Derived From Data Flow Analysis," *Proc. of 2nd Euromicro Conference on Software Maintenance and Reengineering (CSMR'98)*, Deqli Affari, Italy, March 8–11, 1998, pp. 96–103.

[18] Offutt, A. J., and J. H. Hayes, "A Semantic Model of Program Faults," *Proc. of International Symposium on Software Testing and Analysis (ISSTA '96)*, San Diego, CA, January 1996, ACM Press, pp. 195–200.

[19] Robach, C., and Y. Le Traon, "Testability Analysis of Co-Designed Systems," *Proc. of 4th Asian Test Symposium (ATS'95)*, 1995.

PART

II

Validation methods for software components

Because software components are the building blocks for software applications, there is no doubt that their quality has to be ensured. For traditional software systems, great efforts have already carried out to ensure the quality of software modules or classes. Because of the many unique characteristics of software components, in this part we investigate new issues that arise when testing software components, the adjustments that need to be performed for traditional testing strategies, and new software component testing techniques.

As described in Part I, there are usually two different roles that are involved within the component-based software development: *component providers* and *component users*. Each of these roles has different perspectives about how to develop and use software components. Consequently, component users and component developers will test software components in different contexts and with different methods. For component users, third parties deliver the components, and some testing efforts have already been carried out. Component users mainly focus on whether the component matches their specifications and the adequate modifications or customizations that need to be performed. Therefore, the testing should focus on the consistency of the component specification with the user specification and customization testing. As component users usually do not have the access to the source code, black-box testing methods are usually adopted in this context. For component developers, besides the testing for structural coverage to ensure the quality of software components, more efforts should be conducted to ensure the reusability, interoperability, and other related quality features, which are generally overlooked by traditional unit level software testing techniques. In general, component providers are able to access the source code, and therefore, white-box testing techniques are discussed from that perspective in Chapter 7. To efficiently and effectively carry out these tasks, especially for component users, mechanisms for test automation need to be provided. In addition, how to build the automation mechanisms within the software components requires a lot of attention. All these issues are discussed in Chapter 8.

CHAPTER

6

Contents

Black-box testing methods for software components

Black-box testing, which is also known as functional testing, is the testing that ignores the internal mechanism of a system or component and focuses solely on the outputs generated in response to selected inputs and execution conditions. Generally, black-box testing methods are widely used in system level testing and some integration testing. But in the context of software components, when third-party components are adopted into the system and the source code is not available, black-box testing is necessary to make sure these components fulfill the users' requirements.

In this chapter, we first give a brief introduction to black-box testing and different black-box testing techniques. Then we discuss the issues and concerns that arise when applying black-box testing techniques directly to software components. Finally, detailed black-box testing techniques and test adequacy criteria are discussed.

6.1 Introduction

According to IEEE standard [1, 2], black-box testing, also known as functional testing, is:

1. "Testing that ignores the internal mechanism of a system or component and focuses solely on the outputs generated in response to selected inputs and execution conditions;

2. "Testing conducted to evaluate the compliance of a system or component with specified functional requirements."

The primary goal of black-box testing is to assess whether the software does what it is supposed to do. In other words, does the behavior of the software conform to the requirement specification? Black-box testing can be performed at different levels such as unit testing, integration testing, system testing, and acceptance testing. But for different levels there are different requirements and objectives. For instance, black-box testing at the integration level mainly focuses on the functional behavior of interacting modules, while black-box testing at the system level focuses on both functional and nonfunctional requirements, such as performance and reliability. Black-box testing has been widely used for a long time, and different approaches can be used for different situations. Basically, the techniques could be classified into three groups: (1) usage-based black-box testing techniques, such as random testing or statistical testing; (2) error-based black-box testing techniques, which mainly focus on certain error-prone points according to users' experiences regarding how the program behavior usually deviates from the specification (this type of approach includes equivalence partitioning testing, category-partition testing, boundary-value analysis, decision table–based testing); and (3) fault-based black-box testing techniques, which focus on detecting faults in the software. Fault-based testing usually requires a source code to create alternative programs (mutants) for software components, and these mutants can be created by changing the interfaces.

When testing software components from scratch, black-box testing techniques can be directly adopted as the techniques that only rely on specifications. Nevertheless, software components have usually been tested, and very often retesting the component from scratch is usually not acceptable to component users. To adequately apply traditional black-box testing techniques for software components, the following issues need to be taken into consideration:

- *Source code unavailability*: Very often component users will not have access to the source code, and therefore, black-box testing is the most intuitive choice for them to test a software component. Source code unavailability not only prevents the component users from accessing the source code, it also often implies that the component users are not able to access the testing methodologies, test environment, and verification history of the component from component providers. Without this type of information, component users will encounter great difficulties when attempting to determine the test adequacy for black-box testing the software components.

- *Component specifications*: In traditional software systems, for a software module there is only one specification for everybody to follow. Therefore, when the time comes to integrate software modules, people do not need to worry about the functionality of each individual module. Nevertheless, for component-based software, generally two specifications exist. This is because COTS components are developed by a third party and use a generic specification as a result. For reusable in-house developed software components, even though it is possible that the same group of people developed that particular component, very often the component is based on an old specification. When component users need to integrate the component into their system, a user specification for the desirable component needs to be matched. Chances are, these two specifications are similar but different enough to have an impact. A component provider's testing is performed based on one specification, while a component user's testing needs to be based on another specification. How to perform black-box testing within this context is a new challenge to component users.
- *Component customization*: As was discussed above, two specifications could exist for the same component, and they are very often not perfectly matched with each other. Therefore, when adopting a component into the application, customizations are required to reconcile the differences between the two specifications. So when testing the component, component users have two choices: start from scratch or customize the component and make sure the customization will not adversely affect the software component. The major concern for the first option is the cost, which may make it unfeasible according to time or budget constraints. For the second option, it is promising that component users can perform the component testing more efficiently, but new and challenging questions arise regarding how to assess the adequacy, and how to determine the mechanism to reuse the testing efforts from component providers.
- *Component interface:* Unlike in traditional software systems, interactions in component-based software systems can be conducted in a flexible way. For instance, we can access an object through its public methods; or sometimes we can access public data members directly. As to component-based software, interfaces are the only point of contact; that is, the interactions among components have to go through interfaces. Therefore, the specifications of interfaces play a key role in testing software components. If properly conducted, efficient and effective component testing can be achieved through interface-based black-box testing strategies.

6.2 Black-box testing foundations

Before introducing the black-box testing techniques, we need to discuss some fundamental issues, which include how components are specified, the mechanisms that share information between component providers and component users, and how customization will be performed.

6.2.1 Component specification

Component specifications are a base for black-box testing and need to be discussed before getting into any specific testing techniques. For a software component, component interfaces are the point of interaction among components and external entities. Obviously, specification first needs to specify the functionality of each interface and how to interaction with it. Moreover, as defined by C. Szyperski [3], "A software component is a unit of composition with contractually specified interfaces and explicit context dependencies only." Therefore, in addition to what and how the interfaces provide service to the external entities, various constraints and dependency relationships need to be specified as well. Examples of these specifications include sequence constraints, preconditions and postconditions, and dependence constraints.

Naturally, when deploying a component under a certain component model, component models usually have their own way of specifying interfaces. For instance, for CORBA components, there is always an *interface definition language* (IDL) file that specifies the details such as interface name, signature, and error handling mechanism. For EJB components, there is a home interface class and a component interface class that specifies how EJB beans will be created and what services will be provided. For instance, the following is an EJB example to specify interfaces for a banking account component, which provides authentication, withdrawal, deposit, and transfer service.

```
Home interface
public interface BankingAccountHome extends EJBHome {
 public BankingAccount create(int AccountNumber,String
 passwd) throws
  RemoteException, CreateException;
}

Component interface
public interface BankingAccount extends EJBObject {
 public int validatePIN(String pin) throws RemoteException;
 public float getBalance(int bankaccount) throws
 RemoteException;
```

```
    public int deposit(float amount) throws RemoteException;
    public int withdraws(float amount) throws RemoteException;
    public int transfer(int fromBankingAccount, int toBank-
    ingAccount, float amount) throws RemoteException;
}
```

The specification capacity provided by the specific component model is far from enough. For safety-critical systems, more precise specifications are required. The most obvious choice is formal specifications [4, 5]. There are many different types of formal specification languages. Model-based (state-based) formal specification languages, such as Z and VDM, use discrete mathematical entities, such as sets and sequence to build a model for the system. Algebraic specification languages, such as Larch, CLEAR, and OBJ, view a system as a set of equations. Formal specifications can precisely define the functional behavior of the component and can provide an abstract model for software implementation. Therefore, when formal specifications exist for a component, test case generation, test oracle, and test adequacy can be efficiently and effectively carried out. In addition, formal specifications generally improve the quality of the software. Besides these obvious benefits, some inherent overheads may prevent formal methods widely adopted in the software industry. Overhead costs include the cost for staff training, software purchase, and limited tool support.

In the following discussion, we assume component users will have the component interface specifications under their component model and some additional specifications for the constraints and dependence relationships, such as precondition and postcondition.

6.2.2 Component verification history

Component specifications provide the foundations for testing software components for component providers and users. Nevertheless, specifications do not provide the mechanism for interchanging verification information between the two parties. On the one hand, component users could provide feedback of their component testing to the component providers to further improve the component quality. On the other hand, the verification history that is provided by component providers can greatly help the component users to efficiently and effectively perform component testing and determine test adequacy. Without the information, redundant efforts and inappropriate activities may be conducted. For further discussion of this issue, please refer to Chapter 15.

If the component users decide to provide feedback to the component providers, or if software component providers and software certification laboratories are willing to provide component verification history, the exchange information must have the following characteristics:

- Standard and portable;
- Simple, yet powerful;
- Programming language independent;
- Free of proprietary-software requirements for interpreting and running the test;
- Easy to change, and support regression testing.

Fortunately, *Extensible Markup Language* (XML), a general-purpose markup language for representing hierarchical data, can be adopted to fulfill the role. Right now, XML has been widely adopted throughout the software industry, as well as in other industries. XML itself lays out some simple rules; yet it is very powerful for representing data and accommodating changes. In addition, the process of XML can be easily carried out. So far, XML has been used to develop some domain-specific languages such as *XML metadata interchange* (XMI), which supports the interchange of metadata that can be expressed using the *meta-object facility* (MOF) specification [6], *Bean Markup Language* (BML) [7], and XML-based component configuration or wiring language customized for the JavaBean component model. *UML exchange format* (UXF) [8] is an XML-based model interchange format for UML, which is a standard software modeling language by OMG. For instance, the following is a sample model for XML-based test specifications proposed by Morris et al. [9]:

```
<TestSpecification>
 <TestSet Name= "...">
  <TestGroup Name="...">
   <TestGroup>*
   <Invariant DataType="...">*
   <Operation Name = "..." Pre="op_name">
    <invariant>*
    <Constructor>*
    <MethodCall Target="...">*
   <operation>
   <Invariant>
```

Similarly, the output of each test case can be specified in a similar fashion. Given the XML-based description of each test case and test output, component users are able to easily access the verification history of the component, and they are able to make adequate decisions of what to test and when to stop a test.

6.2.3 Component customization

When adopting a software component into a new environment, if the component specification perfectly matches the user specification, then no additional effort is required to change the configuration of the components. In other instances, which are more likely, customizations need to be performed on the component to adjust the component specification to conform to the component users' specification. In general, the customization can be classified into the following three categories, and the effects of these customization activities are discussed along with each individual testing technique.

1. *Generalization customization*: This will release some requirement restrictions in software component specifications. For instance, in the ATM example that is described in Section 6.2.1, the original component specification has a restriction of withdrawing amounts only in multiples of 20. However, in the users' specification, the requirement has no restriction on the withdraw amount. In this case, generalization customization needs to be performed to release the constraints.
2. *Specialization customization:* Contrary to generalization customization, specialization customization applies more restrictions on the software component specification. For example, the ATM example had no restriction on the number of failed PIN inputs that can be conducted. And for the user specification, there is a new restriction that only allows a maximum of three failed PIN inputs. Specialization customization has to be performed to enforce the software component specification.
3. *Reconstruction customization:* Generalization and specialization customizations both assume the input domain for the component remains unchanged. Nevertheless, under certain circumstances customization may change the input domain as a result. For instance, in the ATM example, the original specification only requires a one-time authentication of the user's PIN before carrying out any group of transactions. When that specification changes to a new one because each transaction requires authentication for a PIN, input domain changes as a result.

Given the previous discussion about component specification, verification history, and customization, we now start to discuss individual black-box testing strategies. For each strategy, we first discuss its general principles and then discuss how to make adequate adjustments when testing software components.

6.3 Black box–based testing techniques

For software components, interfaces are the only point of interaction; therefore, when applying black box–based test techniques, we need to follow the specifications of interfaces. Interface specifications usually include how the interface can be invoked, and for certain component models, additional information may also be specified in the interfaces. For instance, and exceptions handling will be specified in the CORBA interface specification.

Traditional black-box testing techniques [10–14] have been extensively discussed for the above situation.

Nevertheless, the component needs to be customized, so even with certification information, a strong test adequacy often does not raise component users' confidence about the software component. To further test the component, different black-box testing techniques are required, which are discussed in the following sections.

6.3.1 Random testing

Random testing is a strategy that requires the "random" selection of test cases from the entire input domain [15, 16]. For random testing, values of each test case are generated randomly, but very often the overall distribution of the test cases has to conform to the distribution of the input domain, or an operational profile. For example, for an ATM system we know that 80% of the transactions are withdrawal transactions, 15% are deposit transactions, and the other 5% are transfer transactions. For random testing, the test case selection has to conform to that distribution. In other words, 85% of the test cases are randomly generated to test the withdrawal transaction, and 15% test deposit functionality, and so on. The advantage of random testing is its efficiency. Test cases can be generated automatically and require no further efforts such as subdomain classification in partition testing. In addition, various research has shown that random testing might be just as effective as many other techniques [15, 17]. On the other hand, to properly test a software system, a distribution for the input domain or an adequate operational profile should be available before test cases are derived. But in reality, it is very difficult to obtain a precise operational profile until the system is up and running in the real environment for a certain period of time.

6.3.3.1 Random testing for software components

Random testing can certainly be utilized in testing software components for component users; and the testing depends on how much component certification information or component verification information is available. If the information is not available, the component users have to treat the component as a newly developed piece to perform the random testing. If the information is available, in order to determine the test adequacy criteria, two questions will be asked: how much of the efforts undertaken by the component providers can be reused and how much additional testing effort is required?

To answer the two questions for random testing, we need to determine the distribution of the test cases that have been carried out by component providers, and the new distribution of the input domain. Again, determining the distribution of the input domain or the operational profile is totally component user specific and can be obtained by traditional approaches. Given the two distributions, new test cases can be generated to reconcile the discrepancy between the two. Furthermore, customizations may be applied to software components. For generalization customization, which releases constraints in the component specifications, part of the input domain that was originally invalid will become valid according to the new requirements. For the specialization customization, of which more constraints are applied onto the component specification, part of the input domain that was originally valid will become invalid according to the new specification. In either case, the portions of the input domain change their characteristics and should be given more attention. Therefore, additional test cases should be randomly generated from those portions. Reconstruction customization is more complex than generalization and specialization customization. As the input domain changes, some of the original test cases may not be applied. Thus, new test cases should be generated for the portion of input domain, which has not been tested.

6.3.2 Partition testing

Random testing can only sample the input domain and will not be able to provide confidence for the entire input domain. In addition, test cases are randomly generated; therefore, extensive redundancy may exist. To overcome these difficulties, equivalence partition testing [18] tries to divide the input domains into k different disjoint partitions $p_1, p_2, \ldots, p_k$, where $p_1 \cup p_2 \cup \ldots \cup p_k$ = input domain, and $p_i \cap p_j = \Phi$ for any i and j where $i \neq j$. Values from each partition have the "all-or-none" property; that is, if an element is selected as the input for the component and that element fails the component, all other elements in that domain will also fail the component. On the other hand, if the component succeeds with that element as input, all other elements in the same partition will succeed as well. Therefore, if partitions, which satisfy this characteristic, can be generated, then test cases can be easily generated by randomly selecting one element from each partition. In addition, the test cases generated by the equivalence partition testing strategies can provide us with the confidence about the entire domain without any redundancy. Unfortunately, there is no systematic way to generate equivalence partitions, and many times it is impossible to develop equivalence partitions. To overcome the difficulties, many systematic partitioning approaches are developed. If formal specifications exist, sometimes partitions can be automatically derived. Given nonformal functional specification, T. J. Ostrand and M. J. Balcer proposed a systematic partition testing strategy called category partition testing [19]. Category partition testing starts with an original functional specification and continues through the individual details of each subprogram to derive a formal test specification. The category partition testing mainly involves seven steps:

1. Decompose functional specifications into functional units;
2. Identify parameters and environment conditions;
3. Find categories of information;
4. Partition each category into choices;
5. Write test specifications for each unit;
6. Produce test frames;
7. Generate test cases.

For instance, according to the category partition testing strategy, the following is a test specification for the withdrawal functionality. In the following test specification, for example, the PIN, the withdraw amount, and the account balance are categories, while a wrong PIN, correct PIN, and multiples of 20 are choices within corresponding categories. To reduce the number of test cases, constraints are also part of the test specification.

```
Parameters:
 PIN
  Wrong PIN [property mismatch]
  Correct PIN [property match]
  Withdraw amount
    Multiple of 20[if match]
                 [property correct]
    Less than 20 [if match]
                 [property wrong]
    Greater than 20 but not multiple of 20
                 [if match]
                 [property wrong]
```

```
Account balance
  Balance > withdraw amount
              [if match and correct]
              [property greater]
  Balance = withdraw amount
              [if match and correct]
              [property equal]
  Balance < withdraw amount
              [if match and correct]
              [property less]
```

6.3.2.1 Partition testing for software components

When component certification information or verification information is not available, partition testing has to be done from the very beginning. Otherwise, when the information is available, we would need to adjust the partition testing in the following ways according to the partitions and user customizations.

1. *Generalization customization:* The effects of the generalization customization are shown in Figure 6.1. Generalization customization could change the characteristic of portion P' as shown in Figure 6.1(b, c), where one partition expands and the other one shrinks. Many times, as shown in Figure 6.1(d), the generalization customization can result in the merge of two partitions to assure the generalization customization will not adversely affect the component. Each changed partition needs to be retested. In other words, p_1 and p_2 need to be rerun. In addition, in Figure 6.1(c), a new test case needs to be generated for the new partition P_2.
2. *Specialization customization*: The effects of the specialization customization are shown in Figure 6.2. Specialization customization will

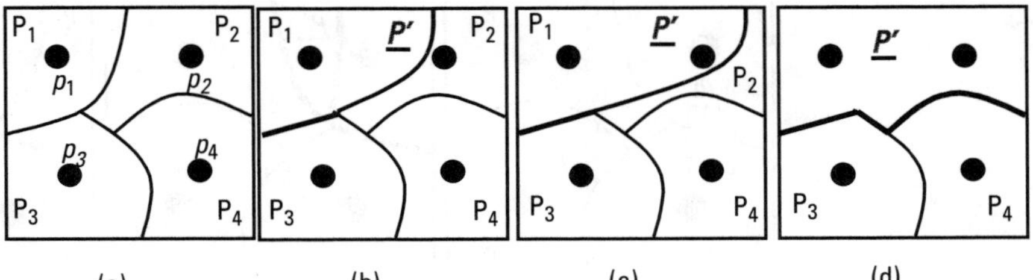

Figure 6.1 Partition testing for generalization customization: (a) before generalization customization; (b) case 1: after generalization customization; (c) case 2: after generalization customization; and (d) case 3: after generalization customization.

introduce more restrictions, and thus, could lead to generate additional partitions, for example, *P′* in Figure 6.2(b, c). Similar to generalization customization, all affected partitions need to be retested, and new test cases need to be generated for newly derived partitions.

3. *Reconstruction customization*: Reconstruction customization is potentially more complex than the first two types of customizations, but it is also less likely to happen. Figure 6.3(b) shows a special case for reconstruction customization, while Figure 6.3(c) demonstrates a more general case. To reassure the customization properly works for the component, a similar strategy can be applied.

6.3.3 Boundary value testing

For the partition testing, input domain will be classified into different disjointed partitions. Ideally, every element in each partition has the same possibility to either reveal or hide a fault. But based on programming experiences, this is usually not true. Values that are close to the boundary of the partition are more likely to expose errors. For instance, take for example a for loop from 0 to 100. Ideally, every integer from 0 to 100 should have the same possibility to cause a failure or not. But based on program experiences, values close to 0 or 100 are more error-prone, as they are more likely to have out-of-boundary errors and other similar errors. Thus, when boundary value testing each partition, not only one element will be selected. Instead, additional elements close to the boundary will be tested as well [20]. Boundary value analysis addresses test effectiveness—focusing on the high-risk areas. Partition testing addresses test *efficiency*—reducing the number of test cases needed to obtain a certain level of confidence in the software under test. Therefore,

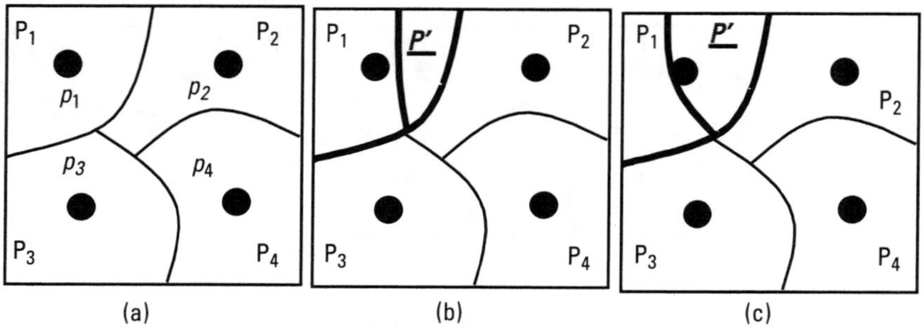

Figure 6.2 Partition testing for specialization customization: (a) before specialization customization; (b) case 1: after specialization customization; and (c) case 2: after specialization customization.

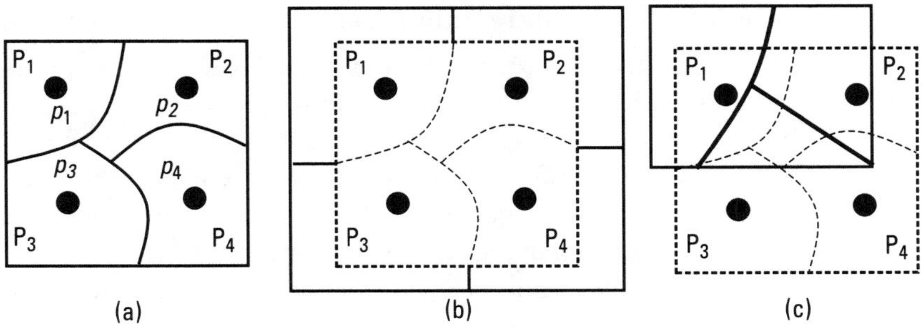

Figure 6.3 Partition testing for reconstruction customization: (a) before reconstruction customization; (b) special case: after reconstruction customization; and (c) general case: after reconstruction customization.

boundary value testing strategies should be used along with partition testing strategies.

Given a one-dimensional space, $x \in [x_{min}, x_{max}]$, we can have the following different types of choices for the boundary values:

1. $x_{min}, x_{min+1}, x_{mid}, x_{max-1}, x_{max}$
2. $x_{min-1}, x_{min}, x_{min+1}, x_{mid}, x_{max-1}, x_{max}, x_{max+1}$

When considering multidimensional space, $x \in [x_{min}, x_{max}], y \in [y_{min}, y_{max}]$ (x and y are independent), different choices can be made for combining the two variables. The following shows some cases of how we chose valid boundary values of x and y. If we want to choose boundary values outside the range, such as x_{min-1} or x_{max+1}, more cases are possible. From Figure 6.4 we can see that the complexity can grow dramatically with the growth of the number of variables involved in the boundary analysis. Moreover, if x depends on y, or vice versa, the selection of boundary values will be more difficult because it may not always be possible to generate all possible boundary values. And it is very difficult to determine the boundary values that can and cannot be reached.

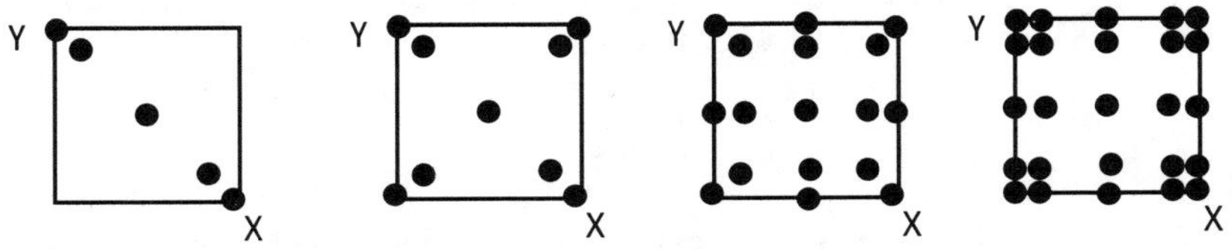

Figure 6.4 Possible test cases for boundary value testing with two variables.

6.3.3.1 Boundary value testing for software components

When customizations are performed on the component, boundary value analysis has to adjust correspondingly.

1. *Generalization customization*: When restrictions are released, the valid value range for the corresponding variable changes accordingly, as shown in Figure 6.5. Assume the customization only affects one variable x, and new_x_{min} and new_x_{max} are new boundaries; so new_x_{min-1}, new_x_{min}, new_x_{min+1}, new_x_{max-1}, new_x_{max}, new_x_{max+1} should be tested. In addition, input values x_{min-1} and x_{max+1} may change their output and thus should be rerun.
2. *Specialization customization:* Figure 6.6 demonstrates the possible effect of specialization customization. Similar to generalization customization, new_x_{min-1}, new_x_{min}, new_x_{min+1}, new_x_{max-1}, new_x_{max}, and new_x_{max+1} should be tested. As to x_{min-1}, x_{min}, x_{min+1}, x_{mid}, x_{max-1}, x_{max}, and x_{max+1}, ideally, they should remain unchanged, for safe critical systems need to be rerun as well.
3. *Reconstruction customization*: Reconstruction customization definitely introduces new boundary values, but it may overlap with existing boundary values, as shown in Figure 6.7. Thus, a general strategy is to test all new boundary values, as well as the boundary values that overlap with new boundary values.

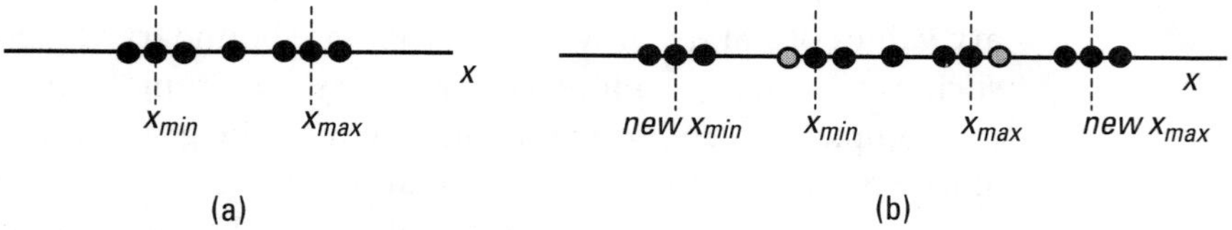

Figure 6.5 Boundary value testing for generalization customization: (a) before and (b) after generalization customization.

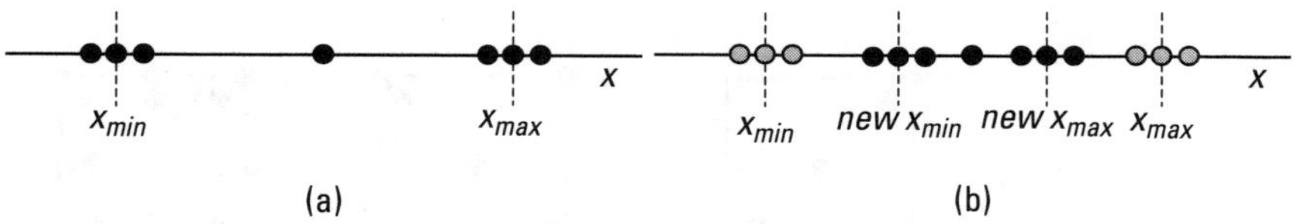

Figure 6.6 Boundary value testing for specialization customization: (a) before and (b) after specialization customization.

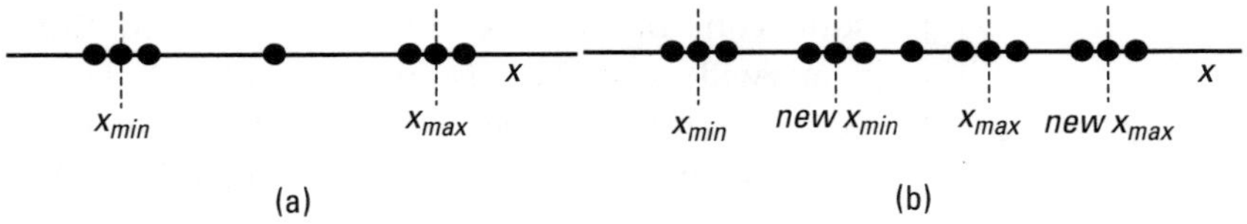

Figure 6.7 Boundary value testing for reconstruction customization: (a) before and (b) after reconstruction customization.

6.3.4 Decision tables–based testing

Partition testing and boundary testing assume variables in the analysis that are mutually independent. In other words, if the variables that are involved in the analysis depend on each other, the partitions or boundary values derived from the analysis may be possible or may not make any sense. To properly analyze complex scenarios where combinations of actions are taken under various logical relationships, decision table–based testing strategies are more appropriate. A decision table includes two parts: (1) all possible conditions and the different combinations, and (2) all possible actions and corresponding responses to different conditions. For instance, the triangle problem to determine whether the three variables *a*, *b*, and *c* can shape a triangle, and if they could, what kind of triangle will be formed. Because the output depends on the relationships among *a*, *b*, and *c*, it is very difficult to divide the input domain into partitions. Consequently it is very difficult to find boundary values. Instead, the decision table–approach, as shown in Table 6.1, can solve this problem nicely.

6.3.4.1 Decision table–based testing for software components

When testing a software component, we can certainly utilize decision-based testing to test individual complex services that are provided by a component. Furthermore, as we have mentioned, there are constraints for interfaces, as well as dependence relationships among different interfaces. With decision table–based strategies, these constraints and dependence relationships will be clearly specified, and therefore, adequately tested.

When customizations are performed for components, provided the verification history information is available, the test adequacy can be determined as follows:

1. *Generalization customization*: When constraints in the specification are generalized, the corresponding change in the decision table is that

some condition will be removed or merged. Consequently, some columns will merge. For instance, in Table 6.1, if we consider isosceles and equilateral triangles as regular triangles, the three middle columns in Table 6.1 will be merged as shown in Table 6.2.

For those merged columns, all corresponding test cases should be rerun to assure no side effects have been brought into the system.

2. *Specialization customization*: Contrary to generalization customizations, specialization will add more constraints as shown in Table 6.3.

   The new constraints often generate one or more new columns. For instance, in Table 6.3, the new constraint C4: $a^2 = b^2 + c^2$ generates a new column. For specialization customization, new test cases are required to cover newly generated columns. Specialization may also split one column into two or more columns. For the same example, if we need to make a distinction between *scalene but not a right triangle* and *scalene and right triangle*, the last column in Table 6.1 will be split into two columns. If that is the case, both of the columns need to be retested.

3. *Reconstruction customization*: As you can imagine, these types of customization activities are more complex then the previous two. The corresponding changes to the decision table can involve adding/removing/modifying rows and columns. The bottom line is for any columns with changed or newly generated values, and for corresponding test cases that need to be rerun or generated.

6.3.5 Mutation testing

Mutation testing, also know as mutation analysis, is a fault-based testing that will determine test adequacy by measuring the ability of a test suite to discriminate the program from some alternative program (mutants). Usually,

Table 6.1 Decision Table–Based Testing Example

Conditions	C1: $a < b + c$	F	T	T	T	T
	C2: $a = b$	–	T	T	F	F
	C3: $b = c$	–	T	F	T	F
Actions	Not a triangle	X				
	Scalene					X
	Isosceles			X	X	
	Equilateral		X			

Assume $a \geq b \geq c > 0$.

Table 6.2 Decision Table for Generalization Customization

Conditions	C1: $a < b + c$	F	T	T
	C2: $a = b$ or $b = c$	—	T	F
Actions	Not a triangle	X		
	Scalene			X
	Regular		X	

Assume $a \geq b \geq c > 0$.

Table 6.3 Partial Decision Table for Specialization Customization

Conditions	C1: $a < b + c$	F	T	T	T	T	—
	C2: $a = b$	—	T	T	F	F	—
	C3: $b = c$	—	T	F	T	F	—
	C4: $a^2 = b^2 + c^2$	—	—	—	—	—	T
Actions	Not a triangle	X					
	Scalene					X	
	Isosceles			X	X		
	Equilateral		X				
	Right triangle						X

Assume $a \geq b \geq c > 0$.

a mutant is created by introducing a single, small, legal syntactic change in the software. Given a program P and corresponding test suite T, the first step of the mutation testing is to construct the alternative programs—that is, mutants, (P). Then each mutant and the original program P have to be executed against each test case. If the output of mutant Q remains the same as the output of the original program P, we consider Q as alive; otherwise, Q is considered to be dead or distinguished. Q remaining alive can be caused from the following two factors:

1. Inadequate test data;
2. The mutant is equivalent to the original program.

If the mutant is equivalent to the original program, then it will not contribute to the adequacy of T and should be eliminated. After equivalent mutants are removed, the number of the remaining mutants that are detected by the test suite can be used to evaluate the corresponding test adequacy. This is formally defined as mutation score:

$$\text{Mutation score} = \frac{|\text{Dead mutants}|}{|\text{Total mutants}| - |\text{Equivalent mutants}|}$$

For early mutating testing techniques, the total number of mutants is usually a very large number, and therefore, it is very expensive to carry out mutation testing. Recent research has shown that by carefully selecting appropriate subsets of mutants, mutation testing can be more efficient and effective, and mutant strength can be equivalent to data-flow testing or domain testing in its ability to determine program correctness [21, 22].

Clearly, even though traditional mutation testing demonstrates high potentials to reveal bugs in a system, the mutation-testing technique typically requires the source code of the program under test, and for complex and large software, it is usually very expensive to be performed. Then, why do we consider mutation testing for software component testing where source codes for components may not available and where efficiency also has a very high priority for software component testing techniques? The answer lies in the fact that component interfaces are the point of interaction, and instead of creating mutants based on the program source code, mutants of software components can be created by syntactically modifying the component interface specification. The interface is the only way to interact with a component, and therefore, this type of mutation testing may provide sufficient confidence to the component. Moreover, for traditional mutation testing, the total number of mutants is often proportional to the square of the number of lines of code. As for mutation testing, the total number of mutations will be proportional to the number of lines of interface specification, which is a much smaller number, and consequently, efficiency could be guaranteed.

To adequately create mutants based on component interfaces, typical interface errors have to be studied carefully. Ghosh and Mathur [23] and Edwards [24] have identified some typical interface-related faults, which include wrong specifications and wrong parameter orders. According to the errors, five types of mutation operators for CORBA-IDL component interfaces are provided, and with proper modifications they can be performed with other component interfaces as well [25–27].

1. *Replace operator*: In CORBA-IDL interface specification, it has to specify the parameter as "in," "out," and "inout," which means input, output or input-and-output parameter. The replace operator will replace one of the parameter types with another one in the list.

2. *Swap operator:* In the interface specification, the order for parameters with the same type requires additional caution. Swap operation will change the order of this type of parameter to test possible specification errors.
3. *Twiddle operator:* Twiddle operation is used for numerical or character variables that substitute x with succ(x) or pre(x).
4. *Set operator:* The set operator will assign a certain value to a parameter. The value usually derived form boundary analysis.
5. *Nullify operator:* Nullify operator nullifies an object reference.

Similar to traditional mutation testing techniques, for interface-based mutation testing an interface-based mutation score can be defined as D/N where D is the number of mutants that are distinguished and N is the number of nonequivalent interface mutants that are generated. Correspondingly, the adequacy criterion requires the interface mutation score to be 1.

In addition to the approaches that are discussed in Sections 6.3.1 through 6.3.5, there are many other black-box testing methodologies—for instance, syntax testing [10, 11], finite-state testing [10, 11], cause-effect graphing [14], and error guessing [14]. Detailed discussion can be found in [10, 11, 14].

6.4 Summary

As discussed in Section 6.3, various complex black box–based testing strategies are critical to the reliability of the target application and have to be carried out. Nevertheless, some concern may arise. First of all, it may not be reasonable for component users to spend considerable effort testing components when they have already spent a fortune purchasing the component. Furthermore, many component users may not be qualified to perform the testing. One of the natural solutions to this problem is for components to provide an automatic testing interface or establish a built-in testing capability. This critical issue will be discussed in Chapter 8 of this book. In addition to black-box testing, when a source is available, white-box testing techniques described in Chapter 7 can be applied.

References

[1] ANSI/IEEE std 1008, "IEEE Standard for Software Unit Testing," 1987.

[2] ANSI/IEEE std 610.12, "IEEE Standard Glossary of Software Engineering Terminology,"1990.

[3] Szyperski, C., *Component Software—Beyond Object-Oriented Programming*, Reading, MA: Addison-Wesley, 1999.

[4] Addy, E. A., and M. Sitaraman, "Formal Specification of COTS-Based Software: A Case Study," *Proc. of 1999 Symposium of Software Reuse*, Los Angeles, CA, 1999, pp. 83–91.

[5] Edwards, S. H., "Toward Reflective Metadata Wrappers for Formally Specified Software Components," *Proc. Workshop on Specification and Verification of Component Based Systems*, held in conjunction with OOPSLA 2001, October 2001.

[6] OMG Meta-Object Facility, version 1.4, http://www.omg.org/technology/documents/formal/mof.htm.

[7] IBM's Bean Markup Language, http://www.alphaWorks.ibm.com/formula/BML.

[8] UML eXchange Format & Pattern Markup Language, http://www.yy.ics.keio.ac.jp/~suzuki/project/uxf/.

[9] Morris, J., et al., "Software Component Certification," *Computer*, Vol. 34, No. 9, September 2001, pp. 30–36.

[10] Beizer, B., *Black-Box Testing: Techniques for Functional Testing of Software Systems*, New York: John Wiley, 1995.

[11] Beizer, B., *Software Testing Techniques*, 2nd ed., New York: Van Nostrand Reinhold, 1990.

[12] Craig, R. D., and S. P. Jaskiel, *Systematic Software Testing*, Norwood, MA: Artech House, 2002.

[13] Jorgensen, P.C., *Software Testing: A Craftsman's Approach*, Boca Raton, FL: CRC Press, 1995.

[14] Myers, G. J., *The Art of Software Testing*, New York: John Wiley, 1979.

[15] Ntafos, S., "On Random and Partition Testing," *Proc. of International Symposium on Software Testing and Analysis (ISSTA)*, 1998, pp. 42–48.

[16] Zhu, H., P. Hall, and J. May, "Software Unit Testing Coverage and Adequacy," *ACM Computing Surveys*, Vol. 29, No. 4, December 1997, pp. 366–427.

[17] Hamlet, D., "Partition Testing Does Not Inspire Confidence," *IEEE Trans. on Software Engineering*, Vol. 16, No. 12, December 1990, pp. 1402–1411.

[18] Weyuker, E., and B. Jeng, "Analyzing Partition Testing Strategies," *IEEE Trans. on Software Engineering*, Vol. 17, No. 7, July 1991, pp. 703–711.

[19] Ostrand, T. J., and M. J. Balcer, "The Category-Partition Method for Specifying and Generating Functional Tests," *Communications of the ACM*, Vol. 31, No. 6, 1988, pp. 676–686.

[20] Hoffman, D., P. Strooper, and L. White, "Boundary Values and Automated Component Testing," *Journal of Software Testing, Verification, and Reliability*, Vol. 9, No. 1, 1999, pp. 3–26.

[21] Offutt, J., et al., "An Experimental Evaluation of Data Flow and Mutation Testing," *Software-Practice and Experience*, Vol. 26, No. 2, 1994, pp. 337–344.

[22] Offutt, J., G. Rothermel, and C. Zapf, "An Experimental Evaluation of Selective Mutation," *International Conference on Software Engineering*, 1993, pp. 100–107.

[23] Ghosh, S., and A. P. Mathur, *Testing for Fault Tolerance*, SERC Technical Report TR-179P, Purdue University, 1997.

[24] Edwards, S. H., et al., "A Framework for Detecting Interface Violations in Component-Based Software," *Proc. of 5th International Conference on Software Reuse*, IEEE CS Press: Los Alamitos, CA, 1998, pp. 46–55.

[25] Delamaro, M., J. Maldonado, and P. Mathur, "Interface Mutation: An Approach for Integration Testing," *IEEE Trans. on Software Engineering*, Vol. 27, No. 3, March 2001, pp. 228–247.

[26] Ghosh, S., and A. P. Mathur, "Interface Mutation," *Journal of Software Testing, Verification, and Reliability*, Vol. 11, No. 4, December 2001, pp. 227–247.

[27] Ghosh, S., and A. P. Mathur, "Interface Mutation to Assess the Adequacy of Tests for Components," *Proc. of TOOLS USA 2000*, IEEE Computer Society Press, Santa Barbara, CA, July 30–August 2, 2000, pp. 37–46.

CHAPTER

7

Contents

White-box testing methods for software components

According to IEEE standards [1], white-box testing, also known as structural testing or glass-box testing, is "testing that takes into account the internal mechanism of a system or component."

White-box testing contrasts with black-box testing in that it requires an internal knowledge of a component. An immediate question is why do we need white-box testing when we have black-box testing to ensure that requirements have been met? Also, if we do need white-box testing, how does white-box testing of software components differ from that of traditional software systems?

The answer to the first question lies in the fact that many program faults can be overlooked by black-box testing [2–4]. For instance, logic errors, which can account for as many as 32% of all possible faults [5], may not be identified effectively by black-box testing; white-box testing is more suitable for that scenario. As for the second question, there is no surprise in the answer: there are not many differences! But this does not mean we do not need to do anything more. Some additional issues need to be addressed when applying white-box testing to software components.

In this chapter, we first investigate typical white-box testing and a hierarchy of different adequacy criteria, and then we look into test-adequacy issues for testing software components.

7.1 Flow graph notation

White-box-testing techniques rely on the structure of a program. Therefore, a representation of the program needs to be defined first. Flow graphs originating from compiler work are often widely used to depict logical control flowing of a program. With proper modification, flow graphs can be used to represent other types of information.

A flow graph G = (V, E) is a direct graph, where V is a set of nodes and each node $v \in V$ represents a basic block. A *basic block* is a maximal sequence of statements, which satisfy all-or-nothing criterion (i.e., if one statement is executed, all the other statements will be executed). The $e \in E \subseteq V \times V$ represents control moving from one node to another node. In a flow graph, there are two special nodes, a begin node and an end node, which represent the entry point and exit point of a program. A *decision node* is a node with two or more outgoing edges that is used to represent branches in a program. For decision nodes, each outgoing edge is associated with a predication, which specifies the conditions under which the control is transferred through this edge. Correspondingly, a *junction node* is a node with two or more incoming edges. For structure programming languages and object-oriented programming languages, a flow graph of a program can be automatically built.

Figure 7.1 demonstrates a pseudocode to determine whether an integer is a composite number. There are a couple of interesting observations. First, some statements need additional processing when constructing a flow graph. For instance, the *for* statement in the example needs to be split into three basic blocks: an initiate block, a loop-control block, and an increment block. All the decision nodes in our example have two outgoing edges. For some statements, such as *switch/case* statements, there could be more than two outgoing edges. We also include S1 in the diagram because it is an initial statement for variable *t*, which will be used later.

It is not only that a flow graph describes the internal structure of a program. More importantly, test cases can be easily tied to a flow graph. Given a test case, there is always a path that corresponds to it. A *path* is a sequence of nodes $(v_1, v_2, \ldots, v_{k-1}, v_k)$ where $(v_i, v_{i+1}) \in E$ for $1 \leq i \leq k$. For a test case, v_1 must be the begin node and v_k must be the end node.

7.2 Path testing

A flow graph provides a means of representing program structures, as well as the test case execution. Therefore, given a flow graph, test cases can be generated accordingly and each test corresponds to a path. Unfortunately, there could be an unlimited number of paths in a flow graph, and we need

```
S1    Boolean Composite(integer t)
S2    if (t > 0) {
S3       for( i = 2; i <= sqrt(t); i++) {
S4          if ( t % i == 0) {
S5             return true;   }
S6       }
S7    }
S8    return false;
```

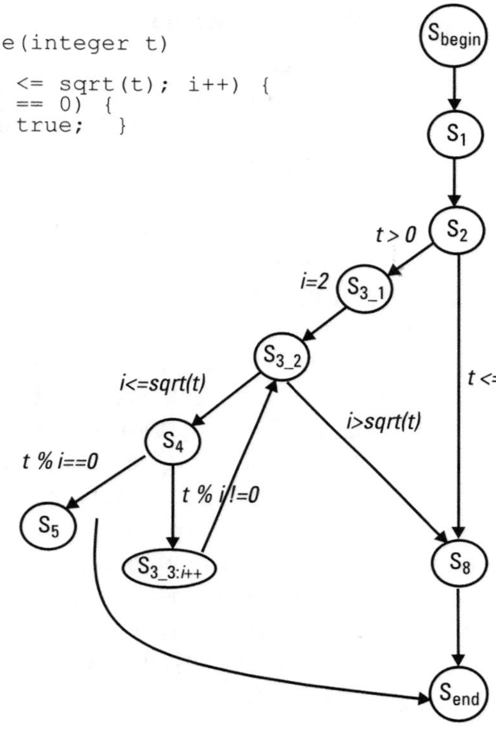

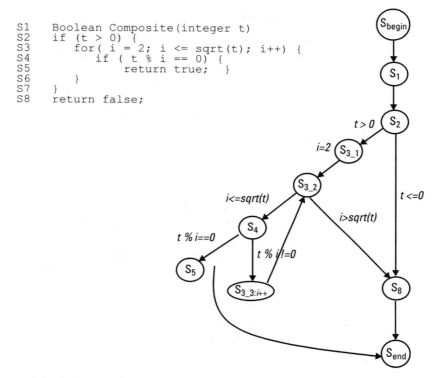

Figure 7.1 A flow graph example.

guidelines to determine how to derive test cases, and when to stop generating test cases. The following sections discuss a group of test-adequacy criteria that can be adopted for different situations.

7.2.1 Statement coverage

When a program is developed, every statement is expected to be executed, and should therefore be run at least once. More formally, a set of test cases T = $\{t_1, t_2, \ldots, t_k\}$ satisfies the *statement coverage criterion* if and only if for any node $v \in V$, there exists a test case t_i and its corresponding execution path P_i, where $v \in P_i$.

For complex statements, such as the statement in Figure 7.1, statement coverage requires coverage of all portions of the statement (i.e., "`i=2`", "`i < sqrt(t)`" and "`i++`" parts). Also, even though statement coverage is a very basic requirement, not all programs can find a test suite that satisfies that criterion. Dead code can occur in a program and will never be reached.

7.2.2 Branch coverage

In flow graphs, decision nodes are special nodes that will determine the next move of the program. For a decision node with a predicate for each of its outgoing associates, to make sure the decision node behaves normally, each of these predicates needs to be tested to be true at least once. In other words, we need to examine all possible transitions from a decision node. More formally, a set of test cases T = $\{t_1, t_2, \ldots, t_k\}$ satisfies *branch coverage criterion* if and only if for any edge $e \in E$, there exists a test case t_i and a corresponding execution path P_i, where $e = (v_i, v_j)$ is in P_i.

Statement-coverage and branch-coverage criteria are very similar. To compare the two adequacy criteria, different mechanisms are proposed. One of the fundamental mechanisms is the subsumed relationship between different adequacy criteria [6]. Test-adequacy criterion C1 *subsume* test adequacy criterion C2, if and only if, given any program P under test and any test suite T, if T satisfies C1, then T must satisfy C2. It can be easily proved that if any test suite T satisfies branch coverage, T satisfies statement coverage because coverage of all edges in a flow graph implies coverage of all nodes in that flow graph. Therefore, branch-coverage criterion subsume statement-coverage criterion. On the other hand, if a test suite T′ satisfies the statement-coverage criterion, it may not satisfy the branch-coverage criterion. For instance, given a test suite T′:

$$t_1 = \{t = 9\}; P_1 = \{S1, S2, S3\_1, S3\_2, S4, S3\_3, S3\_2, S4, S5, Send\}$$
$$t_2 = \{t = -1\}; P_2 = \{S1, S2, S8, Send\}$$

Apparently, test cases in T′ execute all statements in the program. Nevertheless, an outgoing edge S3_2-S8 is not covered, or `i > sqrt(t)` has never been evaluated as true. In order to achieve branch coverage, a third test $t_3 = \{t = 5\}$ needs to be added.

Note, subsume relationships can reflect relative strengths among different adequacy criteria, but there is no guarantee of the fault-revealing abilities of different criteria. This is because subsume relationships measure the structure coverage, not the input domain coverage.

7.2.3 Multiple-condition coverage

Branch-coverage criteria reinforce the testing to cover all outgoing edges of a decision node and consider the predicate as one unit. If the predicate is a Boolean expression that involves multiple atomic predicates or *conditions*, then only two possible combinations are tested. It makes sense to require all combinations of all conditions to be tested at least once. Given a test suite T, if all possible combinations of conditions of any predicates are executed under T, then we can say that T satisfies *multiple condition coverage*.

Similar to statement coverage where 100% statement coverage may not be reached, not all combinations of conditions of a predication can be achieved. For instance, with the predicate "`i > 10 || i < 0`," there are two conditions "`i > 10`" and "`i < 0`" and only three out of four combinations are feasible.

7.2.4 Path coverage

The above three coverage criteria focus on basic elements in a flow graph: statements, branches, and predicates. These criteria cannot reflect the coverage of the paths, which combine nodes and edges. An intuitive approach is to require a test suite to cover all possible paths in a flow graph. But it is not a practical approach because many flow graphs contain a huge number or even an infinite number of paths. For instance, in the example in Figure 7.1, there are $O(t^{\frac{1}{2}})$ different paths, which can be a huge number, depending on t. Instead of a *for* loop in the program, a *while* loop can end up with an infinite number of paths.

To adopt path-coverage criteria in a practical way, we need to select a representative subset of all paths to retest. The selection needs to remove redundant information in a path. There are two straightforward ways. The first is to remove redundant nodes, and the second is to remove redundant edges. In graph theory, an *elementary path* is a path with no repeat occurrences of any node, and a *simple path* is a path with no repeat occurrence of any edge. With these restrictions, given a flow graph, there are usually a very limited number of elementary and simple paths. For instance, in Figure 7.1, there are only three elementary paths:

$$\text{P1} = \{\text{Sbegin}, \text{S1}, \text{S2}, \text{S8}, \text{Send}\}$$

$$\text{P2} = \{\text{Sbegin}, \text{S1}, \text{S2}, \text{S3\_1}, \text{S3\_2}, \text{S8}, \text{Send}\}$$

$$\text{P3} = \{\text{Sbegin}, \text{S1}, \text{S2}, \text{S3\_1}, \text{S3\_2}, \text{S4}, \text{S5}, \text{Send}\}$$

With an additional path P4 = {Sbeing, S1, S2, S3_1, S3_2, S4, S3_3, S3_2, S8, Send}, P1, P2, P3, and P4 are all simple paths.

Simple paths and elementary paths focus on redundant nodes and edges only, and do not indicate the dependence relationships between different paths. McCabe's cyclomatic measurement is an example of specifying dependence relationships between paths [7, 8]. Cyclomatic numbering is based on a theorem of graph theory that for any flow graph, there is a set of paths where all other paths can be represented by linear combination of them. Within the set, no paths can be expressed by linear combinations of other paths, and these paths are called *independent paths*. Therefore, paths within the set are representative paths of the graph and form a base for a new path-testing criterion. More importantly, cyclomatic numbering is a measurement of complexity: It assesses the structural complexity of a program.

The maximum size of a set of independent paths is called the cyclomatic number and can be easily calculated by the following formula:

$$\textit{Cyclomatic number of graph } G = (V, E) = |E| - |V| + 2*|S|$$

In the formula, $|E|$ and $|V|$ are the number of edges and nodes in McCabe's control graph, and $|S|$ is the number of strong connected components. McCabe [9] also provided an algorithm to automatically generate a set of independent paths. In our example in Figure 7.1, the cyclomatic number equals 11 – 9 + 2*1 = 4, and the following set of paths

$$P1 = \{Sbegin, S1, S2, S8, Send\}$$
$$P2 = \{Sbegin, S1, S2, S3\_1, S3\_2, S8, Send\}$$
$$P3 = \{Sbegin, S1, S2, S3\_1, S3\_2, S4, S5, Send\}$$
$$P4 = \{Sbeing, S1, S2, S3\_1, S3\_2, S4, S3\_3, S3\_2, S4, S5, Send\}$$

are a possible set of independent paths. Any other paths—for instance, P′ = {Sbeing, S1, S2, S3_1, S3_2, S4, S3_3, S3_2, S8, Send}—can be represented as a combination of these independent paths—for example, P′ = P2 + P4 – P3.

Based on the cyclomatic number, a test suite T satisfies the *cyclomatic-adequacy* criterion if and only if all independent paths are executed at least once by the test cases in T.

7.2.5 Loop coverage

Actually, the issue of a large number or an infinite number of paths is mainly caused by loop statements in the program, such as for loops and while loops. It is worthwhile to pay a little more attention to how to test loop statements. As with boundary testing in black-box-based approaches, we need to pay more attention to boundary scenarios (i.e., when loops are conducted zero, once, twice, or a maximum number of times, a maximum – 1 number of times). This coverage criterion can be adopted in addition to all the other criteria.

7.3 Data-flow testing

In path testing, a major difficulty is that there are too many feasible paths in a program. Path-testing techniques use only structural information to derive a finite subset of those paths, and often it is very difficult to derive an effective subset of paths.

A program is actually all about manipulating data, which can be generally classified into two categories: data that defines the value of a variable and data that refers to the value of a variable. The following abnormal scenarios may potentially be faults:

- A variable is used before it is defined.
- A variable is defined but never used.
- A variable is defined twice before it is used.

An analysis of the definition and use of a variable can be used to inspect the abnormal scenarios and effectively identify faults in the program.

Within the context of flaw graph, a *definition* occurs in a node. So we claim there is a *definition* of variable *v* in a node *n* if and only if the value of variable *v* is bounded in one of the statements in *n*. Similarly, there is a *use* occurrence of a variable *v* if and only if the value of variable *v* is accessed. The reference of variable *v* can usually be classified into two categories: *computational use* (c-use) and *predicate use* (p-use). If variable *v* is used in defining the value of another variable or is used as an output value, then it is a c-use of variable *v* in node *n*. If the variable is used to determine the true/false value of a predicate, and a predicate associates with an edge e, then this is a p-use of variable *v* in edge *e*. For instance, in Figure 7.1, at node S3_1, there is a definition for variable *i*. For edges (S3_2, S4) and (S3_2, S8), there are p-uses for variables *i* and *t*. If a path $P = (p_1, p_2, \ldots, p_{k-1}, p_k)$, where there is a definition of *v* in p_1 and either p_k has a c-use of variable *v* or (p_{k-1}, p_k) has a p-use of variable *v*, then P is a *definition-use path* (du-path) with respect to variable *v*. Given a du-path with respect to variable *v*, if one of the intermediate nodes also defines *v*, then, the definition at node p_1 will be rewritten, and has no effect on the c-use at node pk or p-use at edge (p_{k-1}, p_k). To distinguish this scenario from other scenarios, a definition-clear path is defined. A $P = (p_1, p_2, \ldots, p_{k-1}, p_k)$ is a *definition-clear path* with respect to variable *v* if and only if P is a du-path with respect to variable *v*, and in addition none of the intermediate nodes, p_2, ..., p_{k-1} have definitions of variable *v*.

Based on the notations of definition, use, c-use, p-use, du-paths, and definition-clear paths, a set of data-flow-based criteria can be used to guide the test-case generation.

A test suite T satisfies the *all-defs* criterion if and only if for every variable *v*, T contains definition-clear paths from every node that has a definition of *v* to either a node that has a c-use of *v* or to an edge that has a p-use of *v*.

A test suite T satisfies the *all-uses* criterion if and only if for every variable *v*, T contains definition-clear paths from every node that has a definition of *v* to all nodes that have c-uses of *v* and to all edges that have p-uses of *v*.

Because uses are further classified into c-uses and p-uses, the all-use criterion can be further classified into four subcategories with different concentration and strength.

A test suite T satisfies the *all-c-uses/some-p-uses* criterion if and only if for every variable *v*, T contains definition-clear paths from every node that has a definition of *v* to every c-use of *v* and if a definition has no c-uses, then a definition-clear path which reaches a p-use of *v* needs to be included in T. On the contrary, a test suite T satisfies the *all-p-uses/some-c-uses* criterion if and only if for every variable *v*, T contains definition-clear paths from every node that has a definition of *v* to every p-use of *v*, and if a definition has no p-uses, then a definition-clear path which reaches a c-use of *v* needs to be included in T. Two other criteria, *all-c-use* and *all-p-use* criteria, just ignore the cases where there are no feasible definition-clear-path scenarios.

For the all-use criteria, all definition and use pairs are tested, but there can be more then one path between the definition and the use of a variable. Therefore, a stronger criterion—all-du-path criterion—needs to be considered. A test suite T satisfies the *all-du-path* criterion if and only if for every variable *v*, T contains all possible definition-clear paths from every defining node of *v* to every use of *v*. The all-du-path criterion has similar issues to those path-based approaches and a number of feasible paths. Similar treatment can be done to obtain a limited-number definition-clear path in this case as well.

Besides these fundamental data-flow criteria, there are several other more complex criteria, such as K-tuples criteria and Laski-korel criteria [10].

Data-flow testing tends to identify import paths among all possible paths. Nevertheless, there are some inherit difficulties. For instance, if array variables or pointer variables are adopted in the program, it is very difficult to conduct precise data-flow analysis. Similar issues arise for interprocedure as well [11]. In addition, the cost of conducting data-flow analysis is much higher than that of path testing.

When testing a program, determining which criterion to choose is all about the balance of the cost and fault-detecting ability. Figure 7.2 shows the subsumed relationships among all the test-adequacy criteria in this chapter [6].

7.4 Object-oriented testing

Nowadays, more and more components are developed by *object-oriented* (OO) programming languages. The above white-box-testing techniques were originally proposed for procedure-based programs. When entering the object-oriented paradigm, adjustments need to be made to adapt some OO features.

7.4.1 Inheritance

In object-oriented programs, a subclass may redefine its inherited functions and other functions may be affected by the redefined functions. When conducting testing for the subclass, how do we treat the inherited functions that

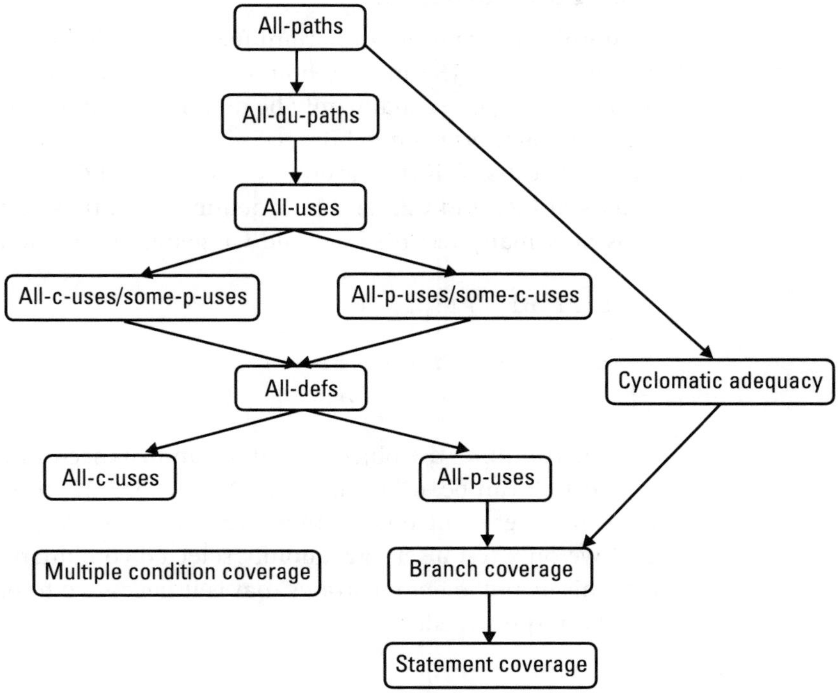

Figure 7.2 Subsumed relationships among test-adequacy criteria.

have been tested? The following pseudocode example shows the fault introduced by the inheritance feature of an object-oriented program.

```
Class foo {
 int v;
 ...
 int f1() { return 1;}
 int f2() { return 1/f1(); }
}

Class foo_child extends foo {
 int f1() { return 0;}
}
```

In the parent class foo, f2 always returns 1 since f1 returns 1. While in the subclass foo_child, f1 has been modified, so, when f2 is invoked in the subclass foo_child, it will fail since f1 returns 0 in the new environment. So even though f2 is inherited from the parent class, it still needs to be retested when we test the subclass foo_child.

7.4.2 Polymorphism

Polymorphism is another important feature of object-oriented programs. With a statement, an object may be bound to different classes during the run time, and the binding can usually only be determined dynamically. The issue introduced by polymorphism is how to treat the dynamic binding issues.

Assume class P1 is the parent class of C11 and C12, and class P2 is the parent class of C21 and C22 and f1 is defined in all these classes. Now the question is how many test cases should we generate for the following code?

```
foo(P1 p1, P2 p2) {
 ...
 return (p1.f1()/p2.f1());
}
```

In the run time, the object p1 can be an instance of P1, C11, or C12, while the object p2 can be an instance of P2, C21, or C22. If the f1 function in C21 returns 0 under some environment, then the function foo will fail. It is likely that if we only test a set of randomly selected combinations, we will miss the fault. This example demonstrates that is it necessary to have a method to test the polymorphism adequately [12].

7.4.3 GUI

GUI is an interesting part of an application. It has many interesting natures that are totally different from the processing parts of the application. The major challenge for GUI programs is that they are very often event driven, and therefore, no clear control flow can be derived from the program source code. Therefore, all the above code-based approaches are not effective.

White-box-testing techniques for traditional programs can be used for testing object-oriented programs. Nevertheless, the test-adequacy criteria for testing object-oriented programs need to be adjusted to accommodate the issues that are raised.

To adequately test OO programs, two types of approaches can be adopted. The first takes OO programs and remodels them so that traditional white-box-testing techniques can be used. The second utilizes OO features to carry out testing.

For the first type of approach, in order to use traditional testing approaches, flow graphs for classes need to be built. As we mentioned before, many OO programs are event driven and do not have a clear control structure. To accommodate the dynamic nature of OO programs, Harrold and Rothermel [13] defined a framework—*class-control flow graphs* (CCFG)—to specify the structure of an OO program. Their framework is based on a call graph of an OO program, and the call graph integrates all possible control flows. Based on the class-control flow graph, all the approaches discussed earlier can be adopted.

The class control flow graph solves the issue of how to apply previous white-box testing techniques on OO programs. But, it does not adequately address the issues caused by inheritance and polymorphism. When determining the adequacy criteria for inherited member functions, statement coverage, branch coverage, and some path coverage techniques only require that nodes, edges, or paths be covered once. As for OO programs, a method in a parent class can be inherited by many child classes (i.e., that method will be invoked in many different scenarios). Simple statement or branch coverage is not sufficient. Similarly, statement, branch-coverage and path-coverage criteria are not sufficient for testing polymorphism. The cause of the above difficulties is the binding issue. *Binding* means an object may be bound to the different classes in the run time in a polymorphic substitution. The following criterion can reinforce the coverage of the binding.

7.4.4 Binding coverage

A test suite T satisfies *all-binding* criteria if and only if every possible binding of each object must be exercised at least once in T when the object is defined or used. If a statement involves multiple objects, then every combination of a possible binding needs to be exercised at least once in T. For instance, in the function foo used above, in the statement "`return (p1.f1()/p2.f1());`"nine different combinations need to be tested.

Many people treat all binding coverage just as a new test adequacy criterion. Actually, it is a new dimension to other adequacy criteria. In other words, it can be combined with other test-adequacy criteria—for instance, statement coverage + all binding coverage, or branch coverage + all binding coverage. When combining binding coverage with other coverage criteria, the number of test cases that need to be generated can quickly go out of bounds. To improve the effectiveness, researchers proposed different methods, such as state-based testing, which is described below, and object flow–based approaches [12, 14, 15].

7.4.5 State-based testing

State-based testing, which can be used at a high level for black-box testing, is very suitable for OO programs because of some OO features like encapsulation. In a state diagram, there are two key elements: states and transitions. Among all states, two are special states: the start state and end state. Any program will start from the start state, and exit at the end state. For classes in an OO program, data members and member functions are encapsulated in one class. Therefore, the values of the data members can be used to determine the states of an instance of the class, and member functions can be used to mange data and can therefore be used to represent the state transitions in the state diagram. In addition, we have noticed that GUI programs are very often event driven. Therefore, state-based techniques work very well for GUI programs.

A state diagram is also a directed graph. Therefore, given a state diagram, a similar coverage hierarchy to that of flow graph can be defined. The basic requirement is to test each state in the state diagram, which is similar to statement coverage. But statement coverage is usually too weak. As with branch coverage, we can further require each of the transitions to be tested at once. State and transition coverage are two fundamental coverage requirements. Many other types of coverage criteria are similar to those of path coverage. More references can be found in [16].

State-based approaches can model the behavior of a program clearly, but obtaining a state diagram from a program is a major issue to overcome. The creation of the state diagram from the program source code usually yields too many states, and the creation of a state diagram based on program specification cannot be fully automated [17].

7.5 Issues in testing software components

Software components are developed either by traditional programming languages or by OO programming languages. Therefore, technically, most of the techniques discussed above can be adopted for testing software components. But there are two major challenges when using white-box-testing techniques to test software components:

- Many traditional software programs are designed and developed in a relatively stable environment, where the specification is well defined and remains stable. On the other hand, software components are usually designed and developed in an isolated environment. When deployed later in another software system, software components may need to be configured or customized in a different environment. Therefore, when testing software components, we can expect to encounter more uncertainties than with traditional software systems.
- The quality requirements for software components are generally higher than the requirements for traditional software modules. This is because white-box testing is often used at unit level. Faults in a program module may have more chances to be identified in a later stage and can be easily fixed internally. For software components, if faults are identified after the delivery, it is harder and more expensive to fix the faults. Moreover, faults in the software component can also hurt the reputation of the component developer.

When performing white-box testing for software components, we need to concentrate on two issues. First, we must identify all possible external and internal scenarios that can occur for a software component, and incorporate them into testing. Second, we must apply stricter adequacy criteria before delivering the component. The first issue is related to both the component specification and the component source code. For instance, for the program in Figure 7.1, without a clear specification, we cannot make any assumptions about the input parameter t. This is a case of an external scenario. For a strong typing programming language like java, t has to be an integer and can be a negative number, a positive number, or 0. More importantly, we know 1 and 2 are two special cases for the composite numbers and they need to be considered in testing as well. No matter what test-adequacy criterion is adopted, all these scenarios need to be considered. For external scenarios, they can be derived without any knowledge of the implementation. For internal scenarios, their derivation is based on the data structure and other information about the source code. For instance, in Figure 7.1, in statement S3, sqrt(t) is used to determine how many times we need to run the loop. Different scenarios can be derived accordingly—for example, what if *t* is 0, what if *t* is a square root of an integer, and what if *t* is not a square root of an integer? If we adopt another way to develop the program, then the "square root" is never an issue. In general, there is no systematic mechanism to derive all possible scenarios; a walk-though along with other careful analysis techniques are possible candidate strategies. Once key scenarios are identified, they will then be taken into consideration of the test adequacy.

7.6 Conclusion

In this chapter, we discuss white-box–based-testing approaches. Many people think white-box testing is the panacea for finding bugs in a program. Unfortunately, white-box testing is a complex process and may not be able find all bugs. In practice, under budget and time limitations, white-box testing very often cannot be carried out as thoroughly as we wish. Subsumed relationships can then be used to determine trade-offs.

Black-box testing and white-box testing target different kinds of bugs in a system. Therefore, to adequately test software components and to assure the best quality, white-box and black-box testing should be combined. More references about this topic can be found at http://www.testingstandards.co.uk.

Testing methodology is only one part of the story. To successfully test software components, test automation, tool support, and many other issues should be addressed. Detailed discussion of these issues can be found in Chapter 8.

References

[1] ANSI/IEEE std 610.12, "IEEE Standard Glossary of Software Engineering Terminology," 1990.

[2] Jones, T. C., *Programming Productivity: Issues for the 80s*, Los Alamitos, CA: IEEE Computer Society Press, 1981.

[3] Pressman, R., *Software Engineering: A Practitioner's Approach*, 5th ed., New York: McGraw-Hill, 2001

[4] Beizer, N., *Software System Testing and Quality Assurance*, 2nd ed., New York: Van Nostrand Reinhold, 1990.

[5] Grady, R. B., *Successful Software Process Improvement*, Englewood Cliffs, NJ: Prentice Hall, 1997.

[6] Rapps, S., and E. J. Weyuker, "Selecting Software Test Data Using Data Flow Information," *IEEE Trans. on Software Engineering,* Vol. 11, No. 4, April 1985, pp. 367–375.

[7] Larsen, L., and M. J. Harrold, "Slicing Object-Oriented Software," *18th International Conference on Software Engineering,* March 1996, pp. 495–505.

[8] McCabe, T. K., *Structured Testing,* Los Alamitos, CA: IEEE Computer Society Press, 1983.

[9] McCabe, T. K., "A Complexity Measure," *IEEE Trans. on Software Engineering,* Vol. 2, No. 4, April 1976, pp. 308–320.

[10] Zhu, H., P. Hall, and J. May, "Software/Unit Test Coverage and Adequacy," *ACM Computing Surveys,* Vol. 29, No. 4, December 1997, pp. 366–427.

[11] Harrold, M. J., and M. L. Soffa, "Inter-Procedural Data Flow Testing," *Proc. of SIGSIFT Symposium on Software Testing, Analysis, and Verification,* December 1990, pp. 58–65.

[12] Chen, M. H., and H. M. Kao, "Testing Object-Oriented Programs—An Integrated Approach," *Proc. of 10th International Symposium on Software Reliability Engineering (ISSRE'99),* 1999, pp. 73–83.

[13] Harrold, M. J., and G. Rothermel, "Performing Data Flow Testing on Class," *ACM SIGSOFT Symposium on the Foundations of Software Engineering,* December 1994.

[14] Alexander, R., and J. Offutt, "Criteria for Testing Polymorphic Relationships," *11th IEEE International Symposium on Software Reliability Engineering (ISSRE '00),* San José, CA, October 2000, pp. 15–23.

[15] Chen, M. H., and H. M. Kao, "Effect of Class Testing on the Reliability of Object-Oriented Programs," *Proc. of 8th International Symposium on Software Reliability Engineering (ISSRE'97),* 1997, pp. 275–282.

[16] Binder, R., *Testing Object-Oriented Systems: Models, Patterns, and Tools,* Reading, MA: Addison-Wesley, 1999.

[17] Whittle, J., and J. Shcumann, "Generating Statechart Designs from Scenarios," *International Conference on Software Engineering,* Limerick, Ireland, 2000, pp. 314–323..

CHAPTER

8

Contents

Test automation and tools for software components

In past decades, people have learned that software testing and maintenance usually costs 40% to 60% of a software project budget [1]. Due to the increase of business competition in the real world, many software workshops are trying their best to shorten the software development life cycle. Hence, software test managers are looking for the solutions to reduce testing costs and time in a test process. In the past decade, software test automation has been approved to be one of the effective solutions.

Until now, many software tool vendors have developed various software tools to help engineers automate a software test process. The basic concepts and methods of software test automation have been discussed in the recently published books [2–5]. These books cover software test automation concepts, issues, solutions, and tools for conventional software projects. However, they do not address the special issues on test automation for modern software components and component-based programs. Test managers and engineers, who work on component-based software construction, are looking for the answers to the following questions:

- What are problems and solutions on the road to test automation for software components?
- What kind of software test tools is needed to support component test automation?
- What are the available software tools for testing software components?

- What are the challenges and concerns in test automation for component-based programs?
- What are the limitations of current software test tools when we apply them on component-based programs?

This chapter discusses test automation of software components and component-based programs. It examines the new issues, challenges, and needs of test automation in component-based software engineering. In addition, it also discusses some examples of component-oriented software test tools. Chapter 12 reports the recent development of component-oriented test frameworks and research efforts to provide systematic test automation solutions for component-based systems and components.

This chapter consists of four sections. The first section reviews the basics of software test automation, including objectives, common issues, and needs. It also provides a test automation process and summarizes the different types of existing test automation solutions and tools. Section 8.2 describes some existing test tools for software components. Section 8.3 discusses the special needs, issues, and challenges in automating a component test process. The discussion focuses on five areas: test management, test execution, test generation, performance validation and measurement, and regression testing. Finally, a summary is given in Section 8.4.

8.1 Software test automation

What is software test automation? *Software test automation* refers to the activities and efforts that intend to automate engineering tasks and operations in a software test process using well-defined strategies and systematic solutions. The major objectives of software test automation are:

- To free engineers from tedious and redundant manual testing operations;
- To speed up a software testing process, and to reduce software testing cost and time during a software life cycle;
- To increase the quality and effectiveness of a software test process by achieving predefined adequate test criteria in a limited schedule. In past years, many businesses and organizations spent much effort on automating software test processes. According to the case studies in [3], a number of them have been able to significantly reduce manual testing efforts and cost. Some of them are able to save as high as 80% of manual testing cost by automating test executions. According to [3], although

some test automation projects did not reduce the testing cost directly, they enhanced the existing manual testing process to deliver better quality software in a limited project schedule. The major key to the success of software automation is to reduce manual testing activities and redundant test operations using a systematic solution to achieve a better testing coverage.

There are various ways to achieve software test automation. Each organization may have its own focuses and concerns in software test automation. Software test automation activities can be performed in three different scopes:

- *Enterprise-oriented test automation,* where the major focus of test automation efforts is to automate an enterprise-oriented test process so that it can be used and reused to support different product lines and projects in an organization;
- *Product-oriented test automation,* where test automation activities are performed to focus on a specific software product line to support its related testing activities;
- *Project-oriented test automation,* where test automation effort is aimed at a specific project and its test process.

Before engaging software test automation, we need to check the current software test process and engineering practices to see where we are on the road to software test automation. We propose the following classification scheme to enable engineers to evaluate the maturity level of software test automation in their test process. As shown in Figure 8.1, there are five basic levels.

- *Level 0: No test tools.* A software test process at this level is a manual process without using any test tools.
- *Level 1: Initial.* A software test process at this level provides engineers with systematic solutions and tools to create, update, and manage all types of software test information, including test requirements, test cases, test data, test procedures, test results, test scripts, and problem reports. No systematic solutions and tools are available to support engineers in test design, test generation, and test executions.
- *Level 2: Repeatable.* A software test process at this level not only provides engineers with tools to manage diverse software testing information, but also provides systematic solutions to execute software tests in a systematic manner. These solutions allow engineers to use a systematic approach to

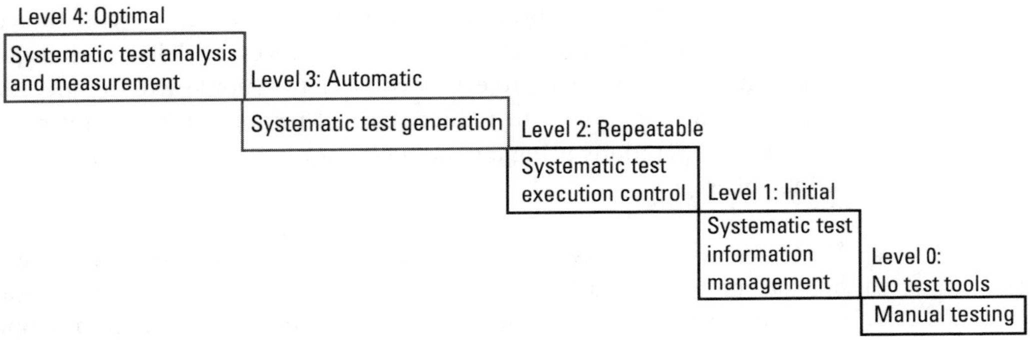

Figure 8.1 Maturity levels of software test automation.

run tests and validate test results. In addition, code-based test coverage analysis tools may be used here to evaluate white-box test coverage. However, no systematic solutions and tools are available to assist test engineers in test design and test generation. There are no systematic tools in place to measure software testability and compare the effectiveness of a software test process according to given test sets and predefined test criteria.

- *Level 3: Automatic.* A test process at this level not only uses test management and test execution tools, but also generates test cases and scripts using systematic solutions. However, the test process has no systematic solutions in place to support software testability measurement and evaluate the effectiveness of software testing.
- *Level 4: Optimal.* This is an optimal level of software test automation. The major advantage of a test process at this level is its systematic solutions for three types of test measurements, including testability measurement, test coverage analysis, and test effectiveness evaluation. The primary benefit of achieving this level is to help engineers understand how well a test process has been conducted, and where software quality issues are.

This maturity evaluation scheme can be used to check the status of test automation for a software test process on a product line or in an organization.

8.1.1 Basics of software test automation

Before spending effort on software test automation, engineers and managers must have a good understanding of software test automation, including its needs, objectives, benefits, issues, and challenges. The essential needs of software test automation are summarized below based on recently published books [2, 3, 5].

- *A dedicated work force for test automation:* Software test automation is not easily carried out unless there is a dedicated team or group who plays as a driving force of software test automation in an organization. This team is responsible for planning, designing, implementing, and deploying test automation solutions by working with developers and test engineers. This group has five major tasks. The first is to identify the essential needs and objectives of test automation. The next is to plan and design test automation solutions. The third is to select and develop necessary tools. The fourth is to introduce, deploy, and evaluate test automation solutions into a project, a production line, and an organization. The last task is to train engineers and maintains the developed tools. Without a dedicated work force, test automation efforts and activities may result in ad hoc ineffective solutions that are poorly reused and highly expensive to maintain.
- *The commitment from senior managers and engineers:* The commitment of senior managers and engineers is essential for software test automation. In addition to the committed budget and schedule for software test automation, they must be committed the efforts into the following areas:
  - Understanding software test automation, including its issues, solutions, difficulties, and complexity;
  - Establishing a well-designed and practice-oriented test automation plan, strategy, and infrastructure for projects, production lines, and organizations;
  - Developing and maintaining test automation solutions;
  - Learning and using the provided test tools to achieve well-defined test criteria.
- *The dedicated budget and project schedule:* Software test automation costs money, although it could shorten a software testing cycle, improve testing quality, and reduce manual testing efforts. A dedicated budget and a well-planed schedule are needed to consider test tool evaluation, development, deployment, and training. In addition, the maintenance cost for automated tests and tools must be included.
- *A well-defined plan and strategy:* It is impossible to achieve the objectives of software test automation without a good plan and a cost-effective

strategy. A good plan usually is practical-oriented with reachable goals and incremental deliverables. An organization-oriented plan should include the following items:

- The ranked test automation objectives and goals;
- A well-defined test tool selection scheme and evaluation policy [2, 6];
- Well-defined quality control, standards test criteria, and test information formats.
- A test automation plan for a software project (or a product line) should be developed as soon as the software requirement specification is available. An ideal plan should address the test automation focuses and needs, specify the required tools, and the implementation schedule.

- *Talent engineers and cost-effective testing tools:* The engineers of a test automation team need a solid knowledge background on software testing and test automation, tool development skills, and working experience on software testing and tool usage. All test tools should be evaluated before they are deployed. In [2], E. Dustin, J. Rashka, and J. Paul provided the details about how to select and evaluation test tools.
- *Maintenance of automated software tests and tools:* In most cases, we easily ignore the fact that all automated software tests and tools (or facilities) must be updated and maintained for the reuse for future product releases and other projects. Sometimes the maintenance cost of automated tests is not cheap. Whenever the maintenance cost is higher than the manual testing cost, these tests should be abandoned.

On the road to software test automation, there are a number of common issues. They are summarized here.

- *Poor manually performed software test process:* A common issue in software test automation is to automate a poor, manually performed software test process. In a poorly managed manual test process, there are no well-defined quality control procedures and test information standards. Engineers use ad hoc test design methods and test coverage criteria. Automating such a test process usually is costly, ineffective, and inefficient for the following reasons:

- Tests from the current test process are not effective in discovering errors. Thus, automating these tests is not effective to uncover errors.
- Engineers are not ready to involve test automation because of their lack of understanding of software test automation.
- Ad hoc quality control process, inconsistent test information formats, and poorly defined test criteria make it impossible to design and develop cost-effective test automation solutions and tools.

- *Late engagement of software test automation in a software product life cycle:* This is another common mistake in the practice of software test automation. In many cases, people do not pay attention to test automation until they found that there is not enough time to conduct software testing manually when software design and implementation has been completed. Engaging software test automation activities in the later phases of a software development process have two serious problems. First, it is too late for engineers to plan and implement a cost-effective test automation plan. For example, it is too late for engineers to build well-structured testable components and develop supporting facilities for component-based software testing after software implementation. Second, it is too late to think about project budgeting and scheduling for test automation.
- *Unrealistic goals and unreasonable expectations:* Engineers and managers involved in test automation frequently set up unrealistic objectives and goals in test automation. This not only affects the proper evaluation of the effectiveness of test automation efforts, but also results in an unrealistic test automation plan that is not easy to be implemented with a limited budget and schedule. In many cases, engineers may come out unrealistic goals that are very hard and costly to achieve. For example, achieving 100% automatic regression tests for functional validation is an unreachable goal. Senior managers may have unreasonable expectations to the effectiveness of software test automation. For example, a senior vice-president may expect to see a big impact of initial test automation efforts on a poorly managed test process for the first product release. This type of unrealistic expectation will affect the objective evaluation of test automation impacts, or even change the committed decisions on a project budget and schedule for test automation.
- *Organization issue:* In many cases, a test automation project fails due to the organization issue. A common problem is the lack of a dedicated work force for test automation. This implies that individual teams with very

limited resources carry out their own test automation efforts and activities without an overall test automation strategy and infrastructure for a project or a product line. These activities focus on a specific part of a project, such as a subsystem, without concerning the future reuse and share of the results of test automation efforts across projects and teams. Hence, the automated tests and developed tools may be poorly maintained, reused, and integrated due to the lack of good collaboration and coordination among different teams.

- *Lack of good understanding and experience of software test automation:* One of the worst situations in which to begin software test automation is one that lacks good engineers and managers. This could result in a poor test automation plan, ineffective test automation solutions, and low-quality test tools because engineers do not understand software test automation issues, problems, and methods. Our suggestion is to find experienced engineers and managers or train them before starting software test automation.

There are a number of essential benefits from test automation. Some of them are:

- Reducing manual software testing operations and eliminating redundant testing efforts;
- Producing more systematic repeatable software tests and generating more consistent testing results;
- Executing many more software tests and achieving a better testing coverage in a very limited schedule.

8.1.2 Software test automation process

To achieve software test automation, a cost-effective process is needed to support engineers to carry on various test automation activities and develop required test automation solutions. As shown in Figure 8.2, a test automation process must consist of the following steps:

- *Step #1: Test automation planning.* This is the initial step in software test automation. The major task here is to develop a plan that specifies the identified test automation focuses, objectives, strategies, requirements, schedule, and budget. In the real world, a test automation plan is usually developed for a specific project or a product line at the earlier phase of a software development process.

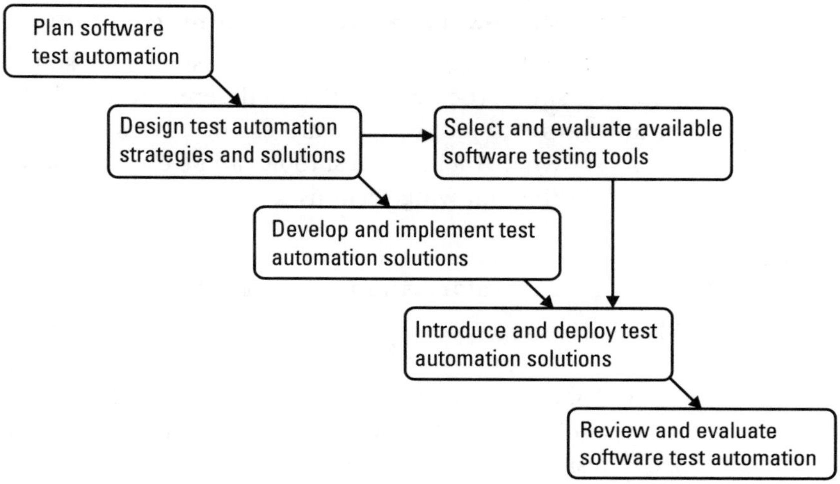

Figure 8.2 A software test automation process.

- *Step #2: Test automation design.* The primary objective of this step is to draw out the detailed test automation solutions to achieve the major objectives and meet the given requirements in a test automation plan. There are two basic tasks. The first is to identify and select the available tools (such as commercial and in-house tools) to support the automation of a test process. To conduct this task, engineers need detailed guidelines and evaluation criteria for selecting test tools. E. Dustin et al. in [2] discussed a set of selection guidelines and criteria. The other task is to design the required automated solutions.
- *Step #3: Test tool development.* At this step, the designed test automation solutions are developed and tested as quality tools and facilities. The key in this step is to make sure that the developed tools are reliable and reusable with good documentation. Many test automation projects failed due to its low quality and poor documentation.
- *Step #4: Test tool deployment.* Similar to commercial tools, the developed test tools and facilities must be introduced and deployed into a project or onto a product line. At this step, basic user training is essential, and proper user support is necessary.
- *Step #5: Review and evaluation.* Whenever a new tool is deployed, a review should be conducted to identify its issues and limitations, and evaluate its provided features. The review results will provide valuable feedback to the test automation group for further improvements and enhancements.

8.1.3 Different types of test automation tools

Effective and efficient test tools are essential to software test automation. Table 8.1 shows the classification of different test tools. In Table 8.2, a number of test tools and tool vendors are listed.

As shown in Table 8.1, test tools can be classified into several categories, as shown in the following sections.

8.1.3.1 Test information management tools

There are three types of test information management tools:

- *Test information management tool:* It supports engineers to create, update, and manage all types of test information, such as testing requirements, test cases, data, procedures, and results. Mercury Interactive's TestDirectory is an example. A Web-based test management system is reported in [7].
- *Test suite management tool:* It enables engineers to create, update, and manage various software test scripts for test execution. TestManger in eValid testing tool (http://www.evalid.com) is a typical example. Other examples are listed in Table 8.2.

Table 8.1 A Classification of Software Test Automation Tools

Test Tool Types	*Basic Descriptions of Different Types of Test Tools*
Test information management	Systematic solutions and tools support test engineers and quality assurance people to create, update, and maintain diverse test information, including test cases, test scripts, test data, test results, and discovered problems
Test execution and control	Systematic solutions and tools help engineers set up and run tests, and collect and validate test results
Test generation	Systematic solutions and tools generate program tests in an automatic way
Test coverage analysis	Systematic solutions and tools analyze the test coverage during a test process based on selected test criteria
Performance testing and measurement	Systematic solutions and tools support program performance testing and performance measurement
Software simulators	Programs are developed to simulate the functions and behaviors of external systems, or dependent subsystems/components for a under test program
Regression testing	Test tools support the automation performance of regression testing and activities, including test recording and replaying

Table 8.2 Different Types of Test Tools and Tools Vendors

Types of Test Tools	*Test Tool Vendors*	*Test Tools*
Problem management tools	Rational Inc.	ClearQust, ClearDDTS
	Microsoft Corp.	PVCS Tracker
	Imbus AG	Imbus Fehlerdatenbank
Test information management tools	Rautional Inc.	TestManager
	Mercury Interactive	TestDirectory
Test suite management tools	Evalid	TestSuiter
	Rational Inc.	TestFactory
	SUN	JavaTest, JavaHarness
White-box test tools	McCabe & Associates	McCabe IQ2
	IBM	IBM COBOL Unit Tester
		IBM ATC
		Coverage Assistant Source Audit Assistant Distillation Assistant Unit Test Assistant
Test execution tools	OC Systems	Aprob
	Softbridge	ATF/TestWright
	AutoTester	AutoTester
	Rational Inc.	Visual Test
	SQA	Robot
	Mercury Interactive	WinRunner
	Sterling Software	Vision TestPro
	Compuware	QARun
	Seque Software	SilkTest
	RSW Software Inc.	e-Test
	Cyrano Gmbh	Cyrano Robot
Code coverage analysis tools	Case Consult Corp.	Analyzer, Analyzer Java
	OC Systems	Aprob
	IPL Software Product Group	Cantata/Cantata++
	ATTOL Testware SA	Coverage
	Compuware NuMega	TruCoverage
	Software Research	TestWorks Coverage
	Rational Inc	PureCoverage
	SUN	JavaScope
	ParaSoft	TCA
	Software Automation Inc.	Panorama
Load test and performance tools	Rational Inc.	Rational Suite PerformanceStudio

Table 8.2 Continued

Types of Test Tools	*Test Tool Vendors*	*Test Tools*
	InterNetwork AG	sma@rtTest
	Compuware	QA-Load
	Mercury Interactive	LoadRunner
	RSW Software Inc.	e-Load
	SUN	JavaLoad
	Seque Software	SilkPerformer
	Client/Server Solutions, Inc.	Benchmark Factory
Regression testing tools	IBM	Regression Testing Tool(ARTT)
		Distillation Assistant
GUI record/replay	Software Research	eValid
	Mercury Interactive	Xrunner
	Astra	Astra QuickTest
	AutoTester	AutoTester, AutoTester One

- *Problem management tool:* It helps engineer bookkeeping and manage the discovered problems during a test process. In [8], J. Gao et al. reported a Web-based problem management system and its deployment experience on a global software production line. Some commercial tools are listed in Table 8.2.

8.1.3.2 Test execution tools

These tools are programs developed to control and execute program tests and test scripts automatically. They usually consist of the capability to set up the selected test scripts and test data, invoke and execute them, and validate the test results based on the expected testing outputs. Mercury Interactive's WinRunner is a typical example. The other samples are listed in Table 8.2.

8.1.3.3 Test generation tools

These tools refer to the programs that generate tests for an under test program using a systematic solution. There are two classes:

- *White-box test generation tools:* They generate white-box tests based on program source code and structures using program-based test models and methods, such as the basis path testing technique.
- *Black-box test generation tools:* They generate black-box tests based on program requirements using black-box test methods, such as random testing and boundary value analysis methods.

8.1.3.4 Test coverage analysis tools

These tools can be used to analyze and monitor the test coverage for a test process based on a selected test criterion. The branch coverage and the statement coverage are typical test coverage criteria for white-box testing. The boundary value coverage is one example of black-box testing. Some commercial test coverage tools are listed in Table 8.2—for example, Software Research's TestWorks Coverage and Rational Inc.'s PureCoverage.

8.1.3.5 Performance measurement tools

These tools support program performance testing and measurement, including performance data tracking and collection, performance evaluation, and analysis. As mentioned in Chapter 11, these tools are developed based on predefined performance models and metrics. Mercury Interactive's LoadRunner is one typical example. It can be used to perform load testing for a server program to check the loading performance of a software product at the system level. Other performance tools are given in Table 8.2.

8.1.3.6 Software simulators

These tools refer to the programs developed to simulate the functions and behaviors of external software/hardware entities, components, or subsystems. Program simulators are necessary and useful for program integration and system testing. OPNET [2], developed by MIL 3 Inc., is one example. It is a network simulation program that is useful to simulate the network behaviors and performance based on defined network models. In general, there are three types of software simulators:

- *Model-driven program simulators:* These simulate the behaviors of a program process based on a specific model, such as a finite state machine.
- *Message-driven program simulators:* These simulate the behaviors of a communication process to generate the protocol-oriented outgoing messages, based on the predefined incoming messages.

8.1.3.7 Regression test tools

These tools support regression testing and consist of the following functions:

- *Software change analysis:* This refers to a systematic facility that identifies various types of software changes and discovers their ripple effects and impacts. These changes include new, updated, and deleted components and elements between product releases [9].

- *Software test selection:* This refers to a systematic facility that assists engineers in selecting reusable tests for regression testing based on software change information. In [10], the authors reviewed the existing test selection methods based on program structural models to find out reusable test cases for object-oriented programs. A requirement-based program test selection method is discussed to select reusable black-box program tests in [11].
- *Test change analysis:* This refers to a systematic solution that identifies the necessary test changes based on the given software changes. The major task of test change analysis is to find reusable tests and identify obsolete tests so that new tests can be added, and existing tests will be updated.
- *Test recorder and replayer:* This refers to a software tool that records the executed tests, and replays them for retesting. Mercury Interactive's WinRunner is a typical retesting tool that is able to record and replay user-system interactive sequences for Windows-based programs using predefined test scripts. Unlike WinRunner, eValider is developed to support the recording and replaying of Web-based system-user interactive sequences.

8.2 Component-oriented software test tools

Recently, a number of software test tools have been developed to support test automation of software components. This section discusses two examples. One of them is a specification-based black-box test tool, known as ADLscope, for software components. The other is a test tool set for Java components that is called Java Compatibility Test Tools.

8.2.1 ADLscope: a specification-based test tool for components

ADLscope is a specification-based automated testing tool that utilizes the formal specification of a program unit as the basis for test coverage measurement. Testers can use ADLscope to validate *application programmatic interfaces* (APIs) written in C programming language based on the given formal API specifications. A formal specification language, known as the *assertion definition language* (ADL), is used to formally specify each API interface for a program component.

According to [12, 13], ADLscope includes the following functional capabilities:

- ADLscope is a formal specification-based testing tool. It supports the unit testing activities for a program component based on its ADL specification. ADL is a formal specification language developed at Sun Microsystems Laboratory to describe behavior of functions in the C programming language. In the ADL specification, the behavior of a C function is specified by a set of assertions. These assertions are based on first-order predicate logic and use the same type system as C. The details descriptions of ADL can be found in the ADL language reference manual [12].
- ADLscope is a test coverage tool. It measures coverage of program units with respect to their ADL specifications. This feature provides testers an estimate of test adequacy based on program specifications. Unlike other program-based test coverage tools, ADLscope measures the test coverage for the program specification instead of program code coverage. The coverage metrics in ADLscope is associated with the expression syntax of the ADL language. Some of the metrics are based on existing test selection strategies. The multiple condition strategy is used on logic expression, and weak mutation testing is used to relational expressions. In ADLscope, various coverage metrics are defined and customized to deal with conditional expressions, implication expressions, relational expressions, equality expressions, normally expressions, group expressions, unchanged expressions, and quantified expressions. The detailed information about the coverage metrics can be found in [13].
- ADLscope facilitates test selection. Based on the reported coverage, testers can select and define additional test data to validate the uncovered conditions and cases specified in program unit specifications.
- ADLscope facilitates unit test activities, including unit test execution, test result reporting, and checking.
- ADLscope facilitates API testing. It is very useful to validate the APIs of a program or components based on the given ADL-based formal specification.

8.2.2 The Java Compatibility Test Tools and technology compatibility kit

To effectively attract and support component users to adopt a specific component technology, a component vendor must provide a set of tools to assist application engineers to develop, test, and deploy newly generated components based on the given technology and its specification. Sun Microsystems has used this approach to support Java-based component specification leads, and its *Java Community Process* (JCP) Members. The JCP is an open organization of international Java developers and licensees whose charter is to develop and revise Java technology specifications, reference implementations, and technology compatibility kits.

8.2.2.1 What are the Java Compatibility Test Tools?

The Java Compatibility Test Tools (Java CTT), recently released by Sun Microsystems, are a set of tools designed to serve this purpose. According to E. Ort in [14, 15], Java CTT are designed to help JCP members create a *technology compatibility kit* (TCK), which is a suite of tests, tools, and documentation that provides a standard way of testing an implementation for compliance with a Java-based technology specification. A TCK is one of the three major deliverables, along with a Java-based component specification and its reference implementation. The specification describes in detail a Java technology, including its APIs. The *reference implementation* (RI) is a working implementation of the Java technology that proves that the specification can be implemented.

The TCK is a test environment for a Java-based component technology. It can be used to test implementations for compliance with the given specification. Each TCK is unique to its associated Java technology; however, all TCKs must have the following elements:

- *TCK test suite:* This includes a set of test cases for testing an implementation compliance with a Java technology specification.
- *Test harness:* This includes a set of applications and tools that support the management functions for the test suite. These management functions include automatic test loading, test execution, and test reporting.
- *Documentation:* This includes the user-level documents that specify how to use the TCK—for example, how to use test harness to run the TCK test suite.

8.2.2.2 The tools in the Java CTT

According to E. Ort [14, 15], the Java CTT includes four tools:

- *JavaTest Harness:* This is a test suite management tool that manages TCK test suites and related test harness. It provides a test bed to assist engineers to control and execute test cases and test scripts in TCK test suites.

- *Spec Trac:* This is a specification-based tool that can be used to check the test coverage for a Java component's APIs. It can be used to identify assertions in a Java component specification, and report on a TCK's test coverage of assertions based on the given specification.
- *Test Generation:* This tool provides a semi-automatic test generation capability. It automates some of the steps required to build TCK test cases and scripts by working with JavaTest Harness. For example, it automatically incorporates commonly used code segments into test script files.
- *Sig Test:* This is a tool that determines if all the necessary API members are in the tested implementation based on a given specification. Moreover, it also checks if there are no additional members that are illegally extend a given API.

8.2.2.3 What is JavaTest Harness?

The JavaTest Harness is a general purpose, fully featured, flexible, and configurable test harness. It manages TCK test suites for unit testing. The JavaTest Harness is a powerful tool for configuring, sequencing, and running test suites that consist of tests for validating component APIs and compilers. It can run tests on a variety of test platforms, including servers, workstations, browsers, and small devices. It provides testers with a variety of test execution models to control and run API compatibility tests, language/compiler tests, and regression tests.

JavaTest Harness provides the following basic functions:

- Create, configure, and TCK test suites;
- Set up and invoke a test execution;
- Monitor a test execution, and collect test results;
- Browse and search diverse test information in test suites;
- Generate and view test reports for a test execution.

JavaTest supports two operation modes:

- *GUI mode:* In this mode, it can be used to browse, configure, and run test cases in test suites, and monitor and report the test results through a graphic user interface.
- *Batch mode:* In the batch mode, it provides a complete test execution function without using a graphic user interface. This allows testers to run tests in test scripts to achieve test automation.

In addition, JavaTest Harness provides the following additional features:

- *HTML reports:* An HTML-based test report can be generated for each test run after test execution.
- *Auditing test runs:* Auditing test runs can be performed in the batch mode and the GUI mode.
- *Web server to monitor batch mode:* The JavaTest Harness provides a small Web server to allow testers to monitor and control test progress while tests are running.
- *Run tests on small systems:* It provides an agent (a separate program that works in conjunction with the JavaTest Harness) to run tests on systems that cannot run the JavaTest Harness.
- *On-line help:* An on-line help facility is provided to assist testers to understand how to use this tool to run test suites and evaluate test results. It also provides context sensitive help, full-text search, and keyword search.

8.3 Component test automation in CBSE

In Section 8.2, we revisited the basics of software test automations. In this section, we discuss the special needs, issues, and challenges of component test automation in the component-based software-engineering paradigm. As shown in Table 8.3, there are a number of challenge questions relating to component test automation. Here, we discuss them in detail.

8.3.1 Systematic management of component testing information

In traditional unit testing, software developers are responsible for performing black-box and white-box testing of software modules before integration. During unit testing, developers create, update, and maintain diverse types of unit test information. Since these software modules are not considered as final products, in many cases, developers are used to managing and maintaining the unit test information in an ad hoc manner. Since components in component-based software engineering become reusable products now, they will be changed and reused later by many component users, it is necessary for component developers and testers to manage and maintain component test information in a systematic way.

Table 8.3 Challenges in Component Test Automation

Areas of Component Test Automation	*Challenge Questions Relating to Component Test Automation*
Component test information management	How does one book keeping and manage component-oriented test information in a systematic way to support component evolution?
Component test execution and control	How does one construct a plug-in-and-play test bed for software components?
	How does one generate component test drivers and scripts in a systematic way?
Component test generation	How does one generate tests systematically for customized software components?
	How does one generate cost-effective black-box tests for software components?
Component test coverage analysis	How does one monitor and analyze component test coverage in a systematic way?
	What are the adequate test coverage criteria for components and customized components?
Component performance measurement	How does one track and measure component performance in a systematic manner?
	How does one create measurable components? How to create plug-in-and-measure environment for software components?
Building testable components	How does one increase component testability in a systematic way?
	How does one create BIT or testable components in a systematic manner?
Component retesting	How does one identify component element changes and their impacts on component tests in a systematic manner?
	How does one select and reuse component tests using a systematic way?

Therefore, on the road to component test automation, the first step is to manage and maintain diverse component test information using systematic solutions. Engineers need three types of systematic solutions:

- *A test information management system:* This is used by engineers to create, update, maintain, and manage diverse component test information. This system allows component developers, test engineers, as well as component users to track and report all types of component test information to support both manual and automatic component tests. The typical component test information includes required test items, test cases, test data, test procedures, and test results.
- *A test suite management system:* This is used by test automation engineers to create, update, maintain, and manage diverse test suites and related test scripts in a systematic manner. This system is essential to support the automatic component test execution. Component developers and test

engineers use this solution to configure, create, and maintain component test suites and test scripts for each component release.

- *A problem management system:* It is used by engineers to create, update, maintain, and manage diverse problem information during component testing. In vendor-oriented component testing, component vendors need this system to track, analyze, and report all types of component problems using a consistent format. During user-oriented component testing, component users need a similar solution for bookkeeping and for reporting all component problems for validated component releases.

Fortunately, many existing tools are available for engineers to manage component test information. However, there are a number of issues and challenges.

- Computerizing all types of component testing information for a component test process is an essential key to the success of component test automation. Achieving this goal requires the support and commitments from engineers and managers.
- Having enterprise-oriented standards and formats for diversified component test information is the necessary foundation in component test automation. This is not only important to vendor-oriented component testing, but also critical to user-oriented component validation, integration, and system testing.
- Because test suite management systems are strongly dependent on component technology, it is not easy for component users to deal with components comprised of different technologies.
- Since current commercial components do not provide enough test information to component users, it is costly and difficult for component users to achieve component test automation due to the lack of component artifacts and testable components.

8.3.2 Automatic test execution for software components

Automatic component test execution not only reduces the number of testing efforts from component developers and test engineers in a component vendor, but also reduces a great deal of efforts in component validation. To achieve this goal, we need a systematic solution for each component to provide the following three functions:

- Set up component tests (including test suite, test data, and procedures) in a component test bed;
- Execute component tests in a component test bed;
- Collect the test results from test executions and validate them based on the expected results.

There are three traditional approaches to establishing a component test bed.

- *Script-based approach:* In this approach, a script-based programming language is defined to allow test engineers to configure component test suites by creating, updating, and maintaining test scripts manually. A script compiler and an execution environment are provided to set up component tests using scripting functions, run these test scripts, and validate the test results. SUN Microsystem's Java Harness is a typical example. It allows engineers to create and execute Java component test scripts to validate component functions in a black-box view.
- *Record and replay approach:* The basic idea of this approach is to implement a systematic solution that allows engineers to record executed tests and results during manual testing, and to replay them later for automatic retesting. This method has been used to develop GUI-based testing tools to perform record and replay black-box tests. Today, many GUI-based testing tools also provide the scripting function to allow users to edit and update GUI-based test scripts. Software Research's eValid is a typical example. Unlike other tools, eValid is developed to test Web-based client software.
- *Component API-based approach:* In this approach, a component test bed is built to interact with components based on component application interfaces. A component test bed usually consists of three parts. The first part plays as a test execution controller that controls test executions of a component. The second part supports component test harness. It assists engineers to create and update component test drivers and stubs manually and/or systematically. The last part works as a test result checker. It collects, monitors, and checks component unit test results. Unlike the first approach, no script languages are used here. Component test drivers and stubs can be generated manually or systematically based on the given component interfaces.

Recently, people have begun to look for new approaches to support component test execution. Creating testable software components with built-in tests is one of them. The basic idea is to insert component unit tests as a part of software components. As mentioned in Chapter 5, the major purpose of this approach is to increase component testability by creating executable and testable components. One typical application of this approach is to insert built-in acceptance tests for software components. Although it is a simple way to create testable components, it increases component complexity and the required system resources. Besides, it increases a lot of programming overhead for component developers unless some automatic solution is provided. Chapter 12 reports the detailed research work on creating testable components.

8.3.2.1 Primary issues and challenges

Based on the current available tools and solutions, engineers encounter the following issues and challenges to achieve automatic test execution for software components:

- The existing script-based test tools usually need a higher scripting cost in preparing test cases and scripts manually. Moreover, component test scripts must be maintained whenever components are changed. The script maintenance cost could be very high if test scripts are not carefully designed. Sometimes, engineers who design test scripts may only focus on how to create executable test scripts without paying attention to test coverage.
- Current solutions for building component test drivers and stubs are highly dependent on component application interfaces. Since each component has its own functions, attributes, and application interfaces, generating component test drivers and stubs is very time consuming and even difficult for component users in user-oriented component testing due to the lack of detailed component artifacts.
- According to our recent industry experience, frequently changing component API interfaces, during a component development process, results in a higher cost and ineffective efforts in automatic test execution when the script-based testing approach is used.

To support component test execution, engineers have the following needs:

- A plug-in-and-play unit test environment to support the test execution of different components. This implies that the unit test environment should be able to support different components.
- Consistent interaction protocols and interfaces between a component and its test bed, test suites, and a unit test repository.
- Systematic solutions to assist engineers to generate component test drivers and stubs based on component API information and available component artifacts, such as a component profile.

8.3.3 Automatic test generation for components

Automatic test generation for components refers to generating component tests (such as test cases, test data, and test scripts) using systematic solutions in component testing.

There are two general approaches for achieving automatic component test generation:

- *White-box component test generation:* In this approach, component tests are created automatically using program-based testing methods to validate component internal structures, logic, data, and functional behaviors. As discussed in Chapter 7, most existing white-box testing methods (such as basis path and state-based testing methods) are still applicable to white-box testing of software components. Hence, existing white-box test tools are still useful for component vendors to generate (or assist to generate) program-based component tests during component testing.
- *Black-box component test generation:* In this approach, component tests are created automatically using specification-based testing methods to validate external visible component functions, behaviors, and interfaces. The essential issue here is how to assist engineers to generate proper test inputs and outputs for each test case. As discussed in Chapter 6, most existing black-box testing methods (such as random test method, equivalence partition, and boundary value analysis) are still applicable to component black-box testing. This implies that the specification-based test tools are still useful for component black-box testing.

With the increase of software component complexity, automatic test generation for components becomes very challenging and difficult for the following reasons:

- Components supporting graphic user interfaces increase the complexity of software test generation. Because each GUI-driven function is performed based on a number of sequential system-user operations involving multiple windows. Each window consists of a number of input and output elements. This really complicates the traditional test generation problem where we used to consider a simple black box with an input vector and an output vector.
- Components involving multimedia inputs/outputs increase an additional complexity factor into the test generation problem.
- Distributed software components and objects bring the distributed communications and interaction issues into component test generation.

Engineers encountered several issues and challenges in component test generation:

- Many existing test tools do not provide test generation features and functions.
- There is a lack of systematic solutions and domain-specific black-box test criteria to support component test generation based on component API specifications.
- Most GUI-based test tools only provide the record-and-replay feature without well-defined test criteria.

The primary needs in automatic component test generation are:

- Well-defined black-box test models and adequate testing criteria for software component to enable engineers to develop systematic test generation solutions;
- Systematic test generation methods and tools to support test generation of domain-specific application functions;
- More white-box test generation methods, criteria, and tools are needed to address the special features of object-oriented software components.

Here, we explain one black-box test model, called flow graphs, for test generation of simple components, such as abstract data types. S. Zweben et al. define a flow graph model to present a software component's specification [16]. According to Zweben et al., a flow graph is a directed graph in which each vertex represents one operation provided by the component, and each directed edge from node A to node B indicates the possibility that the execution control flow inside this component may flow from A to B.

To construct a flow graph based on a component's specification, we can simply follow the procedure given here:

- Identify the component's initial operations from its specification, and use an initial node to represent an initial operation.
- Identify the component's final operations (say, destructors for classes), and use a final node to represent a final operation.
- Identify other component operations given in the specification, create a node for each operation, and label the node with the operation's name.
- Identify all possible control flow from one node to another, and connect them if there is valid control sequence.

Figure 8.3 shows the component specification for Queue, which is written in RESOLVE language. This example is given by S. H. Edwards in [17]. Figure 8.4 displays its flow graph. Similar to the existing white-box test generation methods, a test generation program can be easily developed based on this flow graph. The major function of the test generation program is to generate all legal operation paths, and identify infeasible paths based on the given preconditions and postconditions. For each generated legal path, testers can prepare the right test cases and test data to exercise each path.

8.3.4 Systematic performance testing for components

As pointed out in Chapter 11, component performance testing and measurement is very important for component vendors and users. As discussed in Chapter 11, the scope of component performance validation includes the measurement of the processing speed, throughput, reliability, availability, scalability, and system resource usage. In the real world, engineers must spend a lot of time to prepare performance tests, collect performance measures, and analyze performance test results. To support these activities, systematic solutions are needed in these areas. As mentioned before, most existing software performance test tools are host-centered systems. They can be used to measure system loads, resource utilization, and functional process speed in a black-box view. Many of them do not provide functional features to allow engineers to measure software reliability, availability, and scalability.

There are a number of issues and challenges in automating performance validation and evaluation for components. Here, we list the major challenges.

- Current components (including COTS components) are developed without consideration of how to facilitate component performance validation and measurement. This not only increases the difficulty and complexity in component performance validation and measurement, but also results in a much higher cost.
- There is a lack of cost-effective performance testing tools and evaluation solutions that enable component users to validate and measure component performance in a plug-in-and-measure approach.
- There is a lack of well-defined performance measurement frameworks that allow component developers to use them as a simple and seamless connection interface between components and a systematic component performance evaluation solution.

To validate and evaluate component performance in systematic way, component test engineers and users have the following needs:

```
concept Queue_Template
context
        global conext
            facility Standard_Boolean_Facility
        parametric context
            type Item
interface
        type Queue is modeled by string of math[Item]
            exemplar q
            initialization
                    ensures   q = empty_string
        operation Enqueue (
                    alters    q : Queue
                    consumes  x : Item
            )
            ensures           q = #q * <#x>
        operation Dequeue (
                    alters    q : Queue
                    produces  x : Item
            )
            requires          q /= empty_string
            ensures           <x> * q = #q
        operation Is_Empty (
                    preserves q : Queue
            ) : Boolean
                    ensures   Is_Empty iff q = empty_string
end Queue_Template
```

Figure 8.3 A RESOLVE queue specification. (*From*: [17]. © 2000 John Wiley & Sons, Inc. Reprinted with permission.)

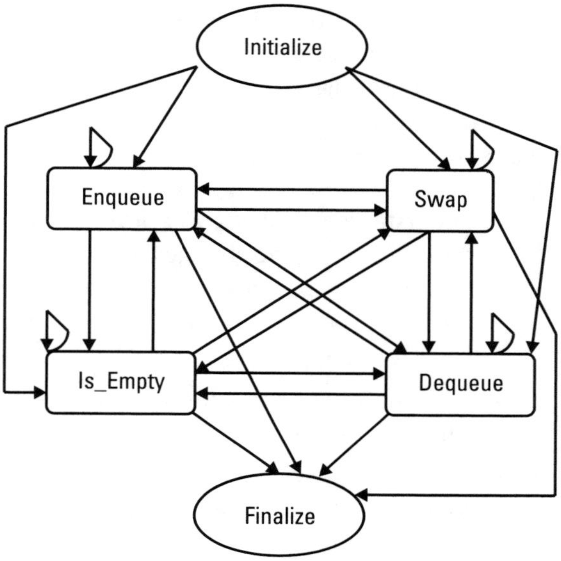

Figure 8.4 The queue flow graph. (*From*: [17]. © 2000 John Wiley & Sons, Inc. Reprinted with permission.)

- Building measurable components that facilitate component performance validation and measurement;
- Well-defined component performance models and measurement metrics, which can be used to implement component performance evaluation solutions;
- Systematic component performance validation and evaluation environments and tools that enable engineers and users to measure component performance in a simple plug-in-and-measure approach.

8.3.5 Systematic regression testing for software components

Software regression testing refers to retesting activities and efforts that validate and confirm the quality of a software product after it is changed. Software regression testing occurs after the first internal/external product release. In Chapter 10, we discuss current issues and solutions in the regression testing of software components and component-based programs. Here, we exam the special test automation needs and issues relating to regression testing of components.

In the component-based software engineering, both component vendors and users must carry on component retesting from one release to another. As mentioned before, reducing redundant testing and manual retesting efforts is a major objective in software test automation. Hence, automating regression-testing operations becomes an important key in the success of test automation. In general, there are four types of engineering activities in component regression testing.

- Understand and identify various types of software changes for a component, including the changes made in the component specification, interface, design, and program code. It is obvious that component users only need to understand component specification changes and component interface changes.
- Understand and analyze component change impacts on component functions, interfaces, structures, and behaviors.
- Examine and select the reusable component tests, and form a new component test suite by adding necessary new tests and deleting obsolete tests.
- Execute component retests and check the test results.

Clearly, automating component retesting needs four types of systematic solutions to support engineering activities. The first is a systematic change identification solution that assists engineers to identify various types of component changes in component function, structure, behavior, and interfaces. The second is a change impact analysis solution that helps engineers figure out what are the change ripple impacts on component elements and tests based on the relationships among its elements. D. Kung et al. [18] discussed software change analysis for object-oriented software. They reported a systematic change impact analysis solution for object-oriented programs to identify *class firewalls*—that is, an enclosure set of the changed and affected classes after software classes are changed. They have implemented a systematic solution and reported its effectiveness in [19]. The third is a systematic retest selection solution that identifies reusable component tests and the affected tests due to software changes. G. Rothermel et al. in [10] presents a good summary and comparison of various test selection techniques based on various program models. They are useful to define systematic test selection methods for software components. T. L. Graves et al. in [20] reported an empirical study of different test selection techniques. The last is the systematic test execution solution for component retests.

Today, engineers encounter the following issues and challenges on the road to automate component regression testing.

- Although current GUI-based testing tools provide a record-and-replay function to support retesting of GUI-based software functions, most of them need engineers to manually identify GUI interface changes and manually update and select reusable GUI-based test scripts. Besides, there are no well-defined adequate test criteria for GUI-based software testing and retesting.
- Existing script-based testing tools do not provide solutions to support engineers to identify and select reusable test scripts in a systematic manner due to the following reasons:
    - No direct mapping relationships between component elements and test scripts are tracked.
    - There is a lack of research results on software change identification and impact analysis solutions to address: (1) component API changes and their impacts to component tests, and (2) internal component changes and their impacts to component APIs.

8.4 Summary

In this chapter, we first review the basic concept of software test automation in terms of objectives, common issues, challenges, and procedural steps. According to our observation, a software test process can be evaluated to check its current maturity level in test automation. In addition, we classify and summarize the existing test tools and list different examples.

Section 8.2 presents two component-oriented test tools. The first is a specification-based black-box test tool, known as ADLScope. The other is JavaTest Compatibility Tools, which are developed by Sun Microsystems. It is developed to assist Java Community Program members to create, implement, and test Java component technology.

Section 8.3 discusses new issues, challenges, and needs in test automation of reusable components. The discussion covered five different areas: test information management, test execution, test generation, regression testing, and performance evaluation. In our view, component test automation is not only useful for component vendors, but also necessary for component users to support component validation and selection.

In the last section, we examine the special test automation issues, challenges, and needs for component-based software. The major difficulty and complexity is caused by the third-party reusable components and the lack of access to their detailed artifacts, as well as nonstandardized component interactions and communication among components. The discussion focuses on three areas: automatic test execution, systematic regression testing, and performance validation and measurement. Because of the lack of research results, systematic test solutions, and tools, it is clear that there is a long way to go to achieve test automation for component-based software.

However, a word caution is needed. Automating a software test process is a very challenge mission for the following reasons:

- It is not easy to find qualified and experienced test automation engineers because software testing and test automation are not taught in many universities.
- Software technologies are constantly changing, so it is hard for test tool vendors to keep the same pace. Therefore, engineers working on projects using new technologies always have problems finding the right test tools.
- Software test automation projects always run into budget and schedule problems because they are not considered the top priorities by senior managers.

Software test automation is one important and effective means to reduce manual testing effort and cost and to speed up and improve the quality of a software test process. In our opinion, successful test automation projects and efforts change a manual test process in the following areas:

- Automate testing operations and reduce manual retest efforts.
- Enforce the establishment of software testing and quality assurance standards, and enhance the quality of tests.
- Replace ad hoc testing practice with systematic testing solutions and well-defined test criteria.

Although a number of successful test automation stories and experience have been reported in the published books and articles [10], many have failed and unsatisfied test automation projects are not yet documented.

References

[1] Brooks, F. P., Jr., *The Mythical Man-Month*, anniversary edition, Reading, MA: Addison-Wesley, 1995.

[2] Dustin, E., J. Rashka, and J. Paul, *Automated Software Testing—Introduction, Management, and Performance*, Reading, MA: Addison-Wesley, 1999.

[3] Fewster, M., and D. Graham, *Software Test Automation—Effective Use of Test Execution Tools*, Reading, MA: Addison-Wesley, 1999.

[4] Hayes, L. G., *The Automated Testing Handbook*, Dallas, TX: Software Testing Institute, 1995.

[5] Poston, R. M., *Automating Specification-Based Software Testing*, Los Alamitos, CA: IEEE Computer Society, 1996.

[6] Kitchenham, B. A., S. G. Linkman, and D. Law, "DESMET: A Methodology for Evaluating Software Engineering Methods and Tools," *IEEE Computing and Control Journal*, June 1996, pp. 120–126.

[7] Gao, J., et al., "Developing an Integrated Testing Environment Using WWW Technology," *Proc. of COMPSAC'97*, IEEE Computer Society Press, 1997.

[8] Gao, J., F. Itura, and Y. Toyoshima, "Managing Problems for Global Software Production—Experience and Lessons," *Journal of Information Technology and Management*, No. 1, January 2002.

[9] Arnold, R. S., and S. A. Bohner, (eds.), *Software Change Impact Analysis*, Los Alamitos, CA: IEEE Computer Society Press, 1996.

[10] Rothermel, G., and M. J. Harrold, "Analyzing Regression Test Selection Techniques," *IEEE Trans. on Software Engineering*, Vol. 22, No. 8, August 1996, pp. 529–551.

[11] Hartmann, J., and D. J. Robson, "Approaches to Regression Testing," *Proc. of Conference on Software Maintenance*, Phoenix, AZ, 1988, pp. 368–372.

[12] Chang, J., and D. J. Richardson, *ADLScope: An Automated Specification-Based Unit Testing Tool*, ICS Technical Report 98-26, Department of Information and Computer Science, University of California, Irvine, CA, 1998.

[13] Chang, J., D. J. Richardson, and S. Sankar, "Structural Specification-Based Testing with ADL," *Proc. of 1996 International Symposium on Software Testing and Analysis (ISSTA'96)*, San Diego, CA, January 1996, pp. 62–70.

[14] Ort, E., "The Java Compatibility Test Tools: JavaTest Harness," October 2001, http://developer.java.sun.com/developer/technicalArticles/JCPtools2/.

[15] Ort, E., "The Java Compatibility Test Tools: Spec Trac," August 2001, http://developer.java.sun.com/developer/technicalArticles/JCPtools/.

[16] Zweben, S., W. Heym, and J. Kimmich, "Systematic Testing of Data Abstractions Based on Software Specifications," *Journal of Software Testing, Verification, and Reliability*, Vol. 1, No. 4, 1992, pp. 39–55.

[17] Edwards, S. H., "Black-Box Testing Using Flowgraphs: An Experimental Assessment of Effectiveness and Automation Potential," *Journal of Software Testing, Verification, and Reliability*, Vol. 10, No. 4, December 2000, pp. 249–262.

[18] Kung, D., et al., "On Regression Testing of Object-Oriented Programs," *Journal of Systems and Software*, Vol. 32, No. 1, January 1996, pp. 21–40.

[19] Kung,D., et al., "Developing an OO Software Testing and Maintenance Environment," *Communications of the ACM,* Vol. 38, No. 10, October 1995, pp. 74–87.

[20] Graves, T. L., et al., "An Empirical Study of Regression Test Selection Techniques," *ACM Trans. on Software Engineering and Methodology,* Vol. 10, No. 2, April 2001, pp. 184–208.

PART

III

Validation methods for component-based software

A vast amount of evidence has shown that the integration of high-quality components may not yield high-quality software systems. For component-based software, on the one hand, the encapsulation or implementation transparency feature improves the integration process. On the other hand, this feature makes it much harder to assess the quality of each component and thus more difficult to assess the quality of the integrated system. When components are added, deleted, or updated during the integration testing and maintenance phases, without source code and other related information, the implementation transparency feature prevents accurate analysis of the component-based software. In Part III of this book, we address this issue from different perspectives of the integration process. When performing integration testing, in Chapter 9, we discuss a model to characterize the interactions among different components. Then, a black box–based approach and a UML-based approach are discussed for obtaining a model from a component-based system. Given the model, subsequent analysis for adequate integration testing is presented. Component-based software systems integrate components from many different sources. This necessitates frequent maintenance activities. In Chapter 10, we first discuss different ways of representing changes that are made to a software component, without guaranteed access to the source code and detailed maintenance activities. Then, strategies for performing regression testing are discussed for different types of maintenance activities. In addition, performance testing and evaluation are the last important steps in the quality control process for software components and systems. To assure the performance of an integrated component-based software system, different approaches are discussed in Chapter 11. To facilitate the entire integration and validation process, frameworks for supporting the validation of component-based systems are presented in Chapter 12.

CHAPTER

9

Contents

Integration testing for component-based software

When developing a component-based software system, we test each component individually. Why should we worry about the assembly of these components? Apparently, components may have been developed by different people, written in different programming languages, and executed in different operating platforms. Therefore, the interfacing among components needs to be tested [1–3]. Integration testing is necessary to ensure communication between components is correct. IEEE defines it as:

> Testing in which software components, hardware components, or both are combined and tested to evaluate the interaction between them.

In this chapter, we explain what integration testing is and describe some typical integration-testing methodologies. We then explore the problems that may be encountered when applying test techniques to component-based software. Finally, a UML-based integration technique is discussed.

9.1 Introduction

When we develop component-based software, we develop and test each component separately. Why can problems be encountered when we integrate them? And how do we perform integration testing to identify them? The following is a fault model that specifies possible faults that are likely to be overlooked during unit-level testing [4]. These interaction-related faults can be classified into *programming-related* faults, which are intercomponent faults, and *nonprogramming-related* faults, which are interoperability faults. Faults that are overlooked during unit testing can be identified during integration testing and are classified as *traditional faults*.

9.1.1 Type I: Intercomponent faults

Even when the individual components have been evaluated separately, there can still be faults in the interactions between them. Programming-related faults that are associated with more than one component are considered to be intercomponent faults. An example is given in the following code. After adding $i = 0$ onto I_1 of component C_1 and testing I_1 and I_2, we cannot detect any problems. But when the component C_1 is deployed together with C_2 and C_3, and I_1 is invoked followed by I_2, a failure will occur in C_3.

```
C1: I1 → i = 0;
    I2 → return i;

C2: I1 → C1::I1();
C3: I4 → j = 1/C1::I2();
```

Many other possible faults can be classified as intercomponent faults—for example, deadlock and livelock. Most of these faults cannot be identified during unit-level testing. Therefore, they are the major targets of integration testing.

9.1.2 Type II: Interoperability faults

Many characteristics of component-based systems, such as heterogeneity, source code unavailability, and reusability, will lead to different types of interoperability problems. These interoperability problems can be classified into system-level, programming-language level, and specification-level interoperability faults.

- *System-level interoperability faults:* In a component-based system, different components can have been built under different infrastructures, and the infrastructures may not be 100% compatible. For example, incompatibility between different CORBA products can affect the interaction between CORBA components.

- *Programming-level interoperability faults:* When components are written in different programming languages, incompatibility between the languages may cause problems. For instance, Microsoft supports the integration of VC and VB components. But the floating-point incompatibility may cause interoperability faults.
- *Specification-level interoperability faults:* Specifications may be misinterpreted by developers, and there are many different ways that specifications may be misunderstood. This type of fault can be caused by the data that pass through the interfaces or the patterns of the component interactions.

Some detailed discussion and examples can be found in Section 3.2.1.

9.1.3 Type III: Traditional faults and other faults

Traditional testing and maintenance techniques can be adopted for those faults that can be isolated within one component. These faults will be identified as traditional faults. Other faults, such as faults related to special-input or special-execution environments, also fall into this category.

9.2 Traditional integration-testing methodologies

Integration testing needs to put a group of components together. To carry out that process adequately, we need to answer two questions: what are the orders we follow to incrementally integrate components, and how do we test the newly integrated components? Some typical approaches follow in the next two sections.

9.2.1 Function decomposition–based integration testing

This approach is based on the function decomposition, which is often expressed in a tree structure. For instance, Figure 9.1 demonstrates a sample functional-decomposition tree for the ATM system we used in previous chapters.

Given the functional-decomposition tree, four different approaches can be used to pursue integration testing.

1. *Big-bang approach:* This approach will integrate all the components at once. The big-bang approach seems simple. Its major drawback is in fault identification. If failures are encountered, all components are

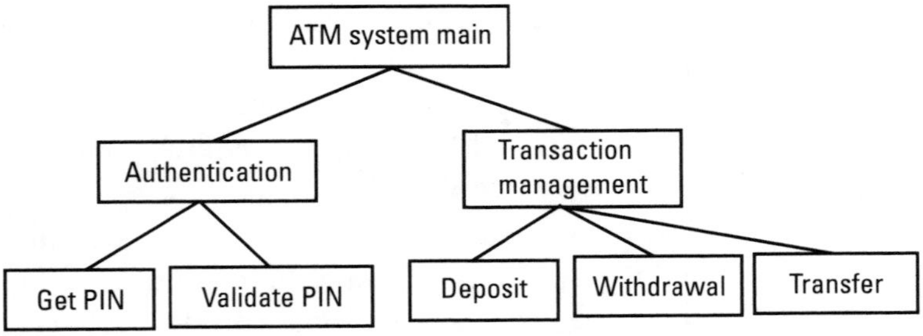

Figure 9.1 ATM system functional-decomposition tree.

subject to the possibility of hosting faults. Therefore, fault identification can be very costly. To avoid the difficulty of identifying faults during integration, the following three incremental approaches can be used. When new components are added and failures are encountered, the faults are most likely to be in the newly added components. Therefore, the faults are often much easier to identify than with the big-bang approach.

2. *Top-down approach*: This approach moves from the top level of the functional-decomposition tree to the leaves of the tree. A breath-first approach is often used to determine the order of integration. For example, in Figure 9.1 we can go after the following order: Main, Authentication, Transaction management, Get PIN, Validate PIN, Deposit, Withdrawal, and Transfer. The difficulty of the top-down approach is that when testing a partial group of components, stubs have to be developed separately for components that are not included.

3. *Bottom-up approach:* In contrast to top-down approach, the bottom-up approach starts from the leaves of the functional-decomposition tree and works up towards the top level. Bottom-up approaches do not require stubs but require drivers instead.

4. *Sandwich approach:* The top-down approach can identify high-level or architectural defects in the systems in the early stage of integration testing, but the development of stubs can be expensive. On the other hand, the bottom-up approach will not be able to identify high-level faults early, but the development of drivers is relatively inexpensive. A combination of the two approaches, called the sandwich approach, tries to use the advantages of the two approaches, while overcoming their disadvantages.

Functional-decomposition approaches do not consider the internal structure of individual components. Therefore, testing is based solely on the specification of the system.

9.2.2 Call graph–based integration testing

The approaches in Section 9.2.1 are based on functional decomposition trees, not the structure of the program. In addition, stubs or drivers need to be developed to fulfill the task. To solve these two difficulties, the call-graph approach has been developed. As the name of the approach implies, it is based on the call graph of a program. Figure 9.2 demonstrates a sample call graph for the ATM system.

Based on the call graph, the internal structure of the program is clearly represented. By following the structure of the call graph, similar incremental approaches such as top-down or bottom-up approaches can be used. But no additional drivers or stubs are required; the original program can be used instead.

Besides the incremental approaches that are used in the functional-decomposition-based methods, other incremental approaches can be incorporated that are based on the characteristics of call graphs—for instance, *neighborhood integration* and *path-based integration*. To speed up the incremental process, neighborhood approaches, instead of incorporating single components, will added all adjacent components (callers and callees) into each integration-testing phase. To take the overall behavior of the system into consideration, paths in the call graph, which corresponds to a single thread of behavior in the system, can serve as the basis for path-based integration testing.

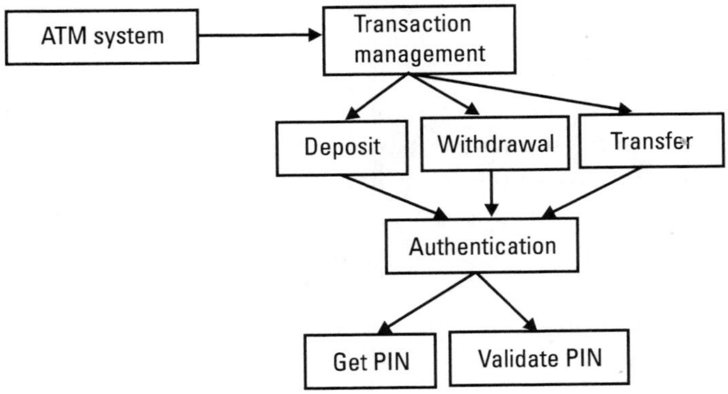

Figure 9.2 ATM system call graph.

Based on call graphs, to test newly integrated components, data flow–based [5, 6] and coupling-based [7] approaches can be adopted. Coupling-based approaches measure the dependence relationships [7] and then the interactions between components. Couplings can be classified into call coupling, parameter coupling, shared-data coupling, and external-device coupling; consequently, coupling def, coupling use, and coupling path are defined and used as the criteria for integration testing.

9.3 A test model for integration testing of component-based software

When performing integration testing for component-based software, some of the approaches mentioned earlier can be used, for instance, functional-decomposition approaches. But the more effective approaches, called graph-based approaches, cannot be used because of the implementation transparent features of component-based software. To overcome this difficulty, M. J. Harrold et al. [8] proposed a testing technique based on analysis of component-based systems from component-provider and component-user perspectives. The technique was adapted from an existing technique [5, 6], which makes use of complete information from components for which source code is available and partial information from those for which source code is not available. The authors of [9, 10] extended their work by proposing a framework that lets component providers prepare various types of metadata, such as program slicing. The metadata were then used to help test and maintain component-based systems. S. Ghosh and A. P. Mathur [11] discussed issues that arise when testing distributed component-based systems and suggested an interface and exception coverage-based testing strategy.

To test component-based software adequately and effectively, we describe a test model, which is an extension of the call graph–based approach. In our model, we define possible test elements that are likely to reveal the integration faults that are described in Section 9.1. Given the test element, a family of test adequacy criteria is then provided. Given the general test model, Section 9.4 discusses a black box–based approach to implementing the test model when source code and other documents are not available to component users. To address the drawbacks of black box–based approaches, which, while capable of overcoming the difficulties arising from the implementation transparent features of component-based software, present heterogeneity as well as other difficulties, an UML-based gray-box testing technique is discussed in Section 9.5.

9.3.1 Test elements

During integration testing, our test model emphasizes interactions between components in component-based software systems and tries to reveal Type I and Type II faults. A component may interact with other components directly through an invocation of the interfaces exposed by those components, an exception, or a user action that triggers an event. A component may also interact with other components indirectly through a sequence of events. The elements that need to be taken into account in component-based testing are:

- *Interfaces:* Interfaces are the most common way to activate components. Therefore, it is necessary to test each interface in the integrated environment at least once.
- *Events:* Every interface that can be invoked after deployment must be tested at least once. This goal is similar to the traditional test criterion that requires every function/procedure to be tested at least once. However, invoking the same interface by different components within different contexts may have different outcomes. Thus, to observe possible behaviors of each interface during runtime, every unique invocation of the interfaces needs to be tested at least once. Moreover, some events that are not triggered by interfaces may have an impact on the components, which need to be tested as well. Therefore, every event in the system, regardless of its type, needs to be covered by some test.

9.3.1.1 Context-dependence relationships

Interface and event testing ensure that every interaction between two components—the client and the server—is exercised. However, when the execution of a component-based software system involves interactions among a group of components, the sequence of event triggering may produce unexpected outcomes. To capture the inter-relationships among events, we define a context-dependence relationship that is similar to the control-flow-dependence relationship in traditional software. An event e_2 has a context-sensitive dependence relationship with event e_1 if there exists an execution path where the triggering of e_1 will directly or indirectly trigger e_2. For a given event, e, it is necessary to test e with every event that has a context-sensitive dependence relationship with e, to observe the possible impact of execution history on the outcome of the execution of e.

Context-sensitive dependence relationships also include indirect collaboration relationships between interfaces and events that occur through other interfaces and events. Therefore, testing context-sensitive dependence relationships may serve to identify interoperability faults that are caused by improper interactions between different components.

9.3.1.2 Content-dependence relationships

The invocation of an interface of a component causes a function inside the component to be invoked. Therefore, when a function declared in an interface v_1 has a data dependence relationship with another function declared in another interface v_2, the order of invocation of v_1 and v_2 could affect the results.

Our concept of a content-dependence relationship between the two interfaces v_1 and v_2 assumes that the two interfaces have a data-dependence relationship. An interface encapsulates one or more signatures, where each signature is a declaration of a function. When an interface is invoked, one or more functions will be executed to perform the requested service. Thus, the interface-dependence relationship can be derived from the function-dependence relationship, which was shown to be useful information in object-oriented class testing [12] and regression testing [13]. More precisely, a function f_2 depends on a function f_1 if and only if the value of a variable that is defined in f_1 is used in f_2. Therefore, a content-dependence relationship is defined as follows: an interface v_2 has a content-dependence relationship with interface v_1 if and only if v_1 contains the signature of f_1, v_2 contains the signature of f_2, and f_2 depends on f_1.

Both the direct interaction among interfaces and events and the context-dependence relationships should be included in the interactions of a component-based system from the control-flow perspective. Content-sensitive dependence, on the other hand, can provide valuable additional information in generating test cases and detecting faults.

9.3.2 Component interaction graph

We define the *component interaction graph* (CIG) to model interactions between components by depicting interaction scenarios. The interactions can take place directly or indirectly. Direct interactions are made through a single event, while indirect interactions are made through multiple events in which execution order and data-dependence relationships may result in different outcomes. A CIG is a directed graph where CIG = (V, E) $V = V_I \cup V_E$ is a set of nodes, where V_I is the set of interface nodes, and V_E is the set of event nodes, and E represents a set of directed edges. Given the CIG, call graph–based approaches can be directly applied.

Moreover, the following two types of entities may require some additional treatment: context-dependence relationships and content-dependence relationships. To identify context-dependence relationships in the CIG, a path-based approach can be used, while to identify content-dependence relationships, data-dependence relationships need to be taken into consideration.

9.3.3 Test-adequacy criteria

Software development costs and schedules are constraints on quality in most software-development processes, often requiring difficult trade-offs [14, 15]. Deciding between budget, schedule, and quality is a practical and difficult software-management task. Testing is a time-consuming process, and is often performed at different levels of detail to satisfy different quality thresholds. For example, white-box approaches provide a variety of test-adequacy criteria, including statement, branch and def-use pairs coverage. Stronger fault-detection criteria often require more complex analysis and test cases.

In our model, the CIG can be used to develop a family of test-adequacy criteria, as depicted in Figure 9.3. The basic levels are *all-interfaces* and *all-events* coverage, which enforce the coverage of all-interface or all-event nodes. Once all interfaces and events have been tested, the direct interactions among components are tested.

To test indirect interactions, *all-context-dependence* coverage will test where the interfaces are invoked in different contexts. To do this, the interface-invocation sequences need to be identified. These sequences can be defined as *context-sensitive paths,* paths between two context-dependent events e_i and e_j in the CIG. Formally, all paths $p = \{e_i, e_{i+1}, e_{i+2}, \ldots, e_j\}$ where $(e_t, e_{t+1}) \in$ CIG. A triggering of e_1 is likely to trigger e_2; therefore, all p are considered viable and need to be tested.

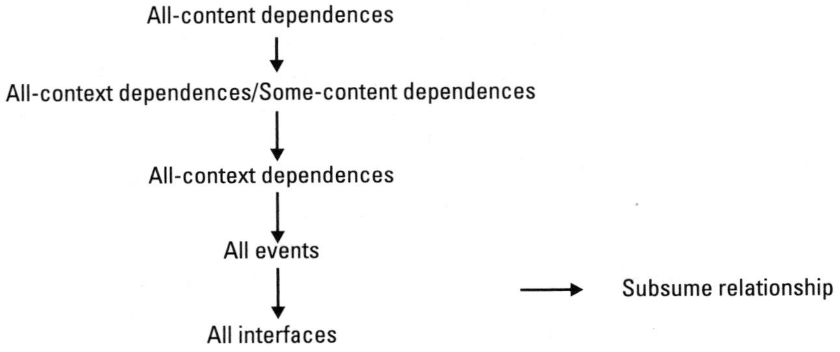

Figure 9.3 Family of the test-adequacy criteria architecture.

All-context dependence will test all the control scenarios in a component-based system. To further test indirect interactions between components, those caused by data interaction must be examined. *All-content-dependence* coverage needs to be satisfied, which requires covering all-content-dependence relationships in the component-based system. Content-sensitive paths can be used to depict these dependence relationships. *Content-sensitive paths* are defined by pairs of context-sensitive paths (p_a, p_b), such that p_a and p_b are context-sensitive paths, and interfaces $I_1 \in p_a$ and $I_2 \in p_b$ have a content-dependence relationship. Every content-sensitive path needs to be exercised, and it is necessary to make sure that the execution of I_1 comes before that of I_2 to observe if a fault could occur due to the content-dependence relationship.

All-content-dependence coverage needs to cover all the context-sensitive-path pairs, which are involved in a content-dependence relationship. Therefore, there will be significantly greater complexity than with the all-context-dependence coverage. Thus, another coverage criterion, *all-context/some-content-dependence* coverage, which requires covering one context-sensitive path for each content-dependence relationship, will adequately test the content dependence and therefore reduce the complexity dramatically.

9.4 Black box–based integration-testing approaches

The central theme of the model is the component interaction graph. Given the graph, all graph-based testing strategies can then be applied. To construct the component-interaction graph for a given component-based software system, the model's test elements need to be identified in that system. The most challenging part of this activity is to obtain necessary information about third-party components for which source code is not available. In constructing the component-interaction graph for a given component-based software, some elements can be identified in straightforward ways by using existing information, while others can be obtained only through preprocessing. As described in the previous section, the elements that must be identified are interface nodes and event nodes, as well as control edges among event nodes and content-dependence edges between interface nodes. CORBA, EJB/J2EE, and COM/DCOM/.NET are the most popular techniques for component-based software development [16]. Each of them has its unique characteristics. Therefore, different approaches are necessary when building different CIGs. In this section, we discuss how to generate a CIG for each test element. We first cover situations where the source code of the components is available. We then cover situations where the source code is not available. In the latter situation, we discuss CORBA, EJB, and COM-based systems separately.

9.4.1 Interface nodes

Regardless of the availability of the source code of components, interfaces are explicitly publicized. Therefore, it is straightforward to identify interfaces and depict them as corresponding interface nodes in the CIG.

- *EJB:* EJB components always include two special classes: home interface and component interface. These specify how EJB beans will be created and what services will be provided. Based on the two classes, interface can be identified.
- *CORBA:* An essential part of CORBA is the OMG interface-definition language. Irrespective of the programming language, platform, and availability of the source code, IDL files must be known, and these files contain all the interfaces provided by the components. So interfaces can easily be acquired.
- *COM:* A central piece of COM is the type library. A type library is a binary description of the COM object or COM interfaces or both, and it is the compiled version of the Microsoft Interface Definition Language (MIDL, which is a Microsoft extension to the OSF-DCE IDL standard). The type library is usually part of the binary image of COM components, or should be deployed along with the binaries. From its type library, it is very easy to acquire a COM component's interfaces.

9.4.2 Event nodes

To identify event nodes, one must know the calling interface that triggers the event and the invoked interface that responds to the event. To identify callers and callees when the source code of the component is available, an instrumented parser of the compiler can be used. When the source code is not available, in general the component is either a server component that provides services only or a server/client component that provides services and also requests services from other components. In the former case, the server will not trigger any events, but in the latter case the server will trigger as well as respond to events. A triggering component should indicate the services that it requires in its specification. Therefore, the event from the calling interface to the called interface can be identified. Our concern is when a server component requests a service. At present, most specifications will provide the name of the called interface but not reveal anything about the calling interface. Under such circumstances, it is necessary to obtain more information from the component provider. One approach is for the provider to support sequence diagrams, which will depict the relationships between callers and callees. An alternative is to take a conservative view where each interface in the component is a candidate caller.

9.4.3 Context-dependence and content-dependence relationships

Context-dependence relationships can be derived using the approaches previously discussed. Content-dependence relationships are more difficult to find. If source code is available, we can apply the algorithm that we developed to identify function-dependence relationships in OO testing [12]. For cases where the source code is not available, we will need to convince component providers to support a class diagram in which each interface is represented as a class and the dependence relationships among classes are depicted. We believe that it should be feasible for providers to supply such high-level design information.

9.5 UML-based integration-testing approaches

Without source code, we can easily obtain interface nodes and event nodes from the component specifications. However, context-dependence relationships and content-dependence relationships would be difficult to derive without a sophisticated mechanism.

To overcome these difficulties, we need to represent the behavior of components without source code precisely. UML [17–20] is a language for specifying, constructing, visualizing, and documenting artifacts of software-intensive systems that can be used. There are several advantages to adopting UML. First, UML provides high-level information that characterizes the internal behavior of components, which can be processed efficiently and used effectively when testing. Second, UML has emerged as the industry standard for software-modeling notations and various diagrams are available from many component providers. Third, UML includes a set of models that provides different levels of capacity and accuracy for component modeling, and it can therefore be used to satisfy various needs in the real world. In UML, collaboration diagrams and sequence diagrams are used to represent interactions between the different objects in a component, which in this research are used to develop a corresponding interaction graph that can be used to evaluate the control flow of a component. Statechart diagrams, on the other hand, are used to characterize internal behavior of objects in a component. Based on the statechart diagram, we further refine the dependence relationships among interfaces and operations that are derived from collaboration diagrams. To characterize the constraints of a class even more rigorously, the formal language that is provided in UML, object constraint language, can be adopted. For more critical scenarios, OCL can be adopted to formally specify the behavior and constraints of a component. Finally, UML is *extensible*, which means that if additional artifacts are found to be important to component-based software systems, they can be added into the UML.

UML is increasingly being used to support the design of component-based systems [17–21]. Some of the UML diagrams have also been used to generate test cases automatically [9, 22–24]. J. Offutt and A. Abdurazik first proposed a mechanism that adapted state specification-based test-data generation criteria to generate test cases from UML statecharts [9]. They subsequently extended their work to generate tests from UML collaboration diagrams [24]. At the same time, L. Briand and Y. Labiche proposed using class diagrams, collaboration diagrams or OCL to derive test requirements [23]. As well as having benefits in testing, these UML-based test elements can also be used in other component-based engineering processes. For example, they can be used in component selection, a process of identifying qualified components for specific component architecture and specification, component customization, and maintenance. Therefore, these elements can become part of the deliverable guidelines for component providers to make sure component-based engineering can be carried out smoothly and adequately.

9.5.1 Context-dependence relationships

When integrating components, programmers typically focus on how to specify component interfaces and events. But how these interfaces and events will interact, and their potential risks to the integrated system, are usually not considered. Context-dependence relationships, which model how interfaces and events interact, can be derived through one of the following approaches.

9.5.1.1 Collaboration/sequence diagram–based approach

UML collaboration diagrams and sequence diagrams focus on interactions between various objects within a use case. (For simplicity, we refer to a use case as a component. If a component covers multiple use cases, we need to develop a consolidated collaboration diagram. If a use case includes more then one component, we need to develop a subset of the collaboration diagram

[20].) In UML sequence diagrams, interactions are ordered by time, while in collaboration diagrams, the same information is presented with the interactions ordered in order of numeric message sequence.

Figure 9.4 describes a partial collaboration diagram and one sequence diagram of an ATM server component. The sequence diagram shows only one of the possible scenarios ordered by time, while the collaboration diagram combines all scenarios in numbered order.

- W1-W2-W3-W4-W5-W6-W7-W8
- W1-W2-W3-W4A-W4A.1
- W1-W2-W3-W4-W5A-W5A.1-W5A.2
- W1-W2-W4-W4-W5B-W5B.1-W5B.2

Also in Figure 9.4, W5, W5A, and W5B demonstrate three alternatives that can occur after the message W4 is passed by the Withdrawal Transaction Manager object to the ATM Account object.

With the collaboration diagrams, we can refine our context-dependence relationships to fit all possible sequences, as shown in Figure 9.4 in a collaboration diagram that could be invoked to precisely model how interfaces and events interact with each other.

9.5.1.2 Statechart-based approach

To model the interactions of components, the collaboration diagram itself is not always sufficient. The behavior of a component can be modeled more precisely by combining the collaboration diagram with statechart diagrams, which are used to describe state-dependent control objects in components.

For example, Figure 9.5 shows a collaboration diagram and statechart. Possible sequences shown in the collaboration diagrams are:

- 1 - 1.1 - 1.2 - 1.3 - 1.4
- 1 - 1.1A
- 2 - 2.1 - 2.2 - 2.3 - 2.4 - 2.5 - 2.6 - 2.7, 2.7a - 2.8
2.6A - 2.6A.1, 2.6A.1a - 2.6A.2
2.6B - 2.6B.1, 2.6B.1a - 2.6B.2
- 2A - 2A.1 - 2A.2, 2A.2a - 2A.3, 2A.2a.1
- 2.6A3 - 2.6A4 - 2.6A5 - 2.6A6 - 2.6A7 - 2.6A8

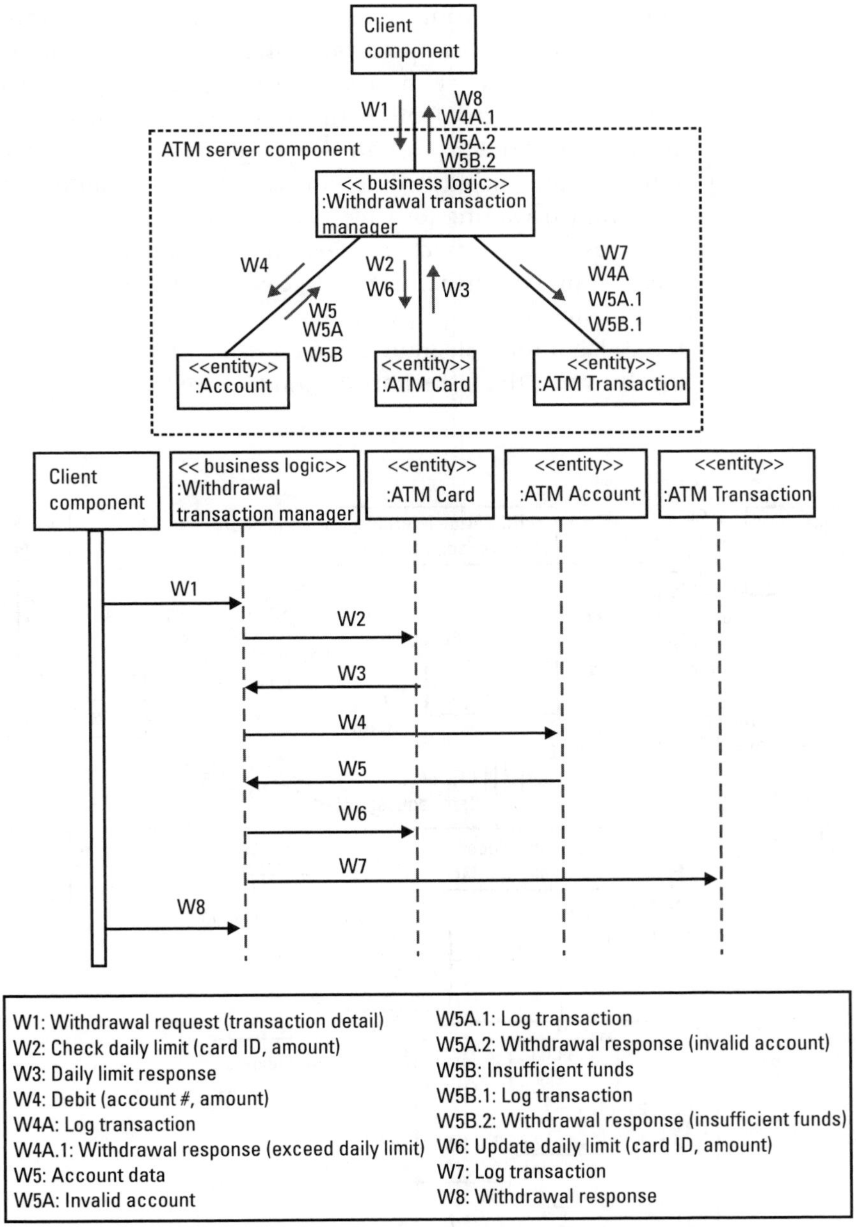

Figure 9.4 Collaboration and sequence diagrams of an ATM server component.

As we can see, the sequence "2A – 2A.1 – 2A.2, 2A.2a – 2A.3, 2A.2a1" (2A.2 and 2A.2a are two concurrent messages) is the only sequence that allows the user to cancel. Nevertheless, this sequence can happen in different contexts, such as the user canceling the current transaction after correctly inputting the PIN, or the user canceling the current transaction after incorrectly inputting the PIN. With the help of the statechart diagram, which is shown in Figure 9.5, we can clearly see that the cancellation sequence needs to be validated in three different scenarios: (1) waiting for PIN, (2) validating PIN, and (3) waiting for Customer Choice.

By using the statechart diagram, the interactions between interfaces and events can be further refined. Given a statechart diagram, our context-dependence relationships will have to include not only all possible sequences in a collaboration diagram, but all possible combinations of the sequences that are shown in the statechart diagram as well.

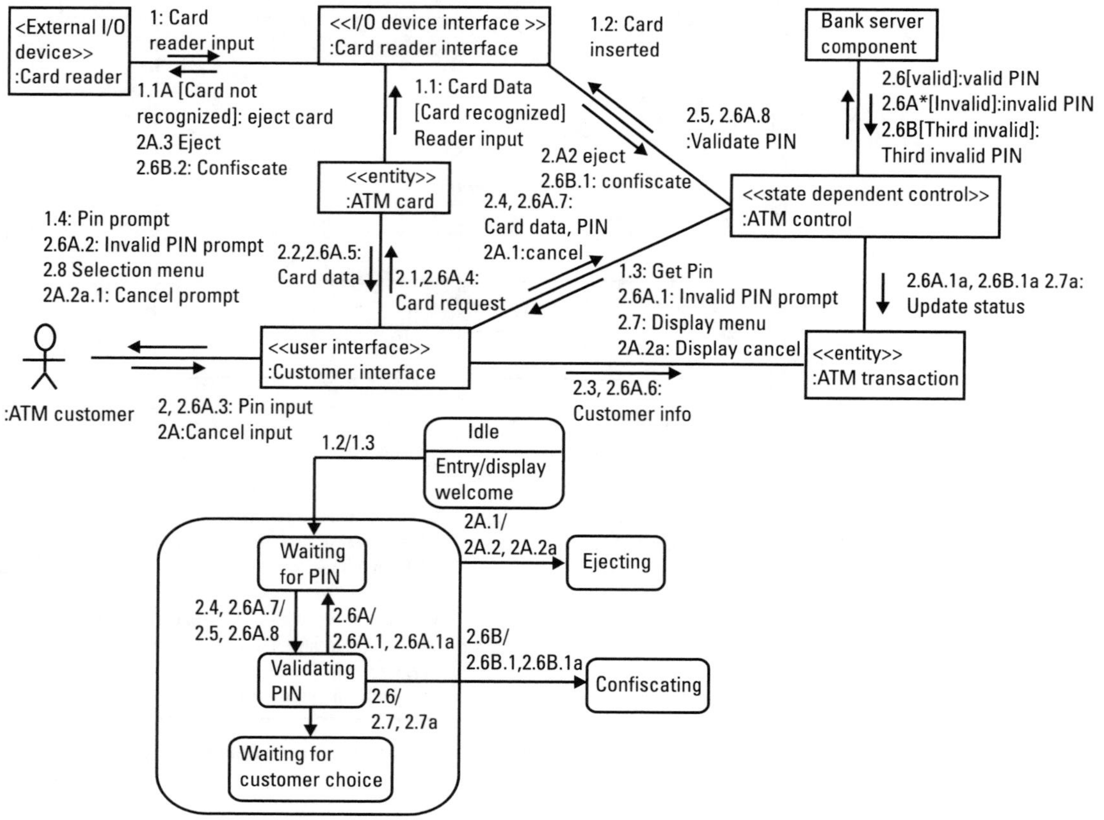

Figure 9.5 Collaboration diagram for validate PIN and statechart for ATM control diagram.

9.5.2 Content-dependence relationships

Content-dependence relationships depict relationships between different interactions. To obtain content-dependence relationships, further processing of UML diagrams is necessary. The following approaches can be adopted.

9.5.2.1 Collaboration-diagram approach

UML collaboration diagrams and sequence diagrams demonstrate interactions between objects within a component. When interactions involve entity classes, collaboration diagrams can demonstrate the dependence relationships between two interactions. For example, Figure 9.6 shows message B_j flowing into entity class *Account*, and no information flows out of *Account*. Generally speaking, message B_j will update information in *Account* objects. We define this type of message as an *update* message. However, messages A_i and A_{i+1} flow into and out of *Account*, which indicates that information about *Account* is retrieved. These types of messages are called *retrieve* messages. Therefore, interactions that include messages A_i and A_{i+1} will depend on sequences that include message B_j. In general, an interface I depends on interface I' if and only if a message sequence invoked by I includes an update message, and another sequence invoked by I' includes a corresponding retrieve message.

9.5.2.2 Statechart diagram approach

Statechart diagrams are able to demonstrate content-dependence relationships from a state-transition point of view. The rationale lies in the fact that if interface I_1 depends on I_2, the state of the component is S_1 after the execution of I_1. When executing interface I_2, the state transitions from S_1 to S_2 depend on

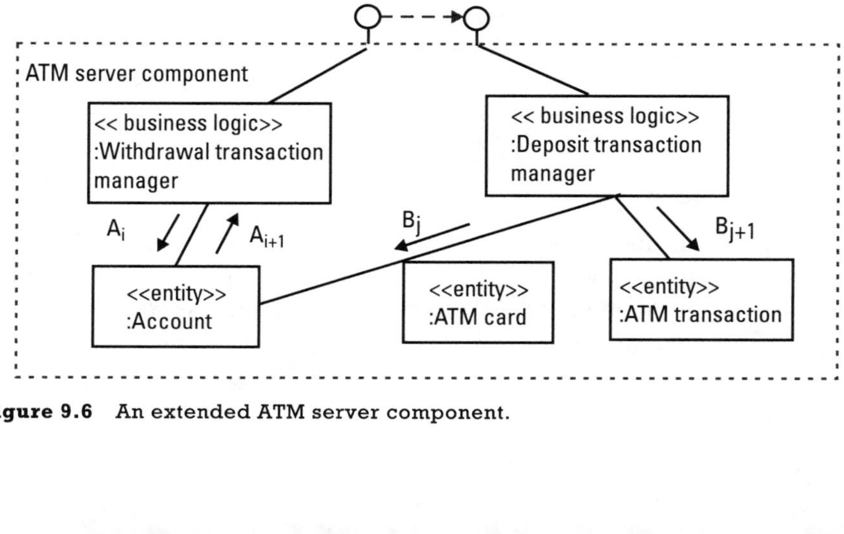

Figure 9.6 An extended ATM server component.

state S_1 and the invocation I_1. To model this type of content-dependence relationships, we eliminate dependence relations that are not effective:

- If the invocation of I_2 remains in the original state, S_1, after the invocation of I_1, the dependence relationship does not affect the behavior of the software; the dependence relationship is not effective.
- From a certain state S, if the invocation of I_2 will always bring the state to S', it will not matter if interface I_1 is invoked before that or not. This indicates that the state transformation is not caused by the dependence relationships.

The advantages of UML-based approaches can be seen from the following: UML diagrams are usually available in the design phase and no extra overhead will be imposed on the component providers. Moreover, component users can derive new pieces of information of their own through the diagrams at an abstract level. This offers great flexibility, extensibility, and probably effectiveness to component-based software development.

Given the UML-based context-dependence and content-dependence relationships, the test criteria provided in our test model have to be modified as follows:

- Each transition in each collaboration diagram has to be tested at least once.
- Each valid sequence in each collaboration diagram has to be tested at least once.
- Each transition in each statechart diagram has to be tested at least once.
- Each content-dependence relationship derived from each collaboration diagram has to be tested at least once.
- Each effective content-dependence relationship derived from each statechart diagram has to be tested at least once.

9.6 Summary

In this chapter, we have given an overview of traditional integration-testing techniques. The difficulties of performing traditional integration testing on component-based components arise because components are from a variety of sources. A lack of knowledge of these components, and a lack of confidence in them, threatens the adequacy of integration testing. In this chapter, we discussed a UML-based approach, which can be done with minimal effort, yet can provide sufficient information to determine how to carry out integration testing and criteria for determining when to stop. This approach can potentially serve as component-deliverable guidelines, which not only facilitate the integration testing for component users, but may be used as guidelines to evaluate the quality of the components.

References

[1] Blackburn, M., et al., "Mars Polar Lander Fault Identification Using Model-Based Testing," *Proc. of 8th International Conference on Engineering of Complex Computer Systems (ICECCS2002)*, IEEE CS Press, 2002.

[2] Garlan, C., R. Allen, and J. Ockerbloom, "Architectural Mismatch or Why It's Hard to Build Systems Out of Existing Parts," *Proc. of 17th International Conference on Software Engineering*, 1995, pp. 179–185.

[3] Weyuker, E. J., "Testing Component-Based Software: A Cautionary Tale," *IEEE Software*, Vol. 15, No. 5, September/October 1998, pp. 54–59.

[4] Wu, Y., D. Pan, and M.-H. Chen, "Techniques for Testing Component-Based Software," *Proc. of 7th International Conference on Engineering of Complex Computer Systems (ICECCS2001)*, IEEE CS Press, 2001, pp. 222–232.

[5] Harrold, M. J., and M. L. Soffa, "Interprocedural Data Flow Testing," *Proc. of 3rd Testing, Analysis, and Verification Symposium*, December 1989, pp. 158–267.

[6] Harrold, M. J., and G. Rothermel, "Performing Dataflow Testing on Classes," *Proc. of 2nd ACM SIGSOFT Symposium on Foundations of Software Engineering*, December 1994, pp. 154–163.

[7] Jin, Z., and J. Offutt, "Coupling-Based Criteria for Integration Testing," *Journal of Software Testing, Verification, and Reliability*, Vol. 8, No. 3, September 1998, pp. 133–154.

[8] Harrold, M. J., D. Liang, and S. Sinha, "An Approach to Analyzing and Testing Component-Based Systems," *1st International ICSE Workshop on Testing Distributed Component-Based Systems*, Los Angeles, CA, 1999.

[9] Offutt, J., and A. Abdurazik, "Generating Test from UML Specifications," *2nd International Conference on the Unified Modeling Language (UML '99)*, October 1999, pp. 416–429.

[10] Orso, A., M. J. Harrold, and D. Rosenblum, "Component Metadata for Software Engineering Tasks," *Proc. of 2nd International Workshop on Engineering Distributed Objects (EDO 2000)*, November 2000, pp. 126–140.

[11] Ghosh, S., and A. P. Mathur, "Issues in Testing Distributed Component-Based Systems," *1st International ICSE Workshop on Testing Distributed Component-Based Systems*, Los Angeles, CA, 1999.

[12] Chen, M., and M. Kao, "Effect of Class Testing on the Reliability of Object-Oriented Programs," *Proc. of 8th International Symposium on Software Reliability Engineering*, May 1997, pp. 73–82.

[13] Wu, Y., M. Chen, and M. Kao, "Regression Testing on Object-Oriented Programs," *Proc. of 10th International Symposium on Software Reliability Engineering*, November 1999, pp. 270–279.

[14] Perry, D. E., and G. E. Kaiser, "Adequate Testing and Object-Oriented Programming," *Journal of Object-Oriented Programming*, January 1990.

[15] Rosenblum, D. S., *Adequate Testing of Component-Based Software*, Technical Report TR 97-34, University of California at Irvine, Irvine, CA, 1999.

[16] Heineman, G. T., and W. T. Councill, *Component-Based Software Engineering: Putting the Pieces Together*, Reading, MA: Addison-Wesley, 2001.

[17] Booch, G., J. Rumabugh, and I. Jacobson, *The Unified Modeling Language Guide*, Reading, MA: Addison-Wesley, 2000.

[18] Cheesman, L., and J. Daniels, *UML Components: A Simple Process for Specifying Component-Based Software*, Reading, MA: Addison-Wesley, 2001.

[19] D'Souza, D. F., and A. C. Wills, *Objects, Components, and Frameworks with UML*, Reading, MA: Addison-Wesley, 1999.

[20] Gomaa, H., *Designing Concurrent, Distributed, and Real-Time Applications with UML*, Reading, MA: Addison-Wesley, 2000.

[21] Fowler, M., and K. Scott, *UML Distilled*, Reading, MA: Addison-Wesley, 2000.

[22] Briand, L., and Y. Labiche, "A UML-Based Approach to System Testing," *4th International Conference on the Unified Modeling Language (UML '01)*, October 2001, pp. 194–208.

[23] Abdurazik, A., and J. Offutt, "Using UML Collaboration Diagrams for Static Checking and Test Generation," *3rd International Conference on the Unified Modeling Language (UML '00)*, October 2000, pp. 383–395.

[24] Yoon, H., B. Choi, and J.-O. Jeon, "A UML Based Test Model for Component Integration Test," *Workshop on Software Architecture and Components, Proc. on WSAC'99*, Japan, December 7, 1999, pp. 63–70.

CHAPTER

10

Contents

Regression testing for component-based software

Component-based software engineering approaches rely on using reusable components as the building blocks for constructing software. On the one hand, this helps improve software quality and productivity; on the other hand, it necessitates frequent maintenance, such as upgrading third-party components or adding new features. The cost of maintenance for conventional software can account for as much as two-thirds of the total cost, and this is likely to be higher for component-based software. *Regression testing,* which is aimed at ensuring that the modified software still meets the specifications validated before the modification activities, is critical to the success of a component-based system. Regression testing involves many different issues, for instance, test-suite management and regression-testing automation. However, the issue of how to selectively retest the modified program has gained most of the researchers' attention. Selective regression testing is also the theme of this chapter.

Regression testing for component-based software is significantly different from regression testing for traditional software [1–5]. This is mainly because traditional software maintenance is usually carried out by the same team; therefore, the team has a full knowledge of the software, including source code, and all the controls to what they can and want to do. For component-based software, maintenance is often carried out by component providers. After that, users have to maintain their systems according to the changes that the component providers have made. Component providers have full control of their components, and can therefore use traditional approaches to maintain their components. For users maintaining their systems with third-party components, the most important questions are how component providers can properly pass the modification information to component users, and how component users can use the information to perform regression testing on their software systems.

In this chapter, we first investigate traditional regression testing techniques and the challenges they are facing when adopted for component-based software. We then focus on a UML-based approach to evaluate modified components properly and so bridge the gap between component providers and component users. Finally, we discuss how to perform regression testing.

10.1 Introduction

Regression testing aims to ensure that a modified software system still meets the specifications validated before modification. To this end, an intuitive approach would involve retesting the modified program using all the test cases in the test pool to make sure the program still preserves the quality that had been validated using those test cases. However, retesting all the test cases is often impractical because of the development cost and the delivery schedule. A selective approach to regression testing is to choose a subset of the test pool that can provide sufficient confidence in the modified system.

For selective regression testing strategies, there are two important issues. The first is identifying portions of the system that will be affected by the modifications. Many approaches have been developed to identify the affected portions. For example, firewall approaches have been developed to identify affected statements and affected functions, and data-flow-based approaches and program-slicing approaches have been developed to identify affected def-use relationships. There are two different ways to apply these approaches: *static* and *dynamic*. A static approach will perform the analysis on the changed program to obtain affected artifacts. A dynamic approach will perform the analysis on the execution history of each test case. Apparently, static approaches are more efficient, because they require only a one-time analysis of the program, but they are less precise. Dynamic approaches have to be performed on the execution history of each individual test case, and are therefore less efficient but more precise. The accurate definition will be discussed later in this chapter.

Once the affected artifacts have been identified, the second question is how well they should be tested. Based on different strategies to determine the adequacy of testing, existing selective regression testing methods can be classified into minimization, coverage, and safe approaches [6]. *Minimization* approaches [7–9] attempt to identify a minimal set of test cases that meets certain structural coverage criteria. An example is the linear-equation approach, which uses the linear equation to select a minimal set of test cases that will achieve segment coverage of the modified program. *Coverage* approaches [10–12] are based on some coverage criteria that are used in minimization approaches. Nevertheless, coverage approaches do not require the minimization that is enforced in minimization approaches. Instead, they seek to select all test cases that exercise changed or affected program components. *Safe* approaches [6, 13–17] place less emphasis on coverage criteria. Instead, they attempt to select every test that will cause the modified program to produce different output than the output from original program. These test cases are called *modification-revealing test cases* [6].

Several models have been developed for evaluating selective regression testing techniques [6, 18–21]. The evaluation has to take input, output, and their relationships into consideration. Inclusiveness, precision and efficiency [6] are the three major considerations in this evaluation. Inclusiveness measures the output of the regression testing process; efficiency measures the input of the regression testing process; and precision considers both input and output.

- *Inclusiveness* measures the ability of a regression technique to include modification-revealing test cases for reexecution. Because modification-revealing test cases generate different outputs when reexecuting, all modification-revealing test cases need to be rerun to guarantee the quality of the modified program. Inclusiveness is a good indicator of how well the system has been reevaluated.
- *Efficiency* assesses the time and space requirements of the regression testing strategy. For instance, the time requirement for a regression testing strategy always includes two parts: *time for test case selection* and *time for test case reexecution*. In fact, the time to rerun the test cases is very easy to forget in the analysis. In general, the more time spent on selection, the fewer test cases selected, and the less time spent reexecuting test cases. The involvement of these two efficiency factors makes the accurate assessment of the efficiency of a regression testing technique very difficult.
- *Precision* is the measurement of the ability of a method to eliminate test cases that do not need to be retested. Precision can be measured by the following formula:

$$\frac{\#\text{Modification\_Revealing\_TestCases\_Selected}}{\#\text{TestCases\_Selected}}$$

When choosing a regression-testing technique for a modified program, inclusiveness, efficiency, and precision all need to be considered. So far, no approach can achieve the best of all three factors. Different approaches try to balance the three factors in different ways.

- The retest-all approach does not require test selection but needs to rerun all test cases. Therefore, this approach is not efficient overall. As all test cases will be rerun, this approach can guarantee inclusiveness.
- Minimization approaches aim to achieve efficiency. So they tend to effectively select a small set of test cases to rerun. However, they often ignore many modification-revealing tests. Therefore, they should not be considered when high reliability is required.
- Coverage approaches select many more test cases to rerun than minimization approaches. Therefore, they are generally less efficient. In terms of inclusiveness, coverage approaches may still omit modification-revealing test cases but they have better inclusiveness than minimization approaches.
- The aim of safe approaches is to select all modification-revealing test cases. Therefore inclusiveness is always 100%. However, efficiency is always a concern for safe approaches. If less time is spent on selection, more test cases will be selected to rerun. Conversely, selecting fewer test cases to rerun takes a lot more time. In general, safe approaches are less efficient than both minimization and coverage approaches.

The choice of regression-testing technique also depends on how the software systems are maintained. Software maintenance activities are often classified into corrective, perfective, adaptive, and preventive maintenance [22].

- *Corrective maintenance* is mainly performed to correct defects that have been discovered. For component-based software, this type of maintenance usually involves modifications to individual components. The overall structure of the component, interfaces, and specifications usually remains unchanged.
- *Perfective maintenance* tries to make changes that will improve the performance of the product, as well as many other quality features. It may also add, delete, or modify functionalities of the product. If it is the first case, the specification of the software remains unchanged and all test cases can be reused. If it is the second or third case, some test cases cannot be reused, and new test cases may need to be generated to meet new requirements.

- *Adaptive maintenance* deals with changes that are responsive to changes in the environment. For instance, an EJB application is upgraded from Weblogic 6.0 to Weblogic 7.0. In most cases, the requirement remains the same.
- *Preventive maintenance*, defined by IEEE standard [23], is "maintenance performed for the purpose of preventing problems before they occur." Preventive maintenance has caused a lot of controversy and has many different interpretations. Regardless of the controversy, regression testing can be performed for preventive maintenance, according to whether the requirement has changed.

Some previous studies [22] have shown that approximately 17.5% of maintenance is corrective maintenance, 60.5% is perfective maintenance, and 18% is adaptive maintenance. A lot of work has been done on regression testing for procedure-based [6, 10, 14, 15, 17, 24] OO software systems [13, 16, 25–28]. Nevertheless, all of these approaches are code-based approaches and apply mainly to corrective maintenance. These approaches have two major drawbacks. First, they cannot always be adopted for component-based software. Second, they very often do not work well for perfective and adaptive maintenance, which make up most of maintenance activities overall.

To perform regression testing adequately on component-based software, we need to solve the two problems raised by the regression testing technique on traditional software systems. To address the implementation transparency characteristic of component-based software, a UML-based technology is developed in Section 10.2—the reasons for using UML-based approaches were explained in Chapter 9. To perform regression testing for perfective and adaptive maintenance, a retest-all approach can be used for scenarios where requirements remain unchanged, and techniques from Chapter 9 can be adopted for the newly introduced requirements. For the cases when retest-all is not feasible, which is pretty common, Section 10.3 introduces a UML-based approach to evaluate the similarity between the new component and the old component. Based on the evaluation, minimization or coverage regression testing approaches can be performed.

10.2 Regression testing for corrective-maintenance activities

After the component has been delivered to users, new faults are likely to be identified, requiring corrective maintenance. Component providers have the source code and can therefore use traditional code-based approaches to define regression tests. However, the code is not available to the component users, so, to create regression testing at the integration and system level, we need a way to represent the changes at a higher level. UML [29–32] provides different diagrams for different purposes. For example, class diagrams are used to specify attributes, operations and constraints of classes, and their inheritance relationships with related classes. Collaboration diagrams are used to specify interactions among classes of components, and the statechart diagram is used to show behaviors of state-dependent objects or the entire component. Different representations of the software can be derived from different diagrams.

Throughout this section, we will use an ATM example to demonstrate our methodology. This ATM component can provide the following services: PIN validation, deposit, withdrawal, and query.

10.2.1 Representing modifications with UML diagrams

If component providers use UML diagrams to design their components, changes to the source code can easily be depicted in the design documents. We consider class diagrams, collaboration diagrams, and statecharts.

- *Class diagrams:* Class diagrams provide a class hierarchy within a component and detailed design information about classes in the component. The importance of the class diagram is that, when modifications are made to one class, the class hierarchy can be used to determine which other classes may be affected. Those changes will then be represented in other diagrams.
- *Collaboration diagrams:* Collaboration diagrams illustrate how objects within a component interact. The collaboration diagram in Figure 10.1 not only shows how objects interact with each other through message description, but also shows the overall control sequence of a component through the numbering mechanism. For instance, the sequence of transitions 1-1.1-1.2-1.3-1.4-2-2.1-2.2-2.3-2.4-2.5-2.6-2.7-2.8 demonstrates a complete PIN-validation process. In collaboration diagrams, capital letters demonstrate alternative edges. For example, 1.1 and 1.1A are to be taken under different conditions. Lowercase letters show messages that will be passed concurrently. For example, 2.7 and 2.7a will be passed from the ATMControl object to the ATMTransaction object and the Customer-Interface object simultaneously.

  Depending on the nature of the modification, different changes to the program can be reflected in collaboration diagrams in two ways:

1. *Localized changes in a specific member function of a class within a component:* In collaboration diagrams, function invocations are represented via messages that are passed among objects. For instance, in Figure 10.1, messages 2.1 and 2.2 (obtain ATMCard data) corresponds to the invocation of the method *getATMCardData*() of an ATMCard object. If the maintenance modifications are to a function that will communicate with other objects, then corresponding messages will be marked as *affected messages*.

2. *Changes that might affect interaction sequences:* When changes are made in a specific function, they can involve adding, removing, or changing function calls. This will consequently change the control sequence of the component. The control sequence is illustrated by numbers and can be managed in one of two ways. First, when adding a new function invocation (i.e., adding a new message into the diagram), we need to add a new sequence after the modified function being called in the collaboration diagram. For example, if we need to add a new message between 2.2 and 2.3, the corresponding number for the new sequence is 2.2.1. Second, when removing a function invocation, corresponding messages need to be deleted from the diagram. This will generate a gap in the message sequence and require renumbering of all subsequent messages. To avoid the renumbering of subsequent messages, we can create a "dummy" message of a self-loop over the modified function, with a range of numbers that correspond to the messages to be deleted.

- *Statechart diagrams:* Statechart diagrams are used to depict state changes for a state-dependent control object or component. Statechart diagrams and collaboration diagrams must be consistent with each other. After modifications are depicted on the collaboration diagrams, the impact of the modifications on the statechart is derived systematically from the collaboration diagrams. The changes in the statechart can add or delete states, or add or delete state transitions. This is illustrated in Figures 10.1 and 10.2. Figure 10.1 shows two changes in boldface: the ability for the user to cancel a transaction (*2A: Cancel Input*); and the ability for the ATM to confiscate a card if the PIN is invalid three consecutive times (*2.6B[ThirdInvalid]*). This is reflected in Figure 10.2 with the two new states Ejecting and Confiscating (in bold).

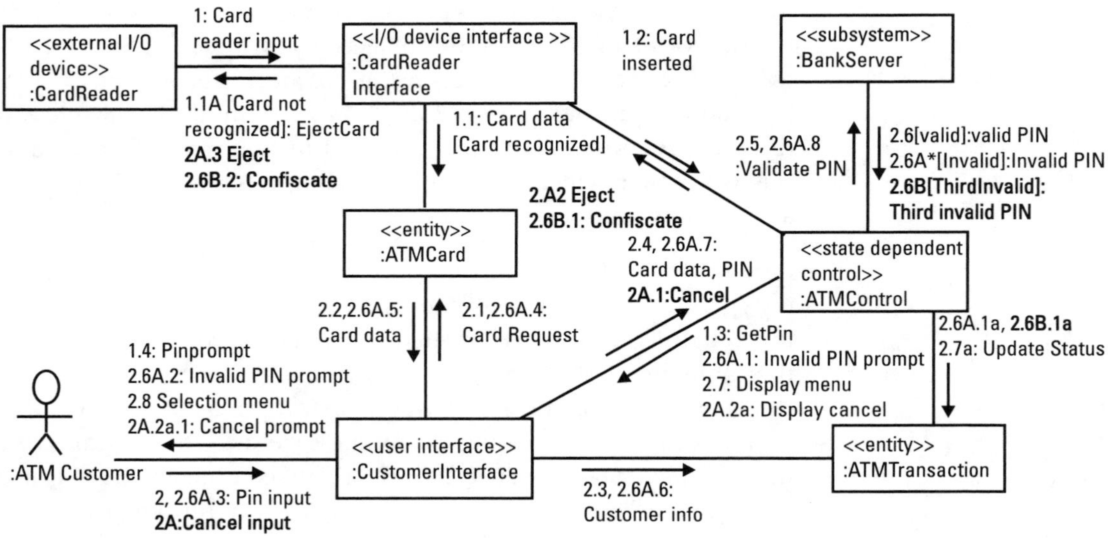

Figure 10.1 Collaboration diagram for validating PINs.

10.2.2 Regression testing for corrective maintenance

The analysis of the UML diagrams in the last section can be used as a basis for developing several regression testing strategies. We assume that the UML diagrams were changed correctly. An obvious minimization approach is to retest each changed artifact in the collaboration diagram at least once. At a further level of detail, a coverage approach can be developed, requiring rerunning of all the test cases that executed any of these affected artifacts. The first approach is cost effective, but may not result in as thorough a test, while the second approach may be much more costly. A middle approach may be developed that relies on change-impact analysis. This has been done at the code level [33], but not at the UML level.

The weakness of the first approach is that only the artifacts that have been directly modified will be tested: indirect effects caused by the modifications will be ignored. With an adequate analysis of all possible side effects that might occur, retesting each individual scenario will be more reliable.

Besides the validation of direct modifications, we need to further identify the effects that these changes may impose on other parts of the component. These effects can be classified into the following two categories:

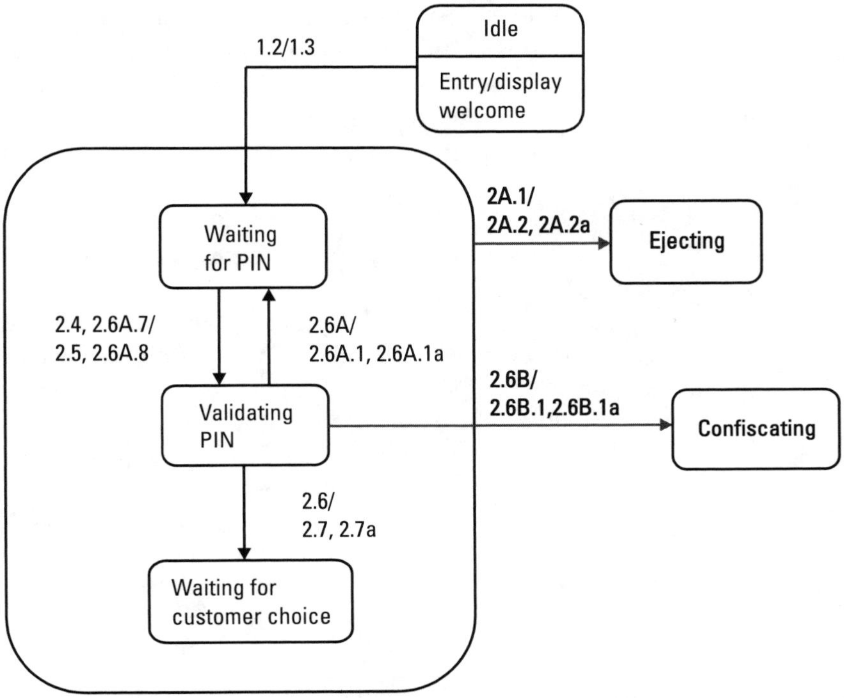

Figure 10.2 Statechart for ATM control diagram.

- *Impacts of changes on control sequences:* In general, artifacts that have been modified in a component can be invoked in different scenarios. For instance, in Figure 10.1, when we add a control sequence 2A, 2A.1, 2A.2, and 2A.3, this sequence can be invoked in many different scenarios. As shown in Figure 10.1, the changed statechart diagram, we need to test that sequence in three different scenarios: (1) cancel while waiting for PIN, (2) cancel while validating PIN, and (3) cancel while waiting for customer choice. Therefore, we not only need to validate the modified artifacts in the collaboration diagram, but also need to retest all possible affected scenarios demonstrated in the statechart diagram. Moreover, with even higher-quality requirements, we may need not only to retest all affected states and state transitions in the statechart diagram, but also to retest all paths that include affected states and transitions.
- *Impacts of changes on data dependencies:* In addition to affected control sequences, changes may also affect the data-dependence relationships

between two control sequences. An invocation of an interface of a component is in fact an invocation of a function implemented by the component. Therefore, when a function declared in an interface v_1 has a data-dependence relationship with another function declared in another interface v_2, the order of invocation of v_1 and v_2 could affect the results.

In a collaboration diagram, messages that flow into an entity object without a corresponding reply message often imply an update of that object. If messages that flow into an object are followed by another message that flows out of that object, it often reflects information retrieval or response after processing from that object. If we change the message that updates an object, we need to retest the control path that retrieves information from the same object. For instance, in Figure 10.3, message W4 enters the Account object, but no message W5 comes out. Therefore, we assume W4 will update information in the Account object. In addition, messages D2 and D3 go into and come out of the Account object, so they retrieve information. Taken in combination, D2 and D3 depend on W4. If changes have been made to W4, then at least one test case that explored this relationship needs to be retested.

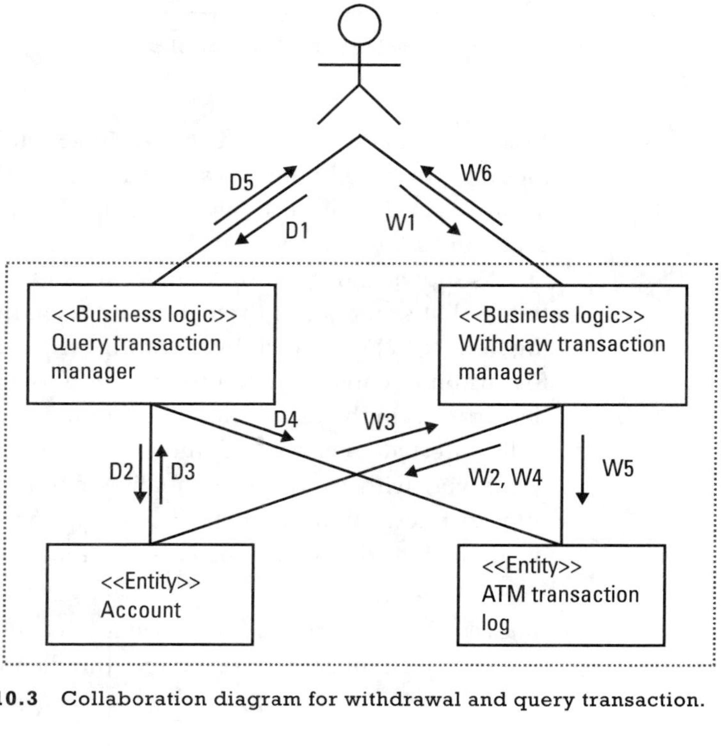

Figure 10.3 Collaboration diagram for withdrawal and query transaction.

10.3 Regression testing for perfective and adaptive maintenance

Perfective and adaptive maintenance may change the requirement of the software component. If that is the case, new test cases need to be generated, and methodologies from Chapter 9 can be applied. In this section, we focus on perfective and adaptive maintenance where requirements do not change. For component-based systems, even maintenance activities may retain the existing interface specifications, but they may require the component to be redesigned. Thus, the internal structure of the new component can be different. To ensure the quality of the system after a new version of the component is integrated, an intuitive approach is to completely retest that component. However, this approach is too costly, particularly when third-party components are expected to be frequently updated. Furthermore, it is against the primary objectives of maintainability and reducing maintenance cost.

How do we determine what tests can be trusted and reused and what tests need to be modified? Based on the observations in the last section, we know that class diagrams, collaboration diagrams, and statechart diagrams can be used to depict the component-control structure and data-dependence relationships among interfaces. This information may be helpful in perfective maintenance as well. Even though the internal structure of a component can be completely different, with new classes and different sequences of control, some artifacts may still be useful for depicting the similarity of the control-sequence and data-dependence relationships.

10.3.1 Control similarity evaluation

Collaboration diagrams represent control sequences by alternative paths, which are annotated by capital letters on the messages. For instance, Figure 10.4 shows the three alternate paths W3, W3A, and W3B. For each alternative path, a guard constraint is defined on the message to determine when the component will execute that path. Given an alternative path, we can partition the executions of the component into different contexts depending on all possible constraints. A *guard* is a Boolean variable or expression that is used to choose alternative paths in a collaboration diagram, while a *context constraint* is a set of guards associated with an execution of an interface of a component. Note that equality guards are compared strictly on the string of characters.

For example, Figure 10.4 shows an execution of the withdrawal interface of the old ATM server component is W1-W2-W3-W4-W5. The execution is constrained by [*Valid account*] and [*Sufficient funds*]. With the new component, the context constraint for the same test case will be [*Within daily limit*], [*Valid account*], and [*Sufficient funds*].

Assume a set of guards $S = \{s_1, s_2, ..., s_p\}$ and a set of context constraints $C = \{c_1, c_2, ..., c_q\}$ of the old component; a set of guards $S' = \{s_1', s_2', ..., s_r'\}$ and a set of context constraints $C' = = \{c_1', c_2', ..., c_s'\}$ of the new component. The following list describes different scenarios that may occur and different strategies that can be adopted:

- The new component's context constraints remain the same after performing the perfective maintenance activities—that is, $S = S'$ and $C = C'$. This means the control structure of the new component remains unchanged. If the new component is adequately tested, we can retest each interface once to make sure the similarity in the control structure will preserve the quality of the component and the component-based system.
- The new component introduces new guards that further partition the execution of the component—that is, $S' \supseteq S$, and for each $c \in C$, there exists a $c' \in C'$, where $c \subseteq c'$. For instance, in Figure 10.4, the old components have two guards and three different context constraints:

  1. [*Valid account*] and [*Sufficient funds*]
  2. [*Valid account*] and [*Insufficient funds*]
  3. [*Invalid account*]

  - The new component introduces one more guard, which is intended to improve the withdrawal transaction by introducing a daily access limit. The modification of the collaboration diagram follows the principles described in the last section; the new diagram is shown in Figure 10.4. Therefore, the new component has four context constraints:

  1. [*Exceed Daily limit*]
  2. [*Within Daily limit*] and [*Valid account*] and [*Sufficient funds*]
  3. [*Within Daily limit*] and [*Valid account*] and [*Insufficient funds*]
  4. [*Within Daily limit*] and [*Invalid account*]

  - For this case, new guards such as [*Exceed Daily Limit*] and [*Within Daily Limit*] need to be retested. The combination of other guards all fall into old contexts. So if the old and new components are properly tested by the same provider, we should have enough confidence that they will be correct. In this case, therefore, we need only to retest context constraint 1 and one of 2, 3, or 4.

- Removing some guards in the new component will merge some partitions in the old component—that is, $S' \subseteq S$, and for each $c \in C$, there exists a $c' \in C'$, where $c \supseteq c'$. The old component had different values for different guards. Thus different outputs are expected. Nevertheless, in the new

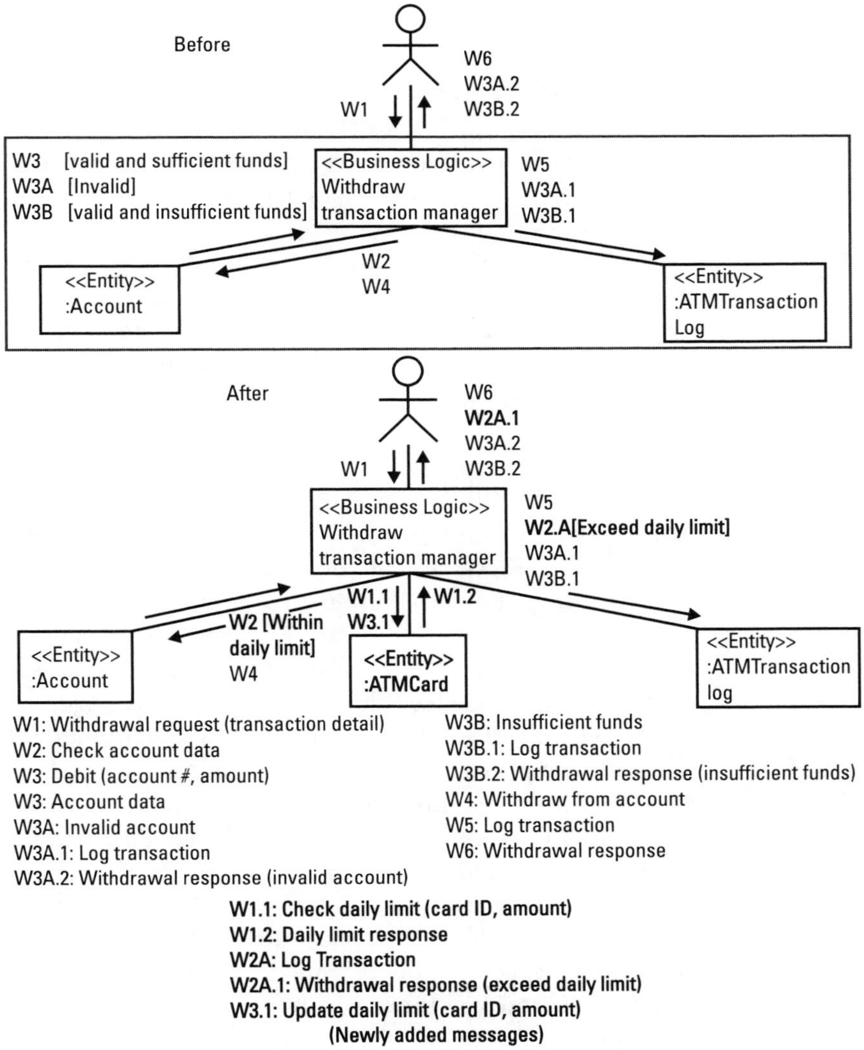

Figure 10.4 Collaboration diagram for withdrawal transaction.

component, two different guards will be treated as one scenario. Thus, it is necessary to properly test the two different values of a removed guard.

- The new component has guards that have been added, removed, or recombined. That is, $S \not\supseteq S'$ and $S' \not\supseteq S$, and there exists $c' \in C'$, where $c' - \{new\ guards\} \not\supseteq$ any $c \in C$ and vice versa. For this case, two different approaches are adopted:

  1. Test all new guards, and retest all context constraints c', where $c' - \{new\ guards\} \not\supseteq$ any $c \in C$. This is a relatively inexpensive strategy, as it ignores potential side effects of merging partitions.

  2. Test all new guards, and retest all context constraints c', where $c' - \{new\ guards\} \neq$ any $c \in C$. This is a safer strategy, as it considers all differences in the partition between the new component and old component, but more test cases will be selected to rerun.

- The order of the guards has some effects. So far, we have assumed that the order of different guards does not adversely affect the integrated system. If that assumption does not hold, the order of guards needs to be taken into consideration. This will incur more overhead than the previous approaches.

10.3.2 Data dependence similarity evaluation

The previous definitions considered the control sequence only. This section considers the effects on data dependences.

Redesigning the old component may change the data dependence relationships among different interfaces. New data dependence relationships can be added, existing dependence relationships can be removed, or the context that enables the dependence relationships may be changed.

If a new data dependence relationship is identified, then either a new test case must be generated or some existing test cases that explore the dependence relationships must be rerun. If a data dependence relationship is removed, then a test case that was originally designated for that relationship needs to be retested. To determine the changes in data dependence relationships, we must identify a context constraint pair (c_1, c_2) or each dependence relationship that defines the scenarios to enable the dependence relationship. To be more specific, under context constraint c_1, an update of an object is issued, and under context constraint c_2, a reference to the same object is issued. To determine whether the scenario varies between the old to the new components, the criteria defined early in this section can be adopted.

Therefore, if either c_1 or c_2 change, the data dependence relationship needs to be retested under the new scenario.

10.4 Summary

We have briefly reviewed traditional regression techniques and discussed how to overcome implementation-transparency problems through a UML-based approach. The UML is now the industry standard for software-modeling notation. UML-based approaches can provide feasible guidelines for component providers to model their behavior precisely and pass through component users. Used properly, the information can efficiently and effectively maintain evolving component-based software. For perfective-maintenance activities, which have not been fully explored by researchers, the UML-based framework can be used to evaluate similarity between old and new components, and this can be the foundation for adequate perfective and other maintenance activities.

References

[1] Cook, J. E., and J. A. Dage, "Highly Reliable Upgrading of Components," *International Conference on Software Engineering*, Los Angeles, CA, 1999, pp. 203–212.

[2] Mezini, M., and K. Lieberherr, "Adaptive Plug-and-Play Components for Evolutionary Software Development," *Proc. of the Conference on Object-Oriented Programming, Systems, Languages, and Applications*, 1998, pp. 97–116.

[3] Voas, J., "Maintaining Component-Based Systems," *IEEE Software*, Vol. 15, No. 4, July/August 1998, pp. 22–27.

[4] Wu, Y., and J. Offutt, "Maintaining Evolving Component-Based Software with UML," *Proc. of the 7th European Conference on Software Maintenance and Reengineering*, March 2003.

[5] Wu, Y., D. Pan, and M. Chen, "Techniques for Maintaining Evolving Component-Based Software," *Proc. of International Conference on Software Maintenance*, 2000, pp. 272–283.

[6] Rothermel, G., and M. J. Harrold, "A Safe, Efficient Regression Test Selection Technique," *ACM Transactions on Software Engineering and Methodology*, Vol. 6, No. 2, April 1997, pp. 173–210.

[7] Hartmann, J., and D. J. Robson, "Techniques for Selective Revalidation," *IEEE Software*, Vol. 16, No. 1, January 1990, pp. 31–38.

[8] Lee, J. A. N., and X. He, "A Methodology for Test Selection," *Journal of System and Software,* Vol. 13, No. 1, September 1990, pp. 177–185.

[9] Sherlund, B., and B. Korel, "Modification Oriented Software Testing," *Software Quality Week Conference,* 1991, pp. 1–17.

[10] Bates, S., and S. Horwitz, "Incremental Program Testing Using Program Dependence Graphs," *Proc. of 20th ACM Symposium of Principles of Programming Languages,* January 1993, pp. 384–396.

[11] Binkley, D., "Reducing the Cost of Regression Testing by Semantics Guided Test Case Selection," *Proc. of International Conference on Software Maintenance,* October 1995, pp. 251–260.

[12] Harrold, M. J., and M. L. Soffa, "Interprocedural Data Flow Testing," *Proc. of 3rd Testing, Analysis, and Verification Symposium,* December 1989, pp. 158–267.

[13] Abdullah, K., and L. White, "A Firewall Approach for the Regression Testing of Object-Oriented Software," *Software Quality Week Conference,* San Francisco, CA, 1997.

[14] Agrawal, H., et al., "Incremental Regression Testing," *International Conference on Software Maintenance,* September 1993, pp. 348–357.

[15] Chen, Y. F., D. Rosenblum, and K. P. Vo, "Testtube: A System for Selective Regression Testing," *Proc. of 16th International Conference on Software Engineering,* May 1994, pp. 211–222.

[16] Rothermel, G., M. J. Harrold, and J. Dedhia, "Regression Test Selection for C++ Software," *Software Testing, Verification, and Reliability,* Vol. 10, No. 2, June 2000.

[17] White, L., and H. K. N. Leung, "A Firewall Concept for Both Control-Flow and Data-Flow in Regression Integration Testing," *Proc. of International Conference on Software Maintenance,* 1992, pp. 262–271.

[18] Graves, T. L., et al., "An Empirical Study of Regression Test Selection Techniques," *Proc. of 1998 International Conference on Software Engineering,* 1998, pp. 188–197.

[19] Leung, H. K. N., and L.White, "A Cost Model to Compare Regression Test Strategies," *Proc. of International Conference on Software Maintenance,* 1991, pp. 201–208.

[20] Rosenblum, D. S., *Adequate Testing of Component-Based Software,* Technical Report TR 97–34, University of California at Irvine, Irvine, CA, 1999.

[21] Rosenblum, D. S., and E. J. Weyuker, "Using Coverage Information to Predict the Cost-Effectiveness of Regression Testing Strategies," *IEEE Trans. on Software Engineering,* Vol. 23, No. 3, 1997, pp. 146–156.

[22] Schach, S., *Object-Oriented and Classical Software Engineering,* 5th ed., Boston, MA: WCB/McGraw-Hill, 2002.

[23] IEEE Standard Glossary of Software Reengineering Terminology, IEEE Std 610.12-1990.

[24] Gupta, R., M. J. Harrold, and M. L. Soa, "Program Slicing Based Regression Testing Techniques," *Journal of Software Testing, Verification, and Reliability*, Vol. 6, No. 2, June 1996, pp. 83–112.

[25] Orso, A., et al., "Using Component Metacontents to Support the Regression Testing of Component-Based Software," *Proc. of IEEE International Conference on Software Maintenance (ICSM2001)*, 2001, pp. 716–725.

[26] Wu, Y., M. Chen, and M. Kao, "Regression Testing on Object-Oriented Programs," *Proc. of 10th International Symposium on Software Reliability Engineering*, November 1999, pp. 270–279.

[27] Harrold, M. J., and G. Rothermel, "Performing Dataflow Testing on Classes," *Proc. of 2nd ACM SIGSOFT Symposium on Foundations of Software Engineering*, December 1994, pp. 154–163.

[28] Kung, D., et al., "On Regression Testing of Object-Oriented Programs," *Journal of Systems and Software*, Vol. 32, No. 1, January 1996, pp. 21–40.

[29] Booch, G., J. Rumabugh, and I. Jacobson, *The Unified Modeling Language Guide*, Reading, MA: Addison-Wesley, 2000.

[30] Cheesman, J., and J. Daniels, *UML Components: A Simple Process for Specifying Component-Based Software*, Reading: MA: Addison-Wesley, 2001.

[31] Fowler, M., and K. Scott, *UML Distilled*, Reading, MA: Addison-Wesley, 2000.

[32] Gomaa, H., *Designing Concurrent, Distributed, and Real-Time Applications with UML*, Reading, MA: Addison-Wesley, 2000.

[33] Lee, M., J. Offutt, and R. Alexander, "Algorithmic Analysis of the Impact of Changes to Object-Oriented Software," *34th International Conference on Technology of Object-Oriented Languages and Systems (TOOLS USA '00)*, August 2000, pp. 61–70.

CHAPTER

11

Contents

Performance testing and measurement

As described in Chapter 2, performance testing and evaluation is the last important step in the quality control process of software components and systems. Since component-based system performance depends on the performance of the involved components, component users must check and evaluate the performance of reusable components (including COTS) before they adopt and deploy them. To provide component customers with performance information, such as component performance boundary and capacity limits, component vendors must evaluate the performance of components to discover the performance issues and capacity limits.

Today, component developers and test engineers have encountered a number of issues and challenges in performance validation and evaluation of software components and component-based programs.

- Most reusable components (including COTS) do not provide the component's performance boundaries and limits as a part of its product specifications and quality report. This increases the complexity of component selection and evaluation on the part of the customers.
- Current component models and technology do not address the needs and issues of component testing, integration, and performance evaluation. Thus, validating the performance of a system, built on third-party reusable components, is very difficult and costly.

- There is a lack of systematic solutions and tools to support component-based software performance testing and measurement. Most existing performance evaluation metrics, solutions, and techniques consider the whole software system (or program) as a single black box. Thus, they are not good enough for test engineers to identify component performance issues and capacity boundaries for component-based programs.

With the advancement of component-based software engineering, we need new evaluation models, metrics, and systematic solutions to support performance validation and evaluation of components and component-based programs. This chapter covers the basic concepts of software performance testing and measurement, including the performance test process, objectives, focuses, problems, and needs. It also discusses various performance evaluation metrics for components and systems. Moreover, different types of performance evaluation approaches are summarized, and several component-oriented performance evaluation models are presented. In addition, the chapter also highlights the essential needs of software performance testing techniques and evaluation tools.

The chapter is structured as follows. Section 11.1 revisits the basic concept of software performance testing and evaluation. Section 11.2 discusses different types of performance evaluation metrics, including utilization, speed, throughput, reliability, availability, and scalability. Section 11.3 covers the different types of performance evaluation approaches, and presents two component-based performance evaluation models. The basic needs of performance testing and evaluation tools are discussed in Section 11.4. Finally, a summary of this chapter is given.

11.1 Basics of software performance testing and measurement

What is software performance testing and measurement? *Software performance testing and measurement* refers to testing activities and evaluation efforts to validate system performance and measure system capacity. As a quality control task, system performance validation and evaluation must be completed before delivering to customers.

There are three major objectives. The first is to validate a product to see if the given component and system performance requirements are satisfied. The next is to find out the product capacity and boundary limits. And, the last is to discover the performance issues, degradations, improvements, and bottlenecks of a software system and its components to support performance tuning and problem fixing.

Conducting cost-effective software performance testing and evaluation requires the following items:

- A well-defined evaluation strategy and performance test process;
- Sound and correct performance evaluation models presenting component and system performance attributes, parameters, and their relationships;
- Well-defined performance measurement metrics useful to compute and evaluate different software performance factors and behaviors;
- A cost-effective software performance testing and evaluation environment equipped with efficient performance monitoring and analysis solutions.

11.1.1 Major focuses in software performance testing and evaluation

During a performance testing and evaluation process, test engineers must identify the major focuses and prioritize them based on a limited budget and schedule. Table 11.1 summarizes the major focuses in performance testing and evaluation.

As shown in Table 11.1, the major focuses of software performance testing and evaluation can be classified as follows:

- *Processing speed, latency, and response time:* Checking the speed of functional tasks of software components (or systems) is always a focus in performance testing and evaluation. The typical examples are checking the processing speed of an e-mail server, and measuring the processing speed of different types of transactions in a bank system. To correctly measure the speed of a functional process, engineers usually need to collect and measure its maximum, minimum, and average processing speed during a performance test process. Validating system-user (or component-user) response time is another common interest in performance testing. Its objective is to confirm that a system (or a component) always generates correct user responses based on users' requests according to the given performance requirements. The other type of performance concern is latency, which is frequently used to measure the delay time of transferring messages, processing events and transactions.

Table 11.1 Classified Focuses in Performance Testing and Evaluation

Focused Area	*Testing and Evaluation Items*	*Examples*
Processing speed, latency, and response time to users (minimum, maximum, and average speed)	System-user (component-user) response time Function processing speed and latency Transaction processing speed and latency Communication speed and latency Data access speed and latency	System/component response time, data query time Application task processing time or functional processing speed Bank transaction speed and latency Network connection time, network latency, message transfer speed Database connection and retrieval speed
Load boundary (maximum, minimum, and average load boundary)	Network traffic load boundary Database load boundary User load boundary Component or server load boundary	Maximum volume of communication traffic Maximum load of concurrent database accesses Maximum number of concurrent user accesses Maximum load for a component or a application server
Throughput (minimum, average, peak, and full-load throughput)	Event processing throughput Task (or process) throughput Transaction throughput Communication message throughput	Domain-specific event throughput Domain-specific task throughput Domain-specific transaction throughput Protocol-oriented message throughput
Reliability (minimum, maximum, and average availability)	Network reliability Computer hardware reliability Computer software reliability Application reliability	Network gateway or switch reliability Client or server machine reliability Component or server reliability Functional process reliability
Availability (minimum, maximum, and average availability)	Network availability Computer hardware availability Computer software availability	Network gateway or switch availability Client or server machine availability Component, server, function, or process availability
Scalability (performance speedup, threshold, and improvement)	Network scalability Data scalability User scalability Transaction scalability	Scalability of network traffic and structure Scalability of data volume and database connectivity Scalability on the number of concurrent users Scalability on the number of concurrent transactions
Utilization (minimum, maximum, and average utilization)	Network utilization Client/server machine utilization	Network bandwidth utilization, such as LAN or Internet utilization CPU, cache, memory, and disk utilization

- *Throughput:* Validating different types of component (or system) throughputs provides useful performance indicators on the product capacity. For example, we can find a component (or system) throughput in processing events, messages, or transactions in a given time period. For a component-based application system, we should pay attention to the minimum, average, and maximum throughputs at the component level as well as the system level.
- *Availability:* System availability analysis helps engineers understand the availability of its hardware and software parts, such as network, computer hardware and software, and application functions and servers. For a component-based application system, it is important to evaluate the availability of its components at the component and system level. Most e-commerce and on-line application systems have very rigorous performance requirements on system availability. For those systems, component and system availability evaluation is very critical.
- *Reliability:* Since software reliability is one of the very important product quality factors, reliability validation and measurement must be performed before a product release. Reliability evaluation focuses on the reliability measurement of a system and its components in delivering functional services during a given time period. Most defense and safety critical software systems have very strict reliability requirements. Using component-based software engineering to develop application systems with rigorous reliability requirements, component engineers and users must check the reliability of all reusable components before they are used and deployed.
- *Scalability:* Checking the scalability of a system and reusable components helps test engineers discover how well a given system (or component) can be scaled-up in terms of network traffic load, input/output throughput, data and transaction volume, and the number of concurrent access clients and users. The focus here is to measure component (or system) performance boundaries and limits, performance improvements, and thresholds under a given platform and system loads.
- *Utilization:* System resource utilization is frequently validated during system performance and evaluation. The goal is to measure the usage of various system resources by components and systems under the given system loads. The common system resources include network bandwidth, CPU time, memory, disk, cache, and so on. Software component users, especially COTS users, must pay special attention to the

component utilization of system resources to make sure that they are used efficiently.

11.1.2 Performance test process

A performance testing process is needed to support the control and management of system performance testing and evaluation. Usually, a performance testing process includes the following working steps (see Figure 11.1):

- *Step 1: Understand and enhance performance requirements.* Before carrying out performance testing and evaluation activities, engineers must understand the system performance requirements. In many cases, the performance requirements are not clearly specified. Hence, performance engineers and testers need to review system specification documents to identify, check, define, and enhance system performance

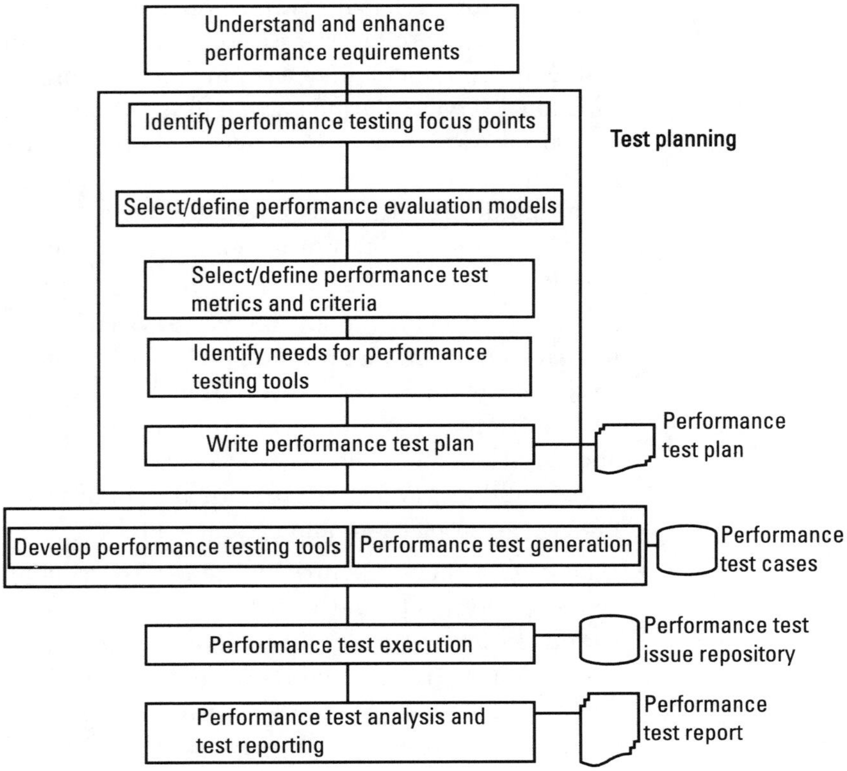

Figure 11.1 A performance test process.

requirements. This step is very important for a performance test process because it is impossible to come out with good performance testing strategy and evaluation solutions without well-defined performance requirements. There are two tasks in this step. The first is to make sure that the necessary performance requirements are specified. The second is to check if the given performance requirements are measurable and technologically achievable.

- *Step 2: Identify the focus areas in performance testing and evaluation*. The major goal of this step is to communicate with system analysts and development managers to identify the major objectives and focus areas for performance validation, and list them according to priority.
- *Step 3(a): Select or define performance evaluation models.* Its major task is to select and/or define performance evaluation models as a basis to define performance evaluation metrics and supporting tools.
- *Step 3(b): Select and/or define performance test metrics and criteria*. In this step, performance test metrics must be selected and/or defined for each specific performance validation objective. For each performance metric, engineers must make sure that it is feasible to collect and monitor the necessary performance data and parameters during performance evaluation.
- *Step 4: Identify needs of performance testing and evaluation tools.* The major task in this step is to discover and define the needs of performance testing and evaluation solutions, including necessary performance data collection and monitoring techniques, performance analysis facilities, and supporting environment and tools.
- *Step 5: Write a performance test plan.* The result of this step is a performance test plan that includes performance validation objectives, focuses, performance models, evaluation metrics, required solutions, and tools. In addition, a performance test schedule must be included to specify tasks, responsible engineers, and timelines.
- *Step 6(a): Develop and deploy performance test tools.*
- *Step 6(b): Design and generate performance test cases, test data, and test suites.*
- *Step 7: Carry out performance tests and evaluation tasks.*
- *Step 8: Analyze and report system performance.* The major task in this step is to come out with a high-quality performance validation and evaluation report based on the collected performance test results. A good report usually presents well-formatted results, and provides a clear picture of system performance issues and boundaries.

11.1.3 Performance testing and evaluation challenges and needs in CBSE

As pointed out in [1, 2], test engineers encountered the following issues during performance validation and evaluation for component-based programs:

- *Lack of component performance information:* Since most reusable components, including COTS components, do not provide component performance information (such as performance boundaries and limits), component users must evaluate their performance to select components. This creates additional cost and effort to component customers.
- *Difficult to collect and monitor component performance data:* Because most reusable components do not provide external accessible built-in facility and interfaces to support component performance testing and evaluation, collecting and monitoring component performance data is costly and difficult.
- *Hard to detect and isolate component performance problems:* Current component technology and models produce components as executable black boxes. Since there are no standard integration solutions available, component users must integrate components to form a system in an ad hoc manner. This causes serious problems in detecting and isolating component performance issues in the testing and evaluation of component-based software systems.

To measure the performance of component-based software systems now, engineers run into two major challenges. The first is the lack of component based performance evaluation models and metrics based on component performance. The next is the lack of systematic solutions and tools to support performance validation and measurement at the component and system levels. Currently, engineers are seeking cost-effective solutions to test and evaluate component and system performance in terms of system resource, utilization, reliability, availability, scalability, speed, and throughputs. Here are several typical questions regarding performance testing and evaluation of component-based software and its parts:

- How can component-based performance evaluation models and measurement metrics be defined and selected?
- How can a product-oriented or enterprise-oriented performance testing and evaluation environment for component-based software be established in a rational way?

Table 11.2 Classification of Performance Metrics

Performance Metrics	*Types of Metrics*	*Typical Performance Metrics*
Performance metrics	System-user response time Database retrieval speed Message processing speed Domain-specific function speed	System response time versus different types of commands System response time versus different types of concurrent users System response time to various types of queries from users Query processing time versus number of concurrent servers Message processing time versus message sizes Call processing time versus number of concurrent agents
Throughput metrics	Transaction throughput Message throughput Task throughput	Total number of processed transactions versus total number of received transactions Total number of processed messages versus total number of received messages Total number of completed tasks versus total number of given tasks
Reliability metrics	Network reliability Computer hardware reliability Computer software reliability Application reliability	Network gateway or switch reliability Client or server machine reliability Component or server reliability Functional process reliability
Availability metrics	Network availability Computer hardware availability Computer software availability	Network gateway or switch availability Client or server machine availability Component, server, function, or process availability
Scalability metrics	Performance improvement Performance boundary and limit	Function speedup Transaction throughput improvement Multithreads processing speed boundary and limit Transaction throughput boundary and limit
Utilization metrics	Network utilization Server machine utilization Client machine utilization	Network traffic (Kbps) versus number of users for each protocol Server utilization versus command rate for CPU, cache, memory, disk utilization Client machine utilization versus command rate for CPU, cache, memory, disk utilization

- How can component performance data be collected and monitored in a systematic way?
- How can the system performance be computed in a gray-box approach based on component performance?

11.2 Performance evaluation metrics

This section discusses various performance evaluation metrics for software components and systems. In the past, many performance metrics were developed to measure software component and system performance. They can be classified into five classes (Table 11.2). The first class refers to the evaluation metrics related to processing speed, such as functional processing speed and response time to users. The second class includes various throughput metrics. They are used to evaluate the throughputs of a system and its components in terms of processing rate of incoming messages, requests, and transactions. The third class consists of reliability metrics. They are used to evaluate system and component reliability. The fourth class refers to availability metrics, which are useful in evaluating the availability of a system and its components. The last class is scalability metrics. They are developed to analyze and/or predict the performance improvements of software systems in processing speed, throughputs, and capacity.

11.2.1 Utilization metrics

Utilization metrics are defined to measure the utilization of system resources in a component (or system) under the given system loads. Although software application systems may involve many different kinds of system resources, engineers usually pay attention to the following types of system resources: network bandwidth, computer CPU, memory, cache, and storage, such as disk. Therefore, they need to define a set of utilization metrics based on different system resources.

Figure 11.2(a) shows the performance testing results on network utilization for an on-line *electronic customer relationship management* (ECRM) system under 7,500 call loads. Here, the network utilization metric is defined based on the ratio of the network traffic (Kbps) over a network to the number of agents during a given performance evaluation time. It is clear that the network utilization of each protocol can be collected, monitored, and analyzed here. Figure 11.2(b) presents the system utilization of single-threaded server in terms of its usage of disk, CPU, RAM, and data inquire server under

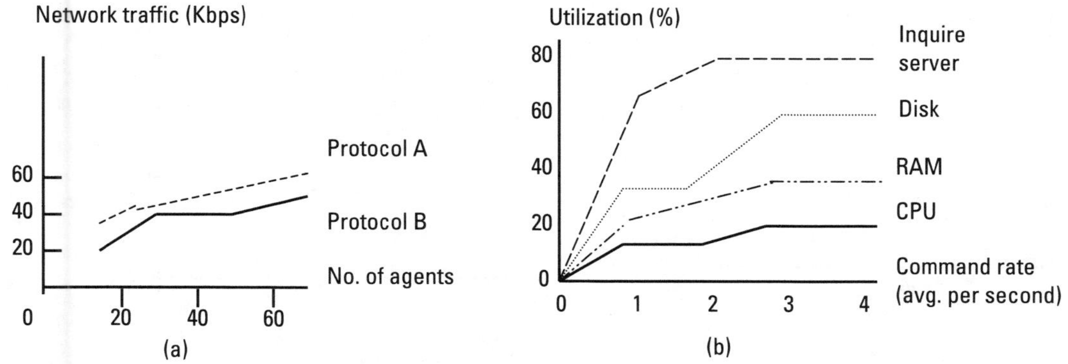

Figure 11.2 System utilization examples: (a) network utilization under 7,500 call loads per time unit; and (b) utilization of single-threaded server.

different system loads. The utilization of a client machine can be analyzed using the same set of utilization metrics.

11.2.2 Speed-related metrics

To measure the processing speed of a component (or system), engineers can define various speed metrics. One of the common speed metrics is the system-user response time. This metric can be used to measure the average, minimum, and maximum user response time for different user groups in a system. For example, Figure 11.3(a) shows the system-user response times of four user groups in an ECRM system based on the number of concurrent access users. They are agents, customers, administrators, and managers. Sometimes, engineers may need to check the system-user response time based on functional features. Figure 11.3(b) displays the system response time of different system commands according to the number of concurrent access users. For components and systems with message communication functions, engineers need to check the processing time of different types of messages. For components and systems with database accesses, engineers must check the database response time of different kinds of database queries, including database connection time. The performance results on database queries are useful in performance tuning to improve data retrieval speed by optimizing database queries. Figure 11.3(d) gives an example, in which the system query response time for each query type is measured according the incoming request rate.

To measure the domain-specific functional speed in components and systems, engineers need to define special performance metrics for each domain-specific function. For example, the call processing is a domain-specific

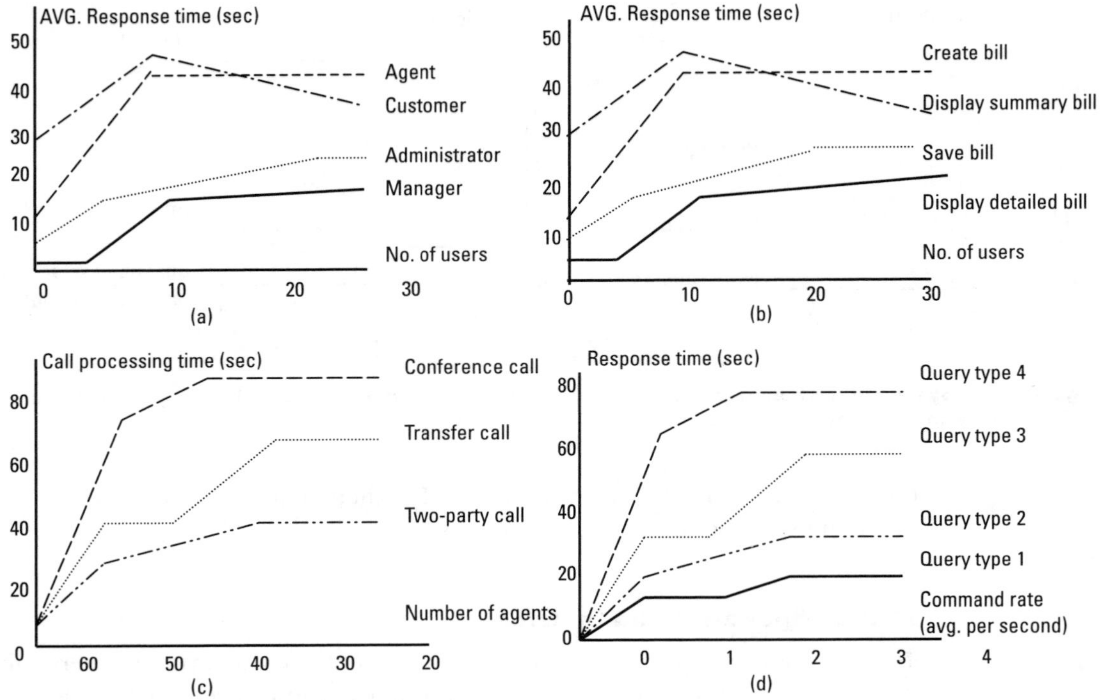

Figure 11.3 Examples of system speed and response time: (a) system-user response time; (b) command-based response time; (c) call processing time (call load: 5,000 per minute); and (d) query response time.

function for a call server in a customer relationship management system. To measure the processing speed for different types of calls, they must evaluate the call processing time for each type of call based on the number of available agents under a specific incoming call load. Figure 11.3(c) displays the call processing speed under a special call load. Readers, who are interested in finding more speed related performance metrics, can read Cahoon's paper in [3]. More examples can be found in [4, 5].

Another type of speed metrics is latency metrics. They can be used to measure various delays of a system (or component). Typical examples are processing delay time, communication delay, and Java Applet download time. Y. Yon et al. [6] presented their processing latency metrics in multiprocessing.

11.2.3 Availability metrics

Component and system availability is a common concern during performance evaluation [7]. To measure system availability, engineers must define availability metrics. In [8], system availability is defined as a ratio of system available time to the total system evaluation time that includes both system available time and unavailable time. Based on this metric, system unavailability can be computed as 1 – availability. Hence, system availability can be formally measured as follows:

$$system\_availability = system\_available\_time \,/\, system\_evaluation\_time$$

where *system_available_time* refers to the time when the system is available to deliver functions and services to its users during the total evaluation time period.

Clearly, this metric is very simple. The basic idea can be applied to measure component availability when software components are considered as a black box. However, this metric has two limitations. First, it does not present any relationship between component availability and component supporting functions. Next, it is not applicable to components with high available capability, such as fault-tolerant features. Today, many software application systems, such as Web-based information systems, e-commerce systems, and defense systems, require high-availability components. How to measure the availability of these components is a challenging issue for engineers. Here we propose two of the component availability metrics to address this issue.

11.2.3.1 Function-based component availability

The function-based component availability metric is developed to measure component availability in terms of component supporting functions.

Definition The function-based availability of a component C_k in a system S refers to the ratio of its total available time T_A (C_k, F_j) of supporting its system function F_j to the performance test period T_p during performance evaluation. Here, T_p actually includes component C_k's available and unavailable time for supporting function F_j. The function-based availability for C_k can be formally computed as follows:

$$FComponent\ Availability\left(C_k, S, T_p\right) = T_A\left(C_k, F_j\right) / T_p$$

Since this metric presents the direct relationship between component availability and supporting functions, component availability measurement can be done through exercising all component functions during performance evaluation. This approach does help engineers to identify the function-oriented performance issue relating to component availability. When component C_k supports m different functions, then its availability can be measured as follows:

$$Component\ Availability = \Sigma_{j=1\ \text{to}\ m}\ FComponent\ Availability(C_k, S, T_p)\ /\ m$$

11.2.3.2 High availability of components

To measure components with high-availability requirements, we first must have a clear understanding of what is a high-available component. Although there are a number of ways to define and develop a high-available component based on different fault-tolerant methods, we have chosen a simple and popular way to define highly available components next.

Definition A high available component is an N-cluster consisting of N redundant components. They are actively running at the same time to support and deliver the same set of functional features as a black box.

Let us assume that $C_{HA} = \{C_0, C_1 \ldots CN\}$ is a cluster of N redundant components. The availability of an N-cluster component C_{HA} in a system S is known as *component high availability*. It can be measured as follows:

$$Component\ High\ Availability = (C_{HA}, S, T_p)$$
$$= Havailable - time(C_{HA})\ /\ [Havailable - time(C_{HA}) + Hunavailable - time(C_{HA})]$$

where $Havailability - time(C_{HA})$ stands for the available time of C_{HA}, and $Hunavailability - time(C_{HA})$ is the unavailable time of C_{HA}, and T_p represents the performance evaluation time period. $Havailability - time(C_{HA})$ can be computed here:

$$HAvailability - time(C_{HA}) = \Sigma_{j=1\ to\ m} T_{j(HA)}$$

where $T_{1\ (HA)}$, $T_{2\ (HA)}$, ..., and $T_{m\ (HA)}$ are the available time slots of T_p.

During each available time slot, at least one component of C_{HA} is active and available to deliver all component functions. Similarly, we can compute *Hunavailable-time* (C_{HA}) here:

$$Hunavailable - time(C_{HA}) = \Sigma_{j=1\ to\ n} T'_{j(HA)}$$

where $T'_{1(HA)}, T'_{2(HA)} \ldots T'_{n(HA)}$ are the unavailable time slots of T_p. During each unavailable time slot, all components of C_{HA} are not available to support and deliver all functions. Figure 11.4 shows an example of three-cluster HA component and its available and unavailable time.

11.2.4 Reliability metrics

Although many different approaches and metrics have been proposed to measure system reliability. The common way is to evaluate the system reliability based on its *reliability of service,* in which a function $R(t)$ is used to present the probability that service survives until time t.

The reliability of the service is often characterized by specifying *mean time to failure* (MTTF) or *mean time between failures* (MTBF). In the computation of the MTTF or the MTBF, it is usually assumed that the exponential distribution best describes the possible occurrence of failures in the service. For different application systems, the reliability of service should be checked to make sure that the above definition is good enough to adequately support the evaluation of system reliability in a given application domain.

When we consider software components as a black box, many system-level reliability evaluation methods are applicable. The details of them can be found in [9–13]. Here, we define component reliability based on component uptime and downtime for services during a performance evaluation time period.

Definition The component reliability of a component C_k in a system S refers to the ratio of the total uptime of the component to the total evaluation time T_p, including both uptime and downtime.

Based on this definition, component reliability can be evaluated easily if there is a systematic solution in place to track component uptime and down-

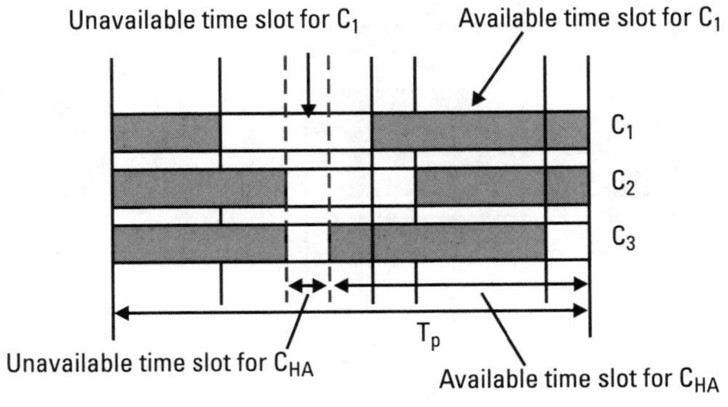

Figure 11.4 A three-cluster HA component and its available and unavailable time.

time. Let us assume that C_k is a non-HA component, and its component reliability during T_p can be computed as follows:

$$Component\ Reliability_{T_P}(C_k) = uptime(C_k) / |T_p|$$

where $|T_p|$ stands for the total performance evaluation time, and *uptime* (C_k) represents the uptime of C_k during T_p, and *downtime* (C_k) includes the downtime and recovery time of C_k.

For HA components, such as an N-cluster component (say, C_{HA}), we need a different metric to evaluate its reliability. Here, we present two different approaches. Figure 11.5 shows the difference between these two approaches.

The first approach is based on a single failure criterion. In this criterion, component C_{HA} is considered down if a single component of C_{HA} has a failure of service in any given time slot during the performance evaluation period. Based on this idea, the reliability of component C_{HA} can be presented as follows:

$$Component\ Reliability_{T_P}(C_{HA})$$
$$= uptime(C_{HA}) / (uptime(C_{HA}) + downtime(C_{HA})) = uptime(C_{HA}) / |T_p|$$

where *uptime* (C_{HA}) represents the total uptime of C_{HA} during T_p, and *downtime* (C_{HA}) stands for the total downtime and recovery time of C_{HA}. To measure *downtime* (C_{HA}), let us assume that $C_{HA} = \{C_{HA1}, \ldots, C_{HAn}\}$ is an N-cluster component. According to the single failure criterion, the downtime of C_{HA} can be measured as follows:

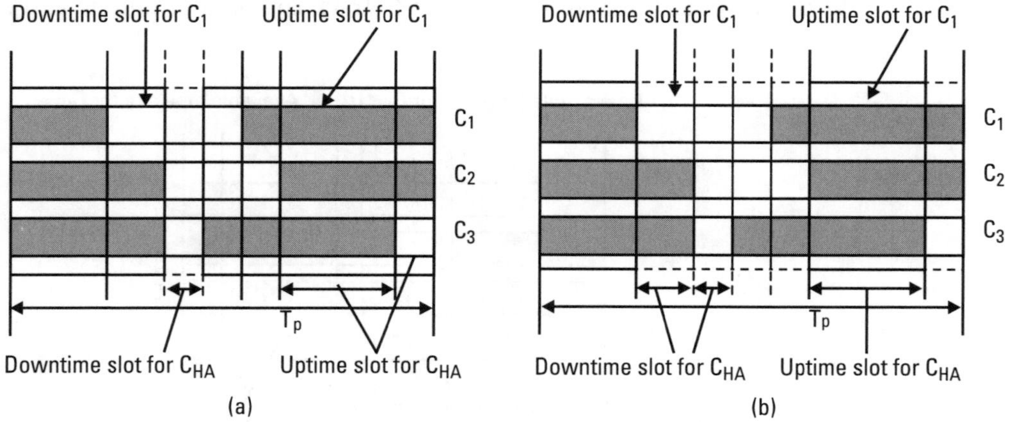

Figure 11.5 A comparison of two approaches.

$$downtime(C_{HA}) = \Sigma T_j \qquad (j = 1,\dots m)$$

where T_j is a time slot in T_p, and at least one of C_{HA} components is down during T_j.

The uptime of C_{HA} can be computed based on its downtime:

$$uptime(C_{HA}) = |T_p| - [downtime(C_{HA})]$$

The other approach evaluates the reliability of component C_{HA} based on its availability for functional services. In this approach, a failure of services of component C_{HA} during a time slot means that all redundant components of C_{HA} have failed to provide their services. This approach uses a different way to compute *uptime* (C_{HA}). For an *N*-cluster component C_{HA} = {C_{HA1}, ..., C_{HAn}}, its *uptime* (C_{HA}) can be computed as follows:

$$uptime(C_{HA}) = \Sigma T_j \qquad (j = 1,\dots m)$$

where T_j is a time slot in T_p, and at least one of C_{HA} components is up and active to provide functional services during T_j.

11.2.5 Throughput metrics

Throughput metrics are usually developed to measure the successful rate of a software component in processing events, messages, and transaction requests in a given evaluation time period. For example, we can define a transaction-oriented throughput metric for software components as follows.

Definition The *transaction throughput* of a component C_k for transaction type *TR* refers to the total number of successfully processed transaction requests in a given performance evaluation period T_p. The formal metric is given here:

$$Throughput(C_k, T_p, TR) = m$$

where *m* is the total number of successfully processed transaction requests.

During component performance testing and evaluation, engineers may apply a number of test sets to find out the maximum, minimum, and average transaction throughput for each type of transaction. Based on the transaction throughput metric, we can define transaction throughput rate next.

Definition The *transaction throughput rate* of a component C_k is the ratio of the total number of successfully processed transaction requests to the total number of received transaction requests during a given performance evaluation period T_p. The formal metric is given here:

$$Throughput\ Rate(C_k, T_p TR) = m / n$$

where m is the total number of successfully processed TR transaction requests during T_p, and n is the total number of received TR transaction requests during T_p.

When the throughput rate reaches 1, component C_k has the highest successful rate in processing transaction requests in TR category. The same idea can be used to define other throughput metrics to measure a component's processing rate of incoming events, messages, and commands.

11.2.6 Scalability metrics

A system (or component) is scalable if it can be deployed effectively and economically to perform specified functions and services over a range of different "sizes," suitably defined. Scalability validation and evaluation is used to check the scalability of a system and its components in the following areas.

- Component and system capacity boundaries in supporting the problem size, data volume, and the number of concurrent access users.
- Performance improvement and degradation under different configurations and settings—for example, process speedup and throughput improvement.

Analyzing system scalability is very interesting and challenging. In the past, a number of measurement models and metrics have been proposed to evaluate system scalability. Most of them focused on the measurement of scalability for parallel systems [14–16]. More research work must be done to study how to measure component and system scalability in component-based systems. Here, we propose several scalability metrics to support the scalability evaluation of components and component-based programs.

11.2.6.1 Component function size boundary

To measure the performance boundary of a software component (say, C_k), we can quantitatively observe the execution performance changes on a fixed-size function problem as the number of concurrent threads in C_k increases. Applying Amdahl's law to software components, we define the execution time of function F_J in component C_k as follows:

$$T(C_k, F_j, n) = T_{seq} + T_{curr} / n$$

where the execution time $T(C_k, F_j, n)$ is divided into two parts: the amount of time spent on serial parts in a component, denoted as T_{seq}, and the execution time spent on n different concurrent program threads in C_k, denoted as T_{curr}. As the number of concurrent threads is increased, the total execution time $T(C_k, F_j, n)$ is bounded by T_{seq}. In practice, the total execution time eventually hits a minimum point (known as the *threshold point*) as more concurrent threads are added. After that, adding more concurrent threads can only consume more system overheads. This increases the execution time T_{seq} for the serial parts.

It is clear that this metric can be used to measure the speedup of the execution time for function F_j in C_k when the number of concurrent threads in C_k is scaled up from n to m.

The speedup of the execution time of F_j in C_k can be computed as follows:

$$T_{speedup}^{Fj}(n,m) = T(C_k, F_j, n) - T(C_k, F_j, m)$$

It is obvious that a negative speedup value indicates a slowdown of the execution, however, a positive speedup value suggests a real speedup. Although $T(C_k, F_j, n)$ can be used to determine the optimal number of concurrent threads to achieve the maximum possible speedup, it has three serious limitations:

- The complex overhead pattern inherent in the component C and the architecture may not be explicitly and precisely evaluated by the term T_{seq}.
- $T(C_k, F_j, n)$ assumes the computation load is balanced.
- $T(C_k, F_j, n)$ cannot be used to evaluate the performance when the function F_j is scaled.

11.2.6.2 Transaction throughput boundary and improvement for component clusters

To scale up the capability of application systems, it is common to use an N-cluster component (which is a cluster of a number of identical components) to support a specific set of functional tasks. In this situation, evaluating system throughput changes of transaction processing needs to find out the system throughput boundary and its threshold points while altering the number of components in the N-cluster component. Let us use a simple metric to quantitatively observe the system throughput changes of *TR* type transactions when altering the number of redundant components in the cluster component.

Here, we measure the system throughput under a fixed test load T_{set} in a given time period T_p as follows:

$$Throughput_{system}\left(T_{set}, n, TR, T_p\right)$$
$$= The\ total\ number\ of\ successfully\ processed\ TR\ transaction\ requests$$

where n is the number of redundant components in the cluster component. It is clear that the system throughput may change as n changes. Using this metric, we can define another metric to measure system throughput improvements:

$$Throughput_{system-imp}(n-m)$$
$$= Throughput_{system}\left(T_{set}, m, TR, T_p\right) - Throughput_{system}\left(T_{set}, n, TR, T_p\right)$$

where $Throughput_{system\text{-}imp}(m, n)$ represents the throughput improvement when the cluster component is scaled up from n identical components to m.

11.3 Performance evaluation approaches

A systematic performance evaluation approach must include the following elements: (1) sound and correct performance evaluation model(s), (2) well-defined performance measurement metrics, (3) cost-effective performance data collection and monitoring techniques, and (4) efficient performance analysis facilities and tools.

11.3.1 Classification of performance evalution approaches

In past decades many performance evaluation models and approaches were proposed. Most of them were developed with a specific focus—for example, evaluating reliability. Many of them can be applicable to software components and component-based programs. Here, we classify them into the following types.

- *State-based performance evaluation:* The state-based performance evaluation approach uses a finite state diagram as a performance evaluation model. It focuses on the performance evaluation of the state-based behaviors in a system and its components. For example, R. C. Cheung in [17] uses a finite state diagram to model the reliability of a system. In his model, a state represents the execution of a single component, and links between states represent the transaction probability from one state to another. The transaction probability data are obtained from the operational profiles of a system. In this model, the reliability of a software system depends on the execution sequence of states and the reliability of each individual state. A state-based matrix can be used to compute the system reliability. The details can be found in [17].

- *User-oriented performance evaluation:* The user-oriented performance evaluation approach uses a model to present the performance of the system-user behaviors. Based on this model, system performance (such as reliability) can be measured using the users' nevigation/operation scenario and usage profiles. C. Wohlin and P. Runeson [18] discussed a user-oriented performance evaluation approach based on a usage model and user profiles. In their approach, the system usage is divided into a herierarchy, where each part represents an aspect of the usage.
  As shown in Figure 11.6, the usage model consists of five levels:
  1. The usage level, which represents the complete usage;
  2. The user type level, which contains all user types;
  3. The user level, which represent the individual users;
  4. The service level, which represents the usage of the available services to the user;
  5. The behavior level, which describes the detailed usage of a single service as a normal Markov chain. The interaction between two different services represents the transition between two different service chains.

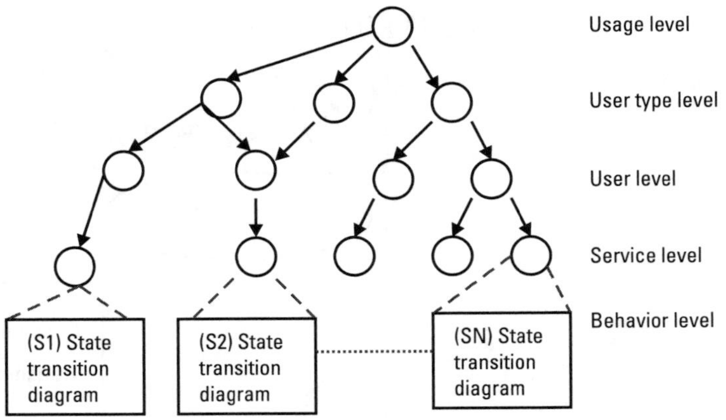

Figure 11.6 An example of the usage model.

- In this model, the system reliability is measured based on a user's perspective usage model and its derived usage profiles. The challenge in applying this approach is how to drive the user's profiles and come up with the correct transition probability parameters in a simple and systematic manner.
- *Scenario-based performance evaluation:* The scenario-based performance evaluation always uses a scenario-based performance model based on system functional scenarios at different levels, including subsystem, component, and class levels. In a scenario-based performance model, system functional scenarios are presented based on component (or class) interaction sequences. The scenario-based approach is very useful to measure scenario-based system function performance, such as system response time, event latency, throughput, reliability, and availability. S. M. Yacoub et al. [19] used a probabilistic model, called *component dependency graph* (CDG), to support reliability analysis for a component-based software. Their approach has three steps. The first estimates the parameters used in the reliability model, including probability of a scenario, *component reliability* (CR), *transition reliability* (TR), *transition probability* (TP), and average component execution time (CE). The second step constructs the component dependency graph (see Figure 11.7). The last step uses a proposed algorithm for reliability and sensitivity analysis.

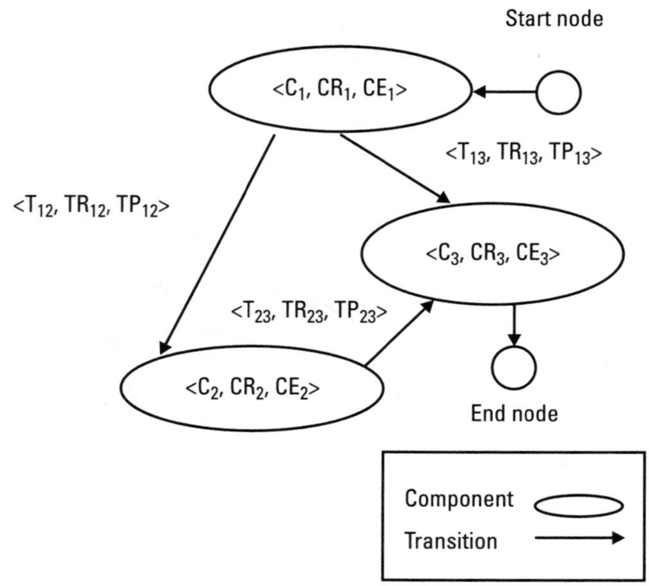

Figure 11.7 An example of CDG.

- *Architecture-based performance evaluation:* In the architecture-based performance evaluation, architecture-based models are developed based on system architectures or architecture styles according to system function and design specifications. W.-L. Wang et al. [20] classify basic architecture styles (or views) in a component-based system into four types: batch-sequential/pipeline, parallel/pipe-filter, fault tolerance, and call-and-return. Using these architecture styles (or views), they transform them into state view(s), and then compute the reliability and transition probability of each state. Later, these state views are integrated to form a global state view of the system. Finally, a state-based transition matrix is constructed based on the global state view to compute the reliability of the system. In fact, this idea also can be used to measure system availability. This approach has two problems. The first problem is how to obtain the accurate transaction probability data in a systematic way. The second problem is the construction of a global finite state that is very difficult for a large system with complex system architecture styles.
- *Transaction-based performance evaluation:* In a transaction-based performance evaluation, a transaction model is used to present the system transactions, sequences, and their relationships. A transaction diagram or a Petri Net are the typical examples. The focus of this approach is the measurement of transaction-oriented system performance, such as domain-specific transaction speed, processing time, latency, throughput, reliability, and availability. They usually are very useful in the performance measurement of transaction driven application systems.

11.3.2 Component-based performance evaluation models

What is a performance evaluation model? For a given software component (or system), a performance evaluation model usually refers to a well-defined model that presents its performance attributes, parameters, and behaviors, as well as their relationships based on system and component architcetures, structures, and behaviors. In order to define and select a perfomance evaluation model, we must check its soundness, correctness, and effectiveness. Well-defined performance evaluation models are very useful to engineers in the following ways:

- They provide a foundation to define and select performance evaluation stratgey and metrics.
- They help engineers define and construct performance testing and evaluation techniques and tools to support performance tracking, data collection, and monitoring, as well as analysis and computation.

- They enhance engineers' understanding of component and system performance issues, and support their performance test design and data prepartion.

11.3.2.1 Component-based scenario model

A *component-based scenario model* is an event-based scenario diagram that represents the interaction sequences among components in a component system. Each diagram consists of three types of elements: (1) a set of component nodes, (2) a set of component interaction events, and (3) component interaction sequences.

Formally, a *component-based scenario diagram* (CSD) can be defined as an ordered triple *CSD*(*N*, *E*, *O*), where *N* is a set of component nodes, and *E* represents a set of interaction events between components. A link $e = (C_r, C_d)$ is a direct edge, which represents an event that is sent from component C_r to component C_d. Each link in *E* has a sequence number. *O* refers to the set of all sequence numbers for links in *E*. Figure 11.8 shows an example that presents the TextChat functional scenario in a component-based *electronic customer relation management system* (eCRM).

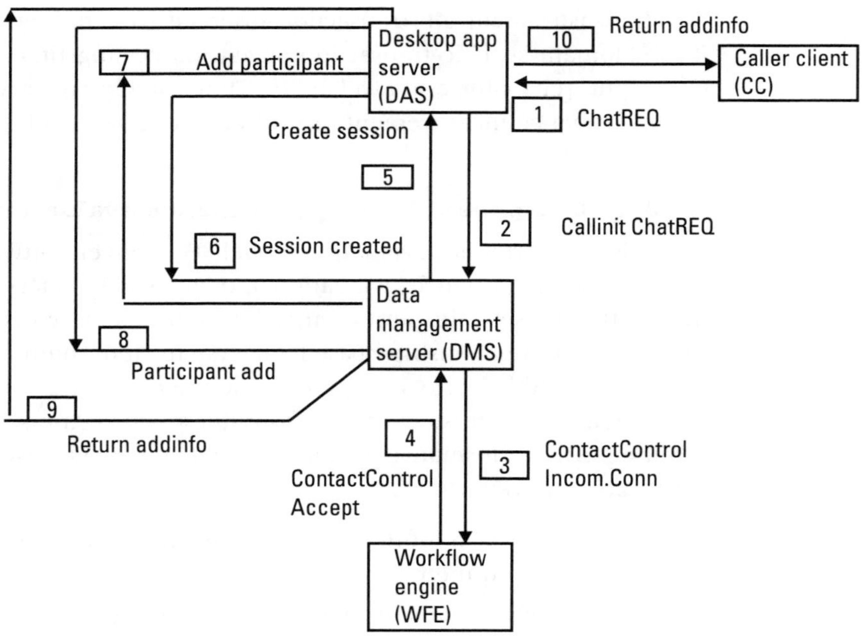

Figure 11.8 An example of a component-based scenario diagram for the TextChat functional scenario in a component-based eCRM system.

The component-based scenario model can be used to measure the performance of event-based functional scenarios at both the component and system levels. Here, we assume that an event-based functional scenario S_q is a sequence of interaction events between components in a system. Clearly, S_q can be represented as a path in a component-based scenario diagram. $S_q = \{E_{i1}, \ldots, E_{im}\}$ is a sequence of events between components, where E_{i1} is the starting event and E_{im} is the ending event.

Clearly, the model assists engineers to understand, measure, and analyze the function-oriented component and system performance based on event-based function scenarios. Engineers can use this model to define scenario-based function performance measurement strategies and metrics in the following ways:

- Validating function or processing speed, event or transaction latency;
- Measuring component and system throughputs for event processing;
- Checking component or system availability and reliability based on the throughput analysis results.

11.3.2.2 Component-based transaction model

A *component-based transaction model* is a transaction-based diagram that represents the transaction flow and interactions among components in a component-based system. Each diagram consists of three types of elements: (1) a set of component nodes, (2) a set of transactions inside components, and (3) data or messages links associated with transactions. Figure 11.9 shows a component-based transaction diagram for a TextChat function scenario in a component-based eCRM system.

Formally, a *component-based transaction diagram* (CTD) can be defined as an ordered quadruple (N, T, NT, L), where N is a finite set of component nodes, T is a finite set of component-level transactions, NT is a set of component-transaction relationships, and L is a set of data or message links between transactions. An element of NT, say, $NT_i = (C_k, T_i)$, represents the association relationship between C_k and T_i. In other words, T_i is a transaction inside component C_k. An element of L, say, $L_i = (T_i, T_j)$, indicates that L_i is a data or message link transferred from T_i to T_j. In other words, L_i is the outgoing data link of T_i and the incoming data link of T_j.

The component-based transaction diagram can be used to measure the performance of a system and its components on transaction-oriented functional features. Based on this model, engineers can define transaction-oriented performance measurement strategy and metrics to support the evaluation of transaction speed, transaction throughput, and transaction-based reliability and availability.

11.4 Performance testing and evaluation tools and techniques

Various tools and techniques are needed for performance testing and evaluation of software components and systems [21, 22]. They can be classified into the following six groups:

- *Performance data generation tools:* Data generation tools are used to generate necessary input data, events, signals, or messages to feed into a targeted system or its components. In many cases, a simple data generator based on a predefined repository (say, a predefined file repository or a database) is enough to generate incoming static (or dynamic) traffic data (or messages). Two common methods are used to generate required performance traffic data. The random data generation approach is the first one. In this method, performance traffic data are generated based on a random function and a seed data value. The other method is known as pattern generation, in which performance traffic data are

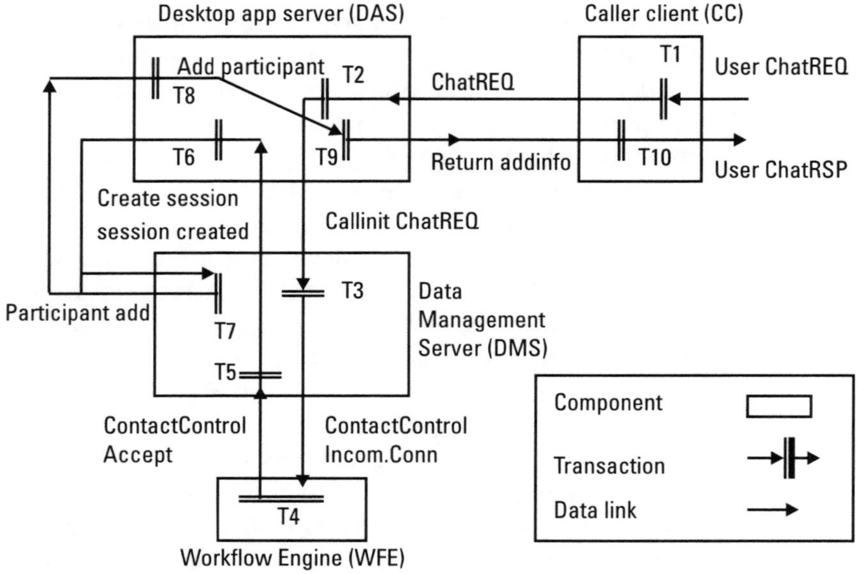

Figure 11.9 An example of a component-based transaction diagram for the TextChat functional scenario in a component-based eCRM system.

produced dynamically based on a set of predefined data patterns. For example, testing the performance of an e-mail server, engineers need an e-mail message generator to send various types of e-mail messages.

- *Component and system simulators:* Simulators are commonly used tools in performance testing and evaluation. They are needed when components and subsystems (such as external parts or third-party subsystems) are not available during system testing and performance evaluation. In many cases, simulators are programs that simulate the behaviors of required components or subsystems. There are two popular ways to construct simulators. The first is known as *state-based simulation,* in which a simulator is developed based on a given finite state machine. It performs and functions based on the state machine model. This method is very useful in coping with a state-driven application process. The other method is known as *message-based simulation,* in which a simulator is created based on predefined protocol messages. It simulates a component (or subsystem) to interact with under-test software components (or systems) by receiving and sending static (or dynamic) protocol-based messages.
- *Performance data collectors:* These are facilities that collect performance data and parameter values during performance testing and evaluation. A typical example is to generate and collect component (or system) error logs and performance trace records. Performance analysis tools will analyze these tracked performance data later.
- *Performance monitoring tools:* They refer to the programs that are used by performance testers to monitor and display component and system performance during system execution and evaluation. In many cases, customers may request a software system to provide some kind of system performance monitoring capability as a part of system functional features. The Microsoft task manager tool on the PC platform is a typical example.
- *Performance analysis tools:* These are necessary to compute and evaluate component and system performance in various aspects based on predefined metrics and collected performance data. These analysis tools are developed to implement the detailed computation of performance metrics. They accept collected performance data and generate performance evaluation results based on the predefined metrics.
- *Performance reporting tools:* These tools are programs that generate different performance reports in various formats based on a performance data repository. The typical formats include tabular, pie chart, line curve, and stack diagram.

11.4.1 Performance tracking and monitoring techniques

To support performance testing and evaluation, engineers need systematic solutions to track, collect, and monitor the performance data of a system and its components. Most reusable components, including COTS components, currently do not support performance testing and evaluation at the component and system levels because today's component models and technologies do not address the issues of component testing, integration, and performance evaluation. In other words, current reusable components do not provide consistent performance tracking capability and interface. Hence, a consistent performance tracking solution is needed for the development of software components and component-based programs. Three types of performance tracking techniques and facilities are needed.

- *System resource tracking techniques:* These techniques refer to the methods that collect and track component usage of system resources, including memory, CPU time, disk space, cache, and network bandwidth. Today, all operating systems (such as UNIX and Windows 98 and 2000) provide basic facilities to allow engineers to use them to collect the real-time system usage data to support the evaluation of component and system utilization of various system resources.
- *Time-based performance tracking techniques:* These techniques collect and track various types of processing time and speed at the component and system levels, for example, the execution time of a component function, the transaction processing time inside a component, and the processing time for an incoming message or event. Based on the collected time-based performance data, different speed-related performance metrics could be used to measure component and system functional process speed, latency, and transaction processing time.
- *Load-based performance tracking techniques:* This technique collects and tracks various types of system loads, such as network traffic load, user access load (e.g., the number of concurrent access users), and transaction load. Based on the collected load-based performance tracking data, various throughput metrics could be used to compute the component and system throughputs in different perspectives. Moreover, these data could be used as samples to find out the necessary operation profiles and probability data for reliability and availability evaluation.

The detailed discussion of various program-tracking methods can be found in [23]. More detailed performance tracking solutions are given in [24].

11.5 Summary

This chapter revisits the basic concepts, issues, and solutions in software performance testing and evaluation for components and component-based programs.

As discussed before, software performance and evaluation have four basic needs: (1) well-defined performance testing strategy, requirements, and focuses, (2) correct and sound and effective performance evaluation models, (3) well-defined performance metrics, and (4) cost-effective performance testing and evaluation tools and techniques.

This chapter first introduced a performance test process and discusses the performance testing objectives and focus areas. Then, it summarized the basic challenges and issues on performance testing and evaluation of component-based programs and components. Next, this chapter presented different types of performance metrics for software components and systems, including processing speed, utilization, throughput, reliability, availability, and scalability metrics. Most of the performance metrics covered here can be considered as the application of existing metrics to software components. New performance metrics are needed to support the performance evaluation of component-based programs.

Later, various existing software performance approaches and models were presented. Most of them can be applied to the performance evaluation of component-based software at the component and system levels. As pointed out before, new component-oriented performance evaluation models and methods are needed in component-based software development.

Finally, this chapter discussed the different types of performance tools and highlighted the required techniques to support performance evaluation in a systematic way.

References

[1] Gao, J., "Testing Component-Based Software," *Proc. of STARWEST'99*, San José, CA, November 1999.

[2] Gao, J., "Challenges and Problems in Testing Software Components," *Proc. of ICSE2000's 3rd International Workshop on Component-Based Software Engineering: Reflects and Practice*, Limerick, Ireland, June 2000.

[3] Cahoon, B., K. S. McKinley, and Z. Lu, "Evaluating the Performance of Distributed Architectures for Information Retrieval Using a Variety of Workloads," *ACM Transactions on Information Systems*, Vol. 18, No. 1, January 2000, pp. 1–43.

[4] Rudolf, A., and R. Pirker, "E-Business Testing: User Perceptions and Performance Issues," *Proc. of 1st Asia-Pacific Conference on Quality Software*, October 2000.

[5] Subraya, B. M., and S. V. Subrahmanya, "Object Driven Performance Testing of Web Applications," *Proc. of 1st Asia-Pacific Conference on Quality Software*, October 2000.

[6] Yan, Y., X. Zhang, and Q. Ma, "Software Support for Multiprocessor Latency Measurement and Evaluation," *IEEE Trans. on Software Engineering*, Vol. 23, No.1, January 1997, pp. 4–16.

[7] Mainkar, V., "Availability Analysis of Transaction Processing Systems Based on User-Perceived Performance," *IEEE 16th Symposium on Reliable Distributed Systems (SRDS '97)*, Durham, NC, October 22–24, 1997, p. 10.

[8] Wood, A., "Predicting Client/Server Availability," *Computer*, Vol. 28, No. 4, April 1995, pp. 41–48.

[9] Gokhale, S. S., M. R. Lyu, and K. S. Trivedi, "Reliability Simulation of Component-Based Software Systems," *Proc. of IEEE 9th International Symposium on Software Reliability Engineering*, Paderborn, Germany, November 4–7, 1998, p. 202.

[10] Krishnamurthy, S., and A. P. Mathur, "On the Estimation of Reliability of a Software System Using Reliabilities of Its Components," *Proc. of IEEE 8th International Symposium on Software Reliability Engineering*, Albuquerque, NM, November 2–5, 1997, p. 146.

[11] Yang, M. C. K., W. E. Wong, and A. Pasquini, "Applying Testability of Reliability Estimation," *Proc. of IEEE 9th International Symposium on Software Reliability Engineering*, Paderborn, Germany, November 4–7, 1998, p. 90.

[12] Tian, J., "Integrating Time Domain and Input Domain Analyses of Software Reliability Using Tree-Based Models," *IEEE Trans. on Software Engineering*, Vol. 21, No. 12, December 1995, pp. 945–958.

[13] Yacoub, S. M., B. Cukic, and H. H. Ammar, "A Component-Based Approach to Reliability Analysis of Distributed Systems," *Proc. of IEEE 9th International Symposium on Software Reliability Engineering*, Paderborn, Germany, November 4–7, 1998, p. 158.

[14] Jogalekar, P., and M. Woodside, "Evaluating the Scalability of Distributed Systems," *IEEE Trans. on Parallel and Distributed Systems*, Vol. 11, No. 6, June 2000, pp. 589–603.

[15] Gokhale, A. S., and D. C. Schmidt, "Measuring and Optimizing CORBA Latency and Scalability over High-Speed Networks," *IEEE Trans. on Computers*, Vol. 47, No. 4, April 1998, pp. 391–413.

[16] Moon, B., and J. H. Saltz, "Scalability Analysis of Declustering Methods for Multidimensional Range Queries," *IEEE Trans. on Knowledge and Data Engineering*, Vol. 10, No. 2, March/April 1988, pp. 310–327.

[17] Cheung, R. C., "A User-Oriented Software Reliability Model," *IEEE Trans. on Software Engineering,* Vol. 6, No. 2, March 1980, pp. 118–125.

[18] Wohlin, C., and P. Runeson, "Certification of Software Components," *IEEE Trans. on Software Engineering,* Vol. 20, No. 6, June 1994, pp. 494–499.

[19] Yacoub, S. M., B. Cuki, and H. H. Ammar, "Scenario-Based Reliability Analysis of Component-Based Software," *IEEE 10th International Symposium on Software Reliability Engineering,* Boca Raton, FL, November 1–4, 1999, p. 22.

[20] Wang, W.-L., Y. Wu, and M.-H. Chen, "An Architecture-Based Software Reliability Model," *Proc. of IEEE 1999 Pacific Rim International Symposium on Dependable Computing,* Hong Kong, December 16–17, 1999.

[21] Dini, P., G. V. Bochmann, and R. Boutaba, "Performance Evaluation for Distributed System Components," *Proc. of 2nd IEEE International Workshop on Systems Management,* 1996.

[22] Everett, W. W., "Software Component Reliability Analysis," *Proc. of IEEE Symposium on Application-Specific Systems and Software Engineering and Technology,* 1999.

[23] Gao, J., et al., "Tracking Software Components," *Journal of Object-Oriented Programming,* August/September 2001, Vol. 14, No. 4, pp. 13–22.

[24] Gao, J., and C. Sun, *Performance Measurement of Software Components—A Systematic Approach,* Technical Report, San José State University, San José, CA, 2002.

CHAPTER

12

Contents

Frameworks for testing component-based software

In recent years, many researchers and software test tool vendors have been spent a lot of time and efforts on developing innovated methods and systematic solutions to address the problems in test automation of components and component-based programs. This chapter focuses on the discussion of a number of selected research efforts on developing new frameworks for supporting the validation of component-based systems and parts.

The chapter consists of four sections. The first section reports the research efforts on BIT components and BIT wrappers to construct components with self-testing and self-checking capability to support automatic component validation. Section 12.2 covers another way to increase component testability by constructing testable components. This section reports the research work on building testable components and the supporting framework. Engineers can use this test framework to build and convert conventional software components into testable components. In addition, a distributed testing environment for testable components is described in Section 12.2. Its goal is to support automatic black-box testing for testable components in a plug-in-and-test approach. Section 12.3 summarizes a systematic solution to monitor component behaviors of component-based software in a distributed environment. As a part of the solution, a component-tracking framework is presented to allow engineers to construct traceable software components. In addition, a distributed component-tracking environment is provided to track and monitor component behaviors in a distributed network. Section 12.4 reports a component performance framework that can be used to build measurable components to facilitate performance testing and evaluation. A distributed performance measurement environment is discussed to demonstrate how to evaluate component performance in component-based programs. Section 12.5 discusses IBM's STCL architecture for the collaboration of test tools to support software projects. This architecture is not only useful to collaborate various test tools for global software productions, but also valuable in providing a rational solution to integrate different test tools to support component-based software development. Finally, a summary is given in Section 12.6.

12.1 BIT components and wrappers

As discussed in Chapter 5, building testable components is very important in component-based software development because it not only increases component testability but also reduces the cost of component validation, integration, and system testing. In the past few years, a number of researchers have been working on seeking different ways in constructing software components to facilitate component testing. One of approaches is known as built-in test components. This section discusses the research work on BIT components and the supporting framework, known as BIT component wrappers.

12.1.1 BIT components

The basic idea of BIT components is to add built-in tests inside components. In the past, the built-in test concept has been widely used in the computer hardware industry. The major purpose is to create self-test computer parts and chips. Recently, this concept has been extended into component software engineering. A number of recently published research papers has addressed this subject. According to Y. Wang et al. [1, 2], a built-in test component is a special type of software component in which special member functions are included as its source code for enhancing software testability and maintain ability. A built-in test component is able to operate in two modes:

- *Normal mode:* In this mode, the component has the same behaviors as a conventional component.
- *Maintenance mode:* In this mode, component users or testers are able to activate its internal built-in tests as its normal function members to support software testing and maintenance.

Figure 12.1 shows a BIT component template as a class object and presents a BIT component sample. It is clear that BIT components include built-in tests as a part of a component. The major advantage of BIT components is to provide a self-test capability to allow component users to easily activate built-in tests without any testing effort. BIT components usually require a higher programming overhead and more system resources than conventional components. In addition, BIT components have two major problems.

- Since component testers and users cannot access the built-in tests inside components, they are costly to update and maintain.
- Only limited component tests can be inserted as built-in tests due to the overhead involved in component programming, compilation, and execution.

12.1.2 BIT component wrappers

To support test execution of built-in tests, component users and engineers need some kind of built-in test capabilities inside components to support test operations and interactions. One approach is to add a simple "hook" interface that can be used in adorning the component with sophisticated BIT capabilities. In other words, adding some built-in parts, known as *BIT wrappers*, into a component to support component self-checking and self-testing by interacting with internal built-in tests. S. H. Edward in [3] presented his strategy for constructing BIT wrappers. These wrappers serve the role of test oracles to check

```
Class class-name {
      // Class interface
      ... // Data declaration
      ... // Constructor declaration;
      ... // Destructor declaration;
      ... // Function declarations;

      // Declarations for built-in tests
      void BIT_1(...);
      ....
      Void BIT_N(...);

      // Normal function implementation
      ... // Constructor's implementation
      ... // Destructor's implementation;
      ... // Function implementations;

      // BIT test functions for built-in tests
      void BIT_1(...) { ... };
      ...
      void BIT_N(...) { ... };
} BITObject;
```

```
Class BIT_INT_Stack {
      // class interface and data declaration

      int stack[N]; // N is the upper limit
      int stack_index;
      BITs_INT_Stack(); // The Stack constructor
      ~BITs_INT_Stack(); // The destructor
      void BITs_Push_INT_Stack(...);
      void BITs_Pop_INT_Stack(...);
      int BITs_Stack_Empty();

      // The built-in-tests declaration
      void BIT_INT_Stack_Test1(...);
      void BIT_INT_Stack_Test2(...);
      ...
      // Normal INT_Stack member functions
      ...
      // Built-in test implementation
      void BIT_INT_Stack_Test1(...) {...};
      void BIT_INT_Stack_Test2(...){...};
      ...
} BITObject;
```

Figure 12.1 A simple BIT object and template. (*From:* [3]. © 2002 John Wiley & Sons, Inc. Reprinted with permission.)

component interface violations and perform dynamic post-condition checking based on preconditions during black-box testing. He claimed the following basic properties of BIT component wrappers [3].

- BIT wrappers are completely transparent to client and component code.
- BIT wrappers can be inserted or removed without changing client code (only a declaration need be modified). This capability does not require a preprocessor, and can be used in most current languages.
- When BIT support is removed, there is no run-time cost to the underlying component.
- Both internal and external assertions about a component's behavior can be checked.
- Precondition, postcondition, and abstract invariant checks can be written in terms of the component's abstract mathematical model, rather than directly in terms of the component's internal representation structure.
- Checking code is completely separated from the underlying component.
- Violations are detected when they occur and before they can propagate to other components; the source of the violation can be reported down to the specific method/operation responsible.
- Routine aspects of the BIT wrappers can be automatically generated.
- The approach works well with formally specified components but does not require formal specification.
- The approach provides full observability of a component's internal state without breaking encapsulation for clients.
- Actions taken in response to detected violations are separated from the BIT wrapper code.

Figure 12.2(a) displays the structure of BIT wrappers and their interactions with a BIT component. According to [3], a component is encased in a two-layer BIT wrapper. The inner layer of the wrapper supports direct and safe access of the component's internals, performs internal consistency checks, then converts the internal state information into a program-manipulable model of the component's abstract state [4]. The outer layer uses the model to check to see if the component maintains any invariant properties it advertises, and also checks the results of each operation to the extent desired for self-testing purpose. Figure 12.2(b) shows a C++ Interface for One_Way List and its BIT wrapper. The detailed implementation method can be found in [3].

S. H. Edward's approach uses a formal component specification language, called RESOLVE [5], to specify components and interfaces. To support this strategy, he highlighted an automated testing framework for BIT components and wrappers. It includes three legs:

- Automatic (or semiautomatic) generation of a component's BIT wrapper;
- Automatic generation of a component's test driver;
- Automatic (or semiautomatic) generation of test cases for the components.

According to [3], a generator has been designed and implemented to process RESOLVE-style component specifications and C++ template interfaces to generate BIT wrappers. Meanwhile, a test driver generator has been designed and is being implemented. The basic idea is based on an interpreter model. In this model, a test driver is viewed as a command interpreter that reads in test cases and translates them into actions on the component under

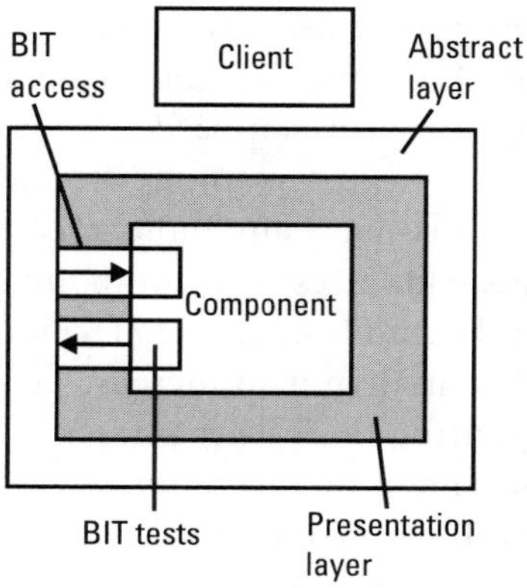

(a)

```
template <class Item>
class One_Way_List {
public
        One_Way_List( );
        virtual ~One_Way_List( );
        // ...
        virtual void Remove_Right(Item& x) = 0;
        // ... other details elided for space
};
```

```
template <class Item, class One_Way_List_Base>
class OWL_BIT_Wrapper;
        public One_Way_List <Item> {
public:
        OWL_BIT_Wrapper( );
        virtual ~OWL_BIT_Wrapper( );
        // ...
        virtual void Remove_Right(Item& x) = 0;
        // ... other details elided for space
private:
        One_Way_List_Base uut; // Wrapper List
};
```

(b)

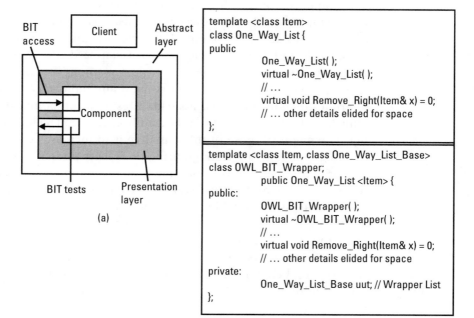

Figure 12.2 A BIT component wrapper example: (a) the structure of a BIT wrapper; and (b) a C++ interface for One_Way List and its BIT wrapper.(*From:* [3]. © 2002 John Wiley & Sons, Inc. Reprinted with permission.)

test. In this approach, a test driver is constructed as an interpreter based on a component's interface definition and the identified operations.

12.2 A framework and distributed environment for testing testable components

Although the built-in test approach and its supporting framework are very useful for self-testing and self-checking of software components, it has three limitations.

- Built-in tests increase the component development cost, including the programming cost for built-tests, and BIT wrappers.
- Built-in tests require extra system overhead and resources for component test execution.
- Built-in tests usually require a higher maintenance cost than normal component tests.

In addition, only limited component tests can be created as built-in tests for a component. To solve these problems, J. Gao et al. [6] introduced a new concept of testable components, known as testable beans. The basic idea is to construct a testable component with a standard test interface to facilitate component testing in a plug-in-and-test manner. Unlike BIT components, their testable components do not contain any built-in component tests as its internal function members. Instead, a testable component has well-defined built-in test interfaces and standard add-on parts to facilitate component testing. The major purpose of this initiative is to increase component testability by design for software testing during the earlier phases of a software project. The primary objective is to provide a new component model that facilitates software testing at the unit level and system level. The basic idea is to establish well-defined testable component architecture, standard component interfaces, and consistent built-in facilities to support component functional testing in a black-box view. The motivation is to regulate the testing interfaces and interactions between a software component under testing and its required test bed, including its test suite and test repository. The detailed concept, requirements, and benefits of testable components have been described in Chapter 5.

12.2.1 Testable component structure

As shown in Figure 12.3, a testable component consists of the following parts:

- A under test conventional component, including all of its entities and elements;
- Static or dynamic built-in parts, which facilities component testing in a systematic manner:
  - *Component profile:* a generated class that provides a built-in capability of accessing the component profile information, such as component interfaces;
  - *Component test entity:* a generated class that provides the built-in common entities that support the setup function for component tests and parameter data during component testing based on the given test case and the component profile;
  - *Component test runner:* a generated class that provides a built-in test control interface between a testable component and external test drivers and test beds;
  - *Component test adapter:* a generated class that interacts with the external visible interface of the under test component. This adapter usually is dependent on the used technology of the component, for example, the underlying programming language.

Besides, each testable component provides a well-defined consistent test interface to facilitate component testing. This test interface provides a set of common test operation interfaces shared by all components, as follows:

- Set up a component test for exercising a component function with given input parameter data and conditions.

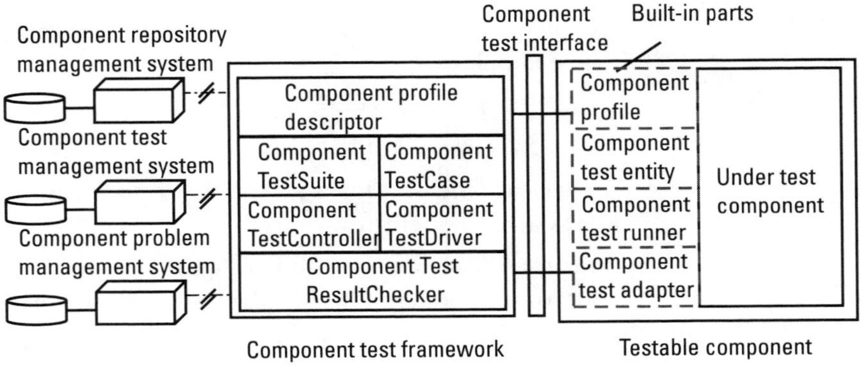

Figure 12.3 Testable component structure and a test framework.

- Invoke the execution of a setup component test.
- Collect the test result of a component test case.

A test interface example is displayed next. A similar test interface example is given in [6].

The detailed testable component examples have been developed and reported in [7]. To support component test automation, a component test framework has been designed and implemented as a middle-tier between testable components and test supporting tools, such as a component test management system, a problem management system, and a component resource management system.

12.2.2 Component test framework

As shown in Figure 12.3, the component test framework, built as a class library, consists of the following parts:

- *Component profile descriptor:* This is a class that is useful to access a predefined component profile of a component and create its component profile descriptor. The detailed examples are given in [7]. A component profile specifies the following information about components:
  - *Component properties:* This includes all basic component information, such as component ID, name, version, platform, OS, programming language, and vendor information.
  - *Component environment:* This consists of the deployed execution information for a component, such as a constructor and configuration profile.
  - *The component interfaces:* This refers to all visible interfaces of a component. The external visible function signatures of a class are typical examples.
- *Component TestSuite:* This is a class that can be instantiated to define and access a component test suite.
- *Component TestCase:* This is a class that can be instantiated to define and access a component test case.
- *Component TestController:* This is a class that can be used to control test execution of component tests based on the given test drivers in a test suite.
- *Component TestDriver:* This is a class that can be used to generate component test drivers based on the given component test case in a test suite.

- *Component Test ResultChecker:* This is a class that is useful for performing the test result checking in a component test driver for a component test.

12.2.3 Generating component test drivers

One of the biggest advantages of a testable component framework is that it provides a base to support the generation of component test drivers during component black-box testing based on given component tests. There are two approaches for generating component test drivers:

- *Manual test driver generation,* where engineers use the provided component test framework to generate black-box test drivers for testable components;
- *Automatic test driver generation,* where a component test bed uses the provided component test framework to generate black-box test drivers for testable components.

Unlike other approaches, the automatic test driver generation is performed dynamically based on the selected component test case and related test data. For each selected component test in a test bed, the dynamic generation procedure consists of four steps:

- Read a component's profile to access component information and generate a class object using *component profile descriptor.*
- Create its component test interface and related built-in wrappers as an instance of *component test entity.*
- Build a dynamic component test driver as an instance of *component runner* by adding the program code that accesses a component test case and data from a component test suite.
- Use a *component adapter* to select and use a predefined component adapter to come out with a component adoption interface that binds the original component interface with a component test driver.

The details of the design and implementation of this test framework can be found in [7].

12.2.4 A distributed component testing environment

As mentioned previously, most existing component-oriented test tools provide engineers with four basic testing support functions:

- *Test suite management function*, which helps engineers to create, maintain, and manage a component test suite for components;
- *Language-based test scripting function*, which enables engineers to write test scripts as program codes;
- *Test script execution function*, which supports test execution and replay;
- *Test result validation function*, which helps engineers to collect, monitor, and check the test results from test executions.

According to our application experience, script-based test tools have three major issues. First, test script generation is an ad hoc manual programming process. During this process, it is not easy for engineers to control test script reusability and quality, and also check test coverage. Next, the test script development for retesting is very costly because the generated test scripts are highly dependent on components' application interfaces. Whenever component application interfaces changed, the corresponding test scripts must be updated. Finally, a script-based test tool usually only supports software components written in a specific programming language; it is impossible to validate components written in different programming languages.

To address these issues, a distributed testing environment, as a research prototype system, has been developed in the Computer Engineering Department at San José State University in 2002. Unlike script-based component test tools, this component test environment focuses on test automation of testable components for black-box tests. It has the following features:

- It provides a systematic solution to support component black-box tests with a plug-in-and-test concept.
- It is developed based on the component test framework (described before), well-defined component profiles, and standard component test interfaces.
- It supports manual and automatic generation of component test drivers based on a component test framework and a given component test suite.
- It supports component test executions for black-box component tests using generated component test drivers.
- It offers a distributed test interaction environment to support engineers to perform Web-based component black-box test operations by accessing a component repository over a client-server network.
- It provides a flexibility to support testable components written in different programming languages. Currently, the system implements two types of test adapters: C++ and Java.

As shown in Figure 12.4, the distributed test environment is structured in four tiers.

- *Presentation layer,* which consists of four basic user interface modules, including test manager GUI, test bed GUI, test reporter, and component viewer.
- *Communication layer,* which supports the communications between a client and a test server. It includes Web server, JSP engine, and a Pipeline module.
- *Test bed layer,* which includes four basic test bed modules, including *test library, test manager, test controller,* and *test adapter.* The test library is the component test framework described before. Test manager is a functional module that supports the management of component test information, such as test suites, test cases, and test drivers. Test controller supports test execution. The test adapter provides the running interface between component test drivers and the test bed. These test bed modules provides a general testing bed for testable components.

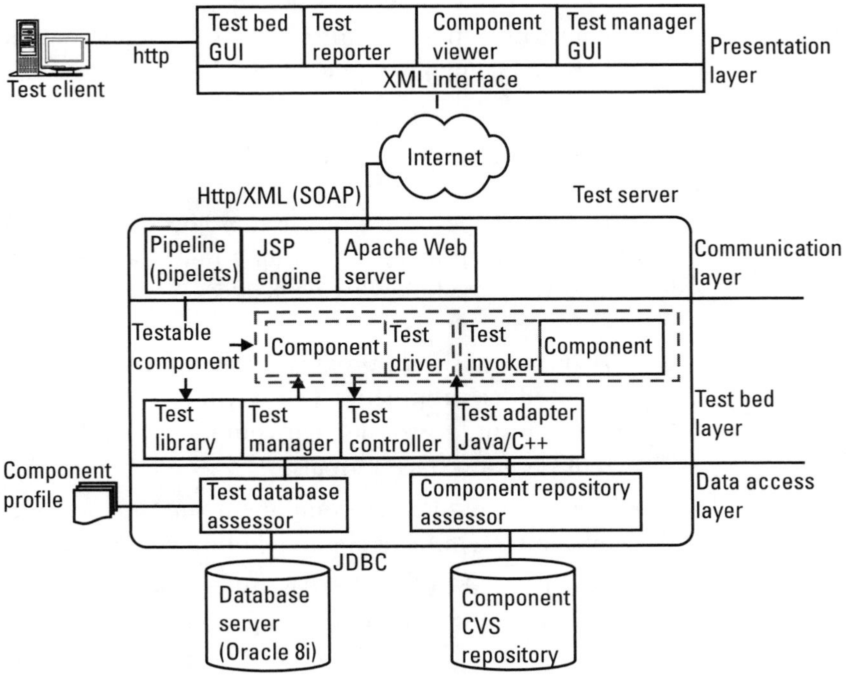

Figure 12.4 A distributed test environment for testable components.

- *Data access layer*, which consists of database access modules for accessing a component repository and component test repository.

The detailed design and implementation can be found in [7]. Although this component test environment is only a research prototype, it provides some useful insights on how to set up a reusable testing environment to support test automation of testable components.

12.3 A framework for tracking component behaviors

As we mentioned before, one of the major issues in validating third-party components is their poor traceability because they do not provide users with transparent program tracking mechanisms. This causes a serious difficulty in understanding and monitoring the program behaviors of software components. There are three major causes.

- Component vendors use ad hoc program tracking mechanisms and inconsistent program trace formats.
- Third-party components do not provide a consistent and transparent component tracking interface to facilitate component users to access external component traceable behaviors.
- Third-party components do not offer a built-in component tracking function to support program tracking of component behaviors.

This section reports on a research project on establishing a systematic solution to solve this problem. According to J. Gao et al. in [8], this solution is developed with the following objectives:

- Define a consistent program tracking solution for components in a distributed environment. This includes a systematic program tracking mechanism, a well-defined component-tracking interface, and consistent program trace formats.
- Create a well-defined component-tracking framework that can be used by component developers to build traceable components.
- Establish a distributed program-tracking environment that can be used to track and monitor component behaviors of component-based software.

12.3.1 Systematic event-based component tracking model

According to [8], an event-based tracking model is developed as a systematic mechanism to monitor and check component behaviors in a component-based program. Its basic idea is influenced by the Java event model for GUI components. All software components and their elements are considered as *tracking event sources*.

In a software component, component engineers or an automatic tracking tool can add built-in tracking code to trigger five types of tracking events. They are performance tracking, operational tracking, error tracking, state tracking, and GUI tracking events. These events are packed as tracking event messages, and added into trace message queues according to their types.

To catch different tracking events, a *tracking listener* is used to receive these events, dispatch them to a *tracking processor* to generate the proper trace, and store them into a specified trace repository. As shown in Figure 12.5, the event-model tracking mechanism relies on a *tracking agent* to support a collection of *traceable components* in a computer. Intuitively, a traceable component is a software component, which is designed to facilitate the observation and

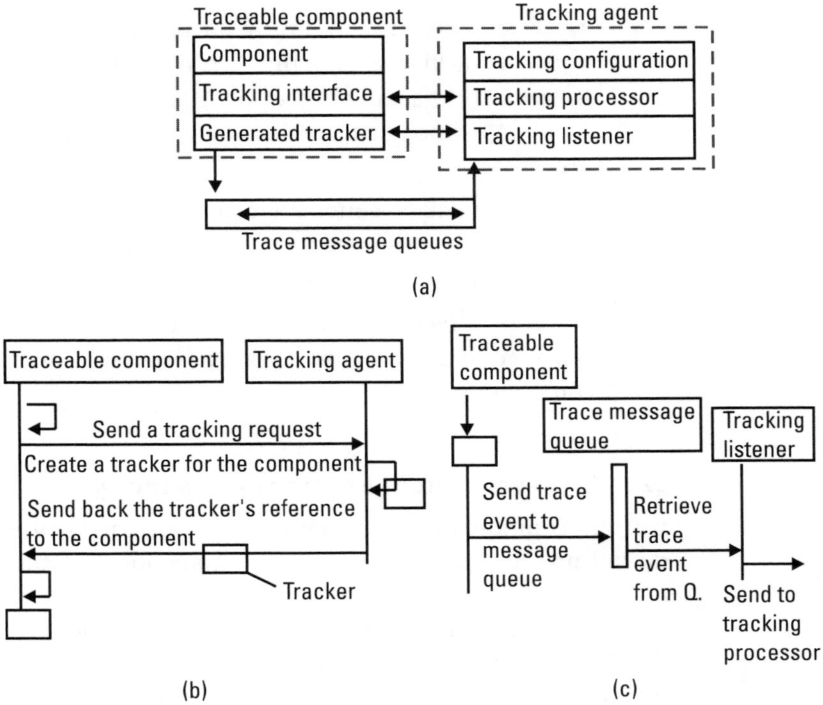

Figure 12.5 A generic component tracking solution: (a) the event-based tracking model for components, (b) establish connection, and (c) exchange trace data.

monitoring of its behaviors, data and object states, function performance, and component interactions to others.

In this solution, a traceable component contains two extra parts in addition to its normal functional parts:

- A *tracking interface*, which is used to set up the connection with the tracking agent. Figure 12.5 shows a procedure to dynamically generate a plug-in tracker for a component by issuing a binding request to the tracking agent through the tracking interface. The detailed implementation example could be found in [8].
- A *dynamic generated tracker*, which is dynamically generated by the tracking agent. Each component receives one *tracker* after it is connected to the agent, since developers can use the generic interfaces to provide various tracking functions for different trace types. The detailed interface template and examples are given in [8].

A tracking agent consists of the following three functional parts:

- *Tracking listener* is a multithread program that listens and receives all types of tracking events through trace message queues, and dispatches them to the tracking processor. Figure 12.5(b, c) shows the interaction sequences among a traceable component, tracking listener, and tracking processor.
- *Tracking processor* generates program traces according to a given trace event based on its trace type, trace message, and data, and stores them in the proper trace repository.
- *Tracking configuration* provides a graphic user interface to allow a user to discover and configure various tracking features for each traceable component.

12.3.2 Java component tracking package

Current component-based technologies (such as JavaBean, EJB, and CORBA) do not provide developers with systematic mechanisms and facilities to track and monitor component behaviors. Thus, developers only can use ad hoc methods to construct traceable components. The Java Tracking Package is developed as a component-tracking framework to provide developers with several generic interfaces for constructing traceable Java components. It includes the following three parts:

- *Component tracker interface:* It provides the generic tracking functional interface between a component tracker and the tracking agent. Various tracking events and requests can be passed to the tracking agent. Using this interface, developers can issue a request to the tracking agent for a component to create a plug-in tracker and set up their connection.
- *Traceable component interface:* It allows developers to bind a plug-in tracker for a component. It supports the binding and discovering functions to facilitate the configuration of trace properties of a component.
- *Tracker adapter interface:* It is provided to support a dummy tracker for the case in which a tracking environment is not set up.

12.3.3 Distributed tracking environment for component software

As shown in Figure 12.6, a distributed component-tracking environment has been developed as a prototype to support the monitoring of component behaviors in component-based software. The system consists of a number of tracking agents and a tracking server. Each machine on the network has a

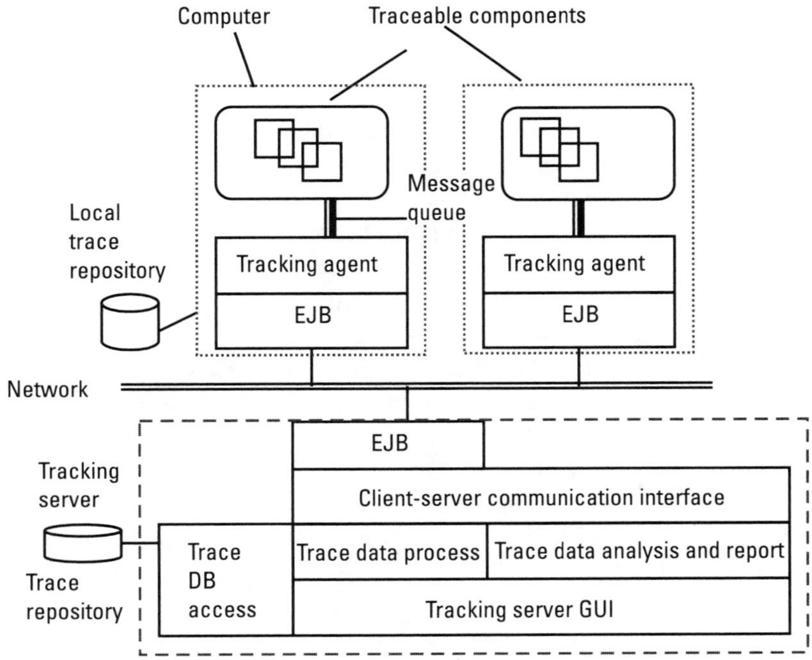

Figure 12.6 A distributed environment for tracking component behaviors.

multiple threading tracking agent based on the EJB technology. It interacts with the plug-in trackers of components on the same machine using trace message queues (in Java Message Queue).

One major function of a tracking agent is to control, record, and monitor diverse component behaviors in an asynchronous mode. Moreover, the tracking agent communicates a tracking server to pass trace data in a given trace format. The tracking server plays the role of a central server to allow engineers to control tracking agents to collect diverse trace data and analyze them.

The tracking server consists of the following parts:

- A communication interface with active program tracking agents to transfer various types of program traces over a network;
- A trace data processor that processes the collected program traces from different tracking gents over a distributed environment;
- A program trace repository that stores and manages all types of program traces from components over a distributed environment;
- A trace analyzer and reporter that enable a user to analyze and report various types of program traces for different components in a distributed system;
- A GUI interface that supports the user interactions between the tracking server and engineers to check and monitor program behaviors in a centralized interface.

According to [8], Java JDK.1.2.2 is used to create the tracking agent and tracking server. A distributed tracking environment is developed to monitor the behaviors of JavaBean components based on an EJB server (JonAS). The Java Application Server (JonAS) is the BullSoft implementation based on the EJB specifications. *Java Message Queue* (JMQ) is used to perform the asynchronized communications between JavaBeans and a tracking agent. In addition, a trace data repository is created and maintained using InstantDB, a 100% Java database server from the *relational database management system* (RDBMS).

Using a standardized trace format is essential to generate consistent trace messages for each component in a distributed program. A well-defined trace format facilitates the generation of understandable trace messages. This increases component understanding and helps error isolation and bug fixing. This distributed tracking environment provides two types of trace information.

- *Trace commands:* Trace commands support the tracking server to control and communicate with tracking agents in the distributed environment.

Each command includes command message ID (00), command code, time stamp, and command parameters.

- *Trace data:* Trace data include message ID (01), trace type, time stamp, component ID, trace data, and trace identifier. Trace data indicates where and when the trace message is generated. A trace identifier includes component identifier, thread identifier, and object identifier. Each trace type has its specific trace data format. Tracking agents generate diverse trace data for each traceable component in a local trace repository.

12.4 A framework for component performance measurement

As mentioned in Chapter 11, performance testing and measurement is important and a challenge for component-based software and its parts for the following two reasons:

- Current third-party components are not developed to facilitate component users to conduct performance testing and measurement at the unit level and system level. Hence, it is impossible to insert component performance probes inside commercial components. In addition, it is costly to collect and conduct performance testing and measurement for components and component-based programs.
- Homegrown components are developed without a consistent performance measurement mechanism, well-defined performance evaluation metrics, and a cost-effective performance evaluation environment.

According to recent reports [9, 10], J. Gao and his students in the Computer Engineering Department at San José State University have been working on establishing a systematic solution to support performance testing and measurement of components and component-based programs in a distributed environment. The major objectives of this systematic solution are to:

- Define standard performance evaluation models, strategies, and metrics for software components and component-based systems;
- Offer a performance measurement framework to support component engineers, application engineers, and system testers to create measurable components or enhance third-party components by adding a performance wrapper;

- Provide a consistent plug-in-and-measure performance evaluation environment for components.

The primary applications of this solution include component performance testing and evaluation and component-based system performance measurement. It allows engineers to evaluate component processing speed, resource utilization, transaction throughput, availability, reliability, and scalability. The solution consists of three basic parts: (1) a component performance library, (2) a distributed component performance evaluation environment, and (3) well-defined component performance tracking methods and formats.

12.4.1 A distributed performance measurement environment for components

Figure 12.7(a) displays the system infrastructure of a distributed component performance measurement environment. It supports performance testing and evaluation of software components and consists of the following parts in a client-server environment.

- *A performance test and evaluation server:* It manages and controls a central performance repository that stores various types of component performance trace records and evaluation results. The performance test server plays a control and management role in supporting performance trace collection, performance analysis and computing, as well as performance monitoring and reporting. The other function of the performance test server is to allow testers to control and configure performance tracking and testing features by communicating with performance agents on different computing stations over the network. Figure 12.7(c) shows the components of a performance test server. Through the GUI interface, engineers can communicate with performance agents on different computing stations to monitor the current component performance, query the agent status, and configure performance-tracking properties. The component performance analysis module analyzes and reports component performance in a graphic format.
- *A number of performance agents:* On each network node (say, a computer station), there is a performance agent that interacts with measurable software components through a dedicated performance message queue. It collects component performance data, and transfers them to the performance test server over the network using dynamic performance messages. The other function of a performance agent is to allow engineers to

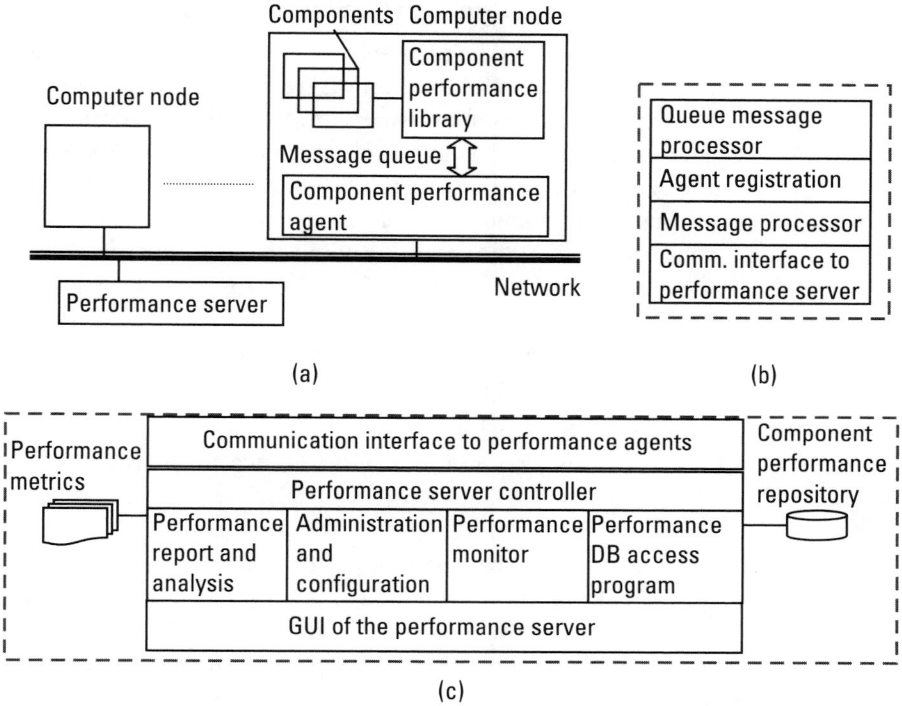

Figure 12.7 A distributed component performance evaluation environment: (a) performance client-server structure; (b) performance agent structure; and (c) performance server structure.

configure and set up the required performance tracking and monitoring features by communicating with the performance test server. Figure 12.7(b) displays the detailed modular structure of a performance agent.

- *A performance library:* It is a well-defined class library that provides component developers and component users a standard component performance framework. This framework contains a set of well-defined performance tracking and analysis classes that support performance measurement of speed, throughput, resource utilization, availability, reliability, and scalability.
- On each computing station over a network, there are two predefined performance tracking message queues. One is used to manage and control the performance tracking messages from different components to the performance library. The other is used to manage and control the performance tracking messages from the performance library to a performance agent.

12.4.2 Component performance library

The component performance library is developed to provide engineers with a standard performance evaluation framework to support performance testing and measurement of software components and component-based programs. One of its purposes is to enable component developers to construct measurable components that facilitate performance measurement. And the other purpose is to enable component users to convert third-party components into measurable components by adding a consistent performance wrapper. The library provides two general functions for each software component:

- Generate and collect different types of component performance traces using well-defined trace formats, and place them into a performance message queue.
- Compute component performance measures based on a set of predefined performance metrics.

Currently, the component performance library includes a set of classes. Six of them are useful for the measurement of component speed, resource utilization, transaction throughput, reliability, availability, and scalability. One of them supports the processing of performance message queues. The library is implemented based on the component performance metrics given in Chapter 11. The details of this library are reported in [10].

12.4.3 Performance tracking techniques

Performance tracking is needed to support performance testing and evaluation. Its major purpose is to help engineers retrieve and monitor component and system performance data. Two different performance-tracking techniques are used. They are the time-based performance tracking method, and the volume-based performance tracking method. Both methods could be used manually or systematically. Here, we only highlight the general idea. The details can be found in [9].

12.4.3.1 Time-based performance-tracking method

The basic idea of this method is to provide a set of programming functions to allow engineers to insert predefined program code to track and analyze the program processing speed from point A to point B on a program control flow inside software components. For instance, to monitor the execution time for a function in a component, a component developer can insert the time-based performance tracking code at the beginning and end point of a component function. The performance library computes the function execution time based on the collected speed performance traces. Typical applications of time-based performance tracking are:

- Check the execution time of a function in a component;
- Check the execution time of a loop or a block of sequential statements in a component;
- Check the execution time of a functional transaction in components;
- Check the waiting-time for receiving a message and/or processing-time of received messages.

In fact, the application of time-based performance tracking is not limited to software components. It can be used for any program code block, loops, repeatable logic, transaction sequences and functional scenarios in component-based programs. Based on time-based performance traces, the performance library computes the maximum/minimum/average processing time for a specific operation, function, or task in a component. The detailed description of this method and its implementation can be found in [10].

12.4.3.2 Volume-based performance tracking

The time-based performance-tracking method is not suitable for stress and load testing because it generates performance traces whenever a component (or a program) exercises the inserted performance tracking code due to its high system overhead. To support engineers in load and stress testing as well as in the measurement of component reliability and availability, the volume-based performance tracking method was developed. Unlike the time-based performance tracking, where the focus is to monitor the processing time for program operations, the volume-based performance tracking focuses on measuring various throughputs in a system and its components. The typical examples are the throughputs of event/transaction/message processing in components. This method is designed to evaluate component and system performance throughput, and support the measurement of component availability and reliability for performance testing and evaluation.

The basic idea of this method is to insert predefined program throughput tracking code into software components to collect and generate throughput performance traces based on a predefined time interval. During the execution of a component (or system), the volume-based performance tracking code counts the occurrence of a targeted event/transaction/message/operation in software components, and generates a performance trace based on a specified time interval. The detailed descriptions about this method and its implementation can be found in [10].

Performance testing engineers can use this method to measure the occurrences of the incoming events (or messages) and outgoing events (or messages) of a component in a specific time slot. This performance data provides useful information for computing component throughputs, the available time and unavailable time for component availability, and the uptime and downtime for component relilability.

12.5 IBM's solution for test tool collaboration

In 1998, IBM formed the Software Test Community Leaders (STCL) group to address issues associated with software testing and quality [11]. The group consists of key technical professionals and managers from testing groups across the various divisions of IBM. One of the first tasks the technical team undertook was to identify and categorize leading tools and practices in use across the divisions. The idea is to enable key tools to be shared by different teams across production lines. As a result of this exercise, the team produced a list of best practices with supporting tools, and discoveredtwo common problems that would inhibit tool and practice sharing.

- Different teams in IBM used diverse test tools to support similar testing processes and common testing practices.
- Many used third-party test tools and homegrown tools, which are not easily integrated and able to interact with one another because they were not designed to interact with other tools.

In mid-1999, the STCL technical leaders decided to take an architectural approach to address these problems. The STCL Architecture Committee was formed in order to exploring high-level solutions to solve test tool interoperation problems based on a new architecture.

12.5.1 The STCL architecture requirements

At the highest level, three requirements were identified for the architecture:

1. It must provide a mechanism for integrating the variety of tools recommended by the STCL. The problem of integrating heterogeneous,

independently developed tools has received attention from other parties, both inside and outside. To address this problem, the STCL Architecture Committee selected a three-phase approach, including data integration, control integration, and interface integration. This led to a more detailed set of integration requirements [11]:

- Data integration must make test-related data available in an open manner regardless of the tool or repository in which the data are stored.
- Data integration must provide a way to maintain associations among related data, even if the data reside in different repositories.
- Control integration should support invocation of externally available functionality on tools that comply with the architecture. This will be important for building highly automated testing environments.
- GUI integration will result in a single user interface for accessing all of the architecturally compliant tools.
- GUI integration should not preclude using the tools as they are currently used today. This is important because different groups will migrate to the new architecture as business conditions permit.

2. It must support site-specific tools and testing processes. To meet this requirement, the STCL listed two detailed requirements [11]:
   - A plug-in-and-play mechanism for incorporating site-specific tools must be included in the architecture.
   - Control integration should allow workflow customization to support different testing processes.
3. It must support legacy tools, while also providing a road map for new tool development within . This requirement identifies two needs: (1) to support and protect the existing legacy test tools, and (2) to provided a road map so that new tools are designed to interoperate. The detailed requirements for achieving this high-level goal are given here [11]:
   - Enhancements to legacy tools must be localized, with no changes to internal tool control structures or data representations.
   - The architecture should provide guidance for the design of new tools in the form of standardized APIs, standardized testing entities, and other reusable items.

- Both legacy and new tools should be supported using reusable components.
- The legacy and new tool requirements should be met with as little variation within the architecture as possible.

These requirements form the foundation for the development of the STCL architecture.

12.5.2 The STCL architecture

The STCL architecture addresses three types of tool integration: data integration, control integration, and GUI integration. The major focus of data integration is to provide a unified view of diverse test information data within the testing organization, regardless of where the data is stored. Control integration allows unrelated tools to invoke one another using a general method. GUI integration focuses on providing a single, consistent interface to support all of architecturally enabled tools. As given in [11], the proposed STCL architecture is a multilayered architecture consisting of the following levels:

1. *Data exchange:* The tool supports the ability to read and write data in the data-exchange format of the architecture.
2. *Data integration:* The tool supports standard functionality on its artifacts (create, read, update, delete) as well as associations with artifacts within other tools.
3. *Control integration:* The tool provides the ability to invoke operations on it via the generic control integration services.
4. *GUI integration:* The tool supports an interface within the standard GUI environment.

Figure 12.8(a) shows an example of the STCL solution. As shown in Figure 12.8(b), the architecture is supported by four basic architectural components, and each component supplies a set of services that can be used by a tool. These four components are:

- *GUI integration component:* It provides a set of common GUI services. The objective of GUI integration is to provide testers with a central user interface to access the tools typically used in the testing process. This interface, when combined with the data and control integration capabilities of the architecture, will provide testers with an *integrated test*

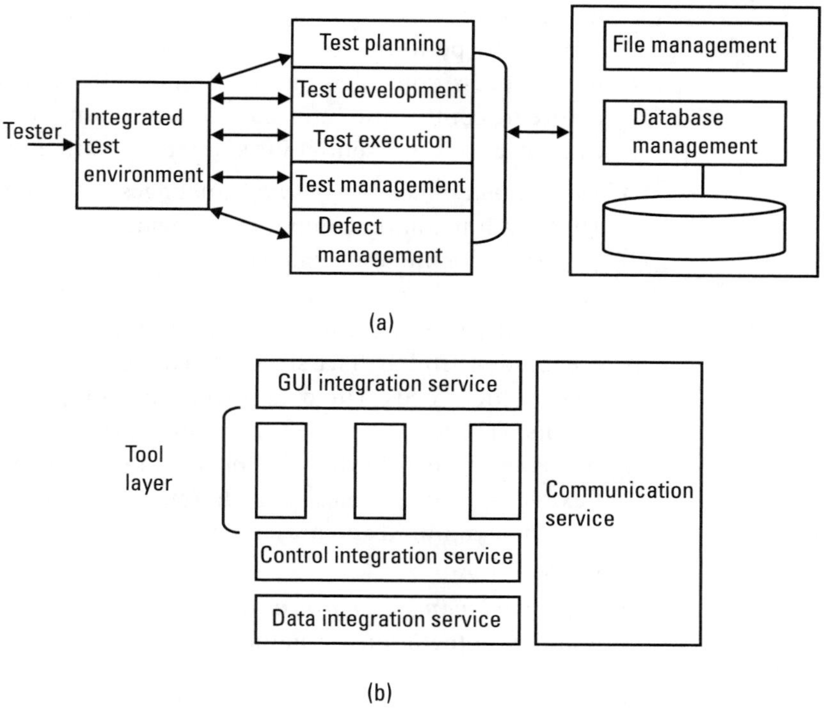

Figure 12.8 The STCL solution and its modified layer architecture: (a) one example of the STCL solution; and (b) the STCL's modified layer architecture. (*From:* [11]. © 2002 IBM. Reprinted with permission.)

environment (ITE). Now IBM uses plug-ins as the extension points to integrate new tools into an ITE.

- *Control integration component:* It offers services for operation invocation and automation. The major purpose of control integration is to provide the ability to distribute and customize the control flow among the various tools integrated in the architecture. To achieve this goal, a flexible, customizable control integration mechanism is essential to allow support of the multitude of testing processes used across the IBM labs. This is especially important for the creation of an ITE. At a high level, a control integration strategy must support two capabilities. First, it should define a tool- and environment-neutral mechanism for coordinating control among the tools used in the testing process. Next, it must provide a tool- and environment-independent way to specify and enforce a software testing process. Two types of control integration are applied: *event management* and *workflow process control*. Event management is essential for control integration. It allows resource providers within the architecture to be made aware of asynchronous events when they occur, and they in turn respond by performing actions specified in their event handlers. Workflow process control is more complex. It involves coordinating and enforcing control and information flow among several distributed applications, and it is essential for supporting varied testing processes.

- *The data integration component:* It manages data content and associations. The data integration component provides a coherent way to access and manage all of the test-related data within an organization. If legacy tools are involved, then they often contain vast amounts of data that need to be shared and integrated with data in other tools. In this case, new artifacts are needed, and artifact structure change may occur as product teams change their focus. The data integration component must consider two factors: (1) the data in the organization is distributed within a heterogeneous set of repositories and tools; and (2) the data artifacts need to be flexible and extensible. In addition, associations among related data should be captured and preserved, even when the data reside in different repositories.
- *The communication component:* It connects the tools to the GUI, control, and data integration components.

12.6 Summary

Recently, component-based software construction has been widely accepted in the software industry. In the past few years, many researchers have devoted their efforts in developing new cost-effective methods and systematic solutions for testing third-party software components and component-based programs. This chapter reports a number of selected research projects on building new frameworks and environments to solve some of the identified problems in testing component-based software and its third-party components.

In Section 12.1, we reviewed the basic concept of BIT components, including its objectives, benefits, and limitations. Meanwhile, we report a supporting framework, known as BIT component wrappers, which is developed by S. Edward's research group. Its major objective is to support component self-testing and self-checking during component black-box testing.

In Section 12.2, we revisited the basic concept of testable components and discuss the structure of a testable component and its test interfaces to support component testing. Moreover, we highlight the research efforts on how to develop a component test framework as a base to support the automatic construction of testable components based on conventional components. In addition, we report the research efforts on establishing a distributed test environment. It implements a plug-in-and-test solution to support black-box component testing for testable components.

Section 12.3 described a component tracking framework and a distributed component-tracking environment that implements a systematic solution to track and monitor component behaviors. The primary objective of this framework is to increase component traceability. Using this framework, engineers are able to construct traceable components that facilitate program tracking in a systematic mechanism and consistent trace formats. The distributed component-tracking environment demonstrates a way to monitor component behaviors in component-based software over a network.

Component performance validation is important in component testing. How to validate and measure third-party components in a cost-effective way is a major issue in component-based software engineering.

In Section 12.4, we covered a component performance framework and a distributed performance measurement environment that supports component performance testing and measurement in component-based software. The component performance framework could be used to develop measurable components that facilitate performance evaluation. The performance measurement environment demonstrates a systematic solution to evaluate component performance for component-based software over a distributed network.

In Section 12.5, we summarized IBM's STCL architecture for collaboration of various third-party tools and homegrown tools to form an integrated test environment. From this architecture, we can learn how to establish a long-term integrated test environment to integrate diverse test tools for various software production lines using a cost-effective solution to enhance the sharing and reuse of test tools in a large company. In component-based software development projects, we expect to see a strong demand for the integration of various test tools to form an integrated test environment to support software validation for component unit testing, integration, and system testing due to the changes and evolutions of third-party components.

As we mentioned in Chapters 2 and 8, there are many open issues and challenges in component-based software testing and test automation. We need more innovative testing methods, systematic solutions, and tools to assist engineers in validating third-party components and component-based programs.

References

[1] Wang, Y., G. King, and H. Wickburg, "A Method for Built-In Tests in Component-Based Software Maintenance," *Proc. of IEEE International Conference on Software Maintenance and Reengineering (CSMR'99)*, March 1999, pp. 186–189.

[2] Wang, Y., et al., "On Built-In Tests Reuse in Object-Oriented Programming," *ACM Software Engineering Notes (SEN)*, Vol. 23, No. 4, July 1998, pp. 60–64.

[3] Edwards, S. H., "A Framework for Practical, Automated Black-Box Testing of Component-Based Software," *Software Testing, Verification, and Reliability*, Vol. 11, No. 2, June 2001.

[4] Edwards, S., et al., "A Framework for Detecting Interface Violations in Component-Based Software," *Proc. of 5th International Conference on Software Reuse*, IEEE Computer Society Press, Los Alamitos, CA, 1998, pp. 46–55.

[5] Sitaraman, M., et al., "Component-Based Software Engineering Using RESOLVE," *ACM SIGSOFT Software Engineering Notes*, Vol. 19, No. 4, 1994, pp. 21–67.

[6] Gao, J., et al., "On Building Testable Software Components," *Proc. of International Conference on COTS-Based Software Systems*, Orlando, FL, February 4–6, 2002.

[7] Cuong, P. Q., *A Framework for Software Component Testing Tool in a Distributed Environment*, Master Project Report, San José State University, San José, CA, December 2002.

[8] Gao, J., E. Zhu, and S. Shim, "Tracking Software Components," *Journal of Object-Oriented Programming*, Vol. 14, No. 4, October/November 2001, pp. 13–22.

[9] Gao, J., and C. Sun, *Performance Measurement of Components and Component-Based Software*, Technical Report, Computer Engineering Department, San José State University, San José, CA, 2002.

[10] Ravi, C. S., and H. Q. Duong, *Performance Testing Framework for Distributed Component-Based Software*, Master Project Report, Computer Engineering Department, San José State University, San José, CA, December 2002.

[11] Williams, C., et al., "The STCL Test Tools Architecture," *IBM System Journals*, Vol. 41, No. 1, 2002.

PART

IV

Quality assurance for software components and component-based software

In recent years, product and service quality has become a principal means for many corporations to distinguish themselves from their competitors. IEEE defines software quality as "the degree to which software meets customer or user needs or expectations." The vast majority of the modern quality assurance methodology has been motivated to improve hardware or service quality. Although *software quality assurance* (SQA) is very different from *hardware quality assurance* (HQA), much of the modern quality assurance methodology is applicable to software quality. There exists a large amount of published literature on software quality assurance in general, but quality assurance for software components or software systems assembled with such components has received little attention so far. Part IV is devoted to quality assurance for software components and component-based software systems.

This is a big subject, and we can only provide an overview in the space available. In our attempt to weave a coherent story about this big subject, it is inevitable that our personal views permeate the whole discussion. However, we have tried to place a balanced emphasis on the published literature, particularly peer-reviewed journal articles as well as books written by recognized experts in the related fields, rather than relying merely on our personal experience or anecdotes. Many references are provided as a result.

Our approach to introducing quality assurance for software components begins by briefly summarizing the modern quality assurance methodology for hardware products and highlighting the main differences between HQA and SQA. These differences turn out to play an important role in our entire approach, and they are discussed in both Chapters 13 and 16.

A major reason why quality assurance has contributed significantly to hardware quality is that HQA involves not only control of the development and production processes at the management level but also precise definition and accurate measurement of product quality. This balance has earned HQA the recognition as a distinct and worthy discipline. However, a common complaint about the current status of SQA is its predominant process orientation. We have attempted to achieve a balance between the process and product orientations. For example, in addition to the process-oriented software life cycle models and the process-oriented life cycle standards of IEEE 12207 and IEEE 1517, we discuss Dromey's product quality models and other product-oriented approaches. We first summarize briefly and extend the life-cycle processes in Chapter 13 and then discuss the more recent concept of product quality models in Chapter 14. Software standards and certification will be discussed in Chapter 15. The discussions of Chapters 13 and 14 on issues and solutions associated with software components and component-based software, respectively, pave the way for the discussions of Chapter 15 on standards and certification. Chapter 15, in turn, paves the way for some detailed and more focused discussions in Chapters 16 and 17 on the verification of quality.

The practice of developing and using software components is a relatively new phenomenon, and new issues continue to emerge. Some of these issues can be resolved or alleviated using methods developed for conventional software, while the others require innovations. We have attempted a balance between generic SQA methods and some recently proposed approaches motivated explicitly by the development or reuse of software components. For example, we address the seminal quality factors/criteria/metrics framework proposed in 1977 by McCall et al. for SQA in general (in Chapter 13), as well as IEEE 1517, which is the *IEEE Standard for Information Technology—Life Cycle Process—Reuse Processes*, approved and published in 1999 (in Chapter 17).

Many software metrics have been proposed, and many claims and conjectures about their relationships with software quality have emerged. Some of the conjectures are based on theoretical reasoning, or even intuition, and have not yet been empirically validated. We certainly understand that decisions are routinely made based on subjective beliefs about these claims and conjectures, and we also understand the necessity of such practices in real-world project management. However, we have attempted to strike a balance between theoretical reasoning and empirical validation. For example, in Chapter 16 we discuss not only many software metrics that have been used to help measure or predict software quality but also their empirical validation. Other examples include a discussion in Chapter 17 on the success and failure factors in the reuse of software components at the program-management level and a discussion on failure modes and failure causes at a technical level, both based on empirical results reported in the literature.

In our weaving of a coherent story, we have also attempted to fill gaps where space allows. For example, IEEE 1517 contains only a brief discussion of the development and quality assurance processes for software components, and as a result, we propose an integrated process in Chapter 13.

CHAPTER

13

Contents

Quality assurance for software components

In recent years, product and service quality has become a principal means for many corporations to distinguish themselves from their competitors. Many definitions of quality exist; we adopt AT&T's simple definition of quality: customer satisfaction. This definition is equivalent to the one adopted by IEEE (for software) [1], which defines software quality as "the degree to which software meets customer or user needs or expectations." According to this definition, quality assurance can simply be interpreted as assuring customer satisfaction. The focus of this chapter is quality assurance for software components. Quality assurance for component-based software systems will be the focus of Chapter 14.

Although the vast majority of recent developments in quality assurance has been motivated to improve hardware or service quality, much of the modern quality assurance methodology is applicable to software quality. There is a significant amount of literature on software quality assurance, but quality assurance for software components or for component-based software systems has received little attention.

Our approach for introducing quality assurance for software components is as follows. We begin by briefly summarizing the modern quality assurance methodology for hardware products and then point out several fundamental differences between hardware quality assurance and software quality assurance. These differences turn out to play an important role in our entire approach. More differences, particularly those related to quality measurement and quality metrics, will be discussed in Chapter 16.

HQA has clearly been a recognized and important academic discipline for decades. SQA, however, seems to overlap significantly with *software development* (SD) and seems not to have been embraced by software developers the same way the hardware quality assurance has been embraced by hardware developers. The differences between HQA and SQA may actually offer a good explanation; they also seem to indicate that the distinction between the discipline of SQA and that of SD is much weaker than the distinction between the discipline of HQA and hardware development. This may be evidenced by the fact that both SQA books and SD books cover subjects like software configuration management, test and evaluation, and verification and validation. Although SQA may not constitute a distinctive discipline from SD, it is a critical function of any SD organization. To establish credibility and to win embrace from SD personnel, SQA must be conducted with systematic rigor and not just with ad hoc activities. All this has motivated our emphasis on a rational SQA process, with the relationships among all the steps clearly stated and perhaps even derived. Despite the significant amount of overlap between SD and SQA, few books (e.g., [2]) address the two as an integrated activity. This has motivated us to develop an integrated process for both SD and SQA for software components.

This chapter is focused on the classical approach to SQA, which is predominantly process oriented. Although the classical SQA approach addresses high-level process issues well, a good high-level SQA process does not guarantee software quality. The more modern product-oriented approaches to SQA and the companion software product quality models (e.g., [3, 4]) will be discussed in Chapter 14. Software measurement plays an important role in both SQA and project management. Because of its quantitative nature, it requires a rigorous approach. Chapter 16 discusses, among other subjects, a framework for software measurement developed by N. E. Fenton [5], and proposes an extension.

This chapter is organized as follows. Section 13.1 addresses the modern methodology for HQA and the main differences between HQA and SQA. Section 13.2 briefly reviews the subject of SQA in the way that is commonly treated in the SQA literature. After discussing the main issues involved in quality assurance for software components in Section 13.3, we provide an integrated development and quality assurance process for software components in Section 13.4. Concluding remarks are given in Section 13.5.

13.1 The main differences between hardware and software quality assurance

This section begins with a brief review of the modern methodology for hardware quality assurance. It then addresses some key differences between hardware quality assurance and its software counterpart. A major goal of these discussions is to learn what portions of the tremendous amount of effort already invested in hardware quality assurance are applicable to software quality assurance and to understand key unique features of software quality assurance.

13.1.1 Modern methodology for assuring quality of hardware

The modern methodology for HQA can be characterized by the following phases and steps. The italicized phrases are the major concepts or techniques developed for and used in modern methodology for HQA.

13.1.1.1 Requirements definition phase

The requirements definition phase includes the following major tasks:

- Identify and define a complete set of *external quality characteristics*, together with requirements and targets. These characteristics are typically solicited from the customer through user surveys or focus groups, and are usually vague or abstract (e.g., reliable). They capture only customer expectations. This is the rationale for the qualifier "external." The qualifier "complete" refers to the necessity that the set encompasses all aspects of customer expectation. The characteristics may be qualitative attributes or quantitative variables.
- Based on the external quality characteristics identified, defined, and targeted in the previous step, identify and define a complete set of *internal quality characteristics* and determine a detailed and complete set of *requirements and specifications*. Vague or abstract external quality characteristics must be clarified, refined, and translated into internal quality characteristics in engineering terms (e.g., mean time between failures). In addition to such internal quality characteristics, there also exist quality characteristics that are not of direct concern to the customer but are of direct concern to the producer for reasons such as cost reduction. Such characteristics include the capacity for self-diagnosis for more efficient repair and the reuse of components previously developed by the producer. These two types of internal quality characteristics and the

resulting requirements and specifications drive the entire engineering process internal to the hardware developer. This explains the qualifier "internal." The qualifier "complete" in this step refers to the necessity that these requirements and specifications meet or exceed the requirements and targets translated from the external quality characteristics developed in the previous step as well as those set for internal engineering purposes.

13.1.1.2 Design phase

The design phase includes the following major tasks:

- Design a product that *satisfies the internally set requirements and specifications* developed in the previous phase and, in turn, satisfies the requirements and targets associated with the external quality characteristics.
- Due to the wear-and-tear nature of most hardware products, anticipate possible failure modes, estimate life distributions of the product and its components, and design reliability into the product (*design for reliability*).
- Due to the innate imperfection of most manufacturing processes, design the product and a "capable" manufacturing process with anticipation of possible manufacturing issues (*design for manufacturability*).
- Due to the wear-and-tear nature of most hardware products and the concomitant maintenance requirements, anticipate maintenance issues and consider these issues in product design (e.g., self-diagnosis) (*design for maintainability*).

The last three bullets may be included in the first one, particularly if reliability, manufacturability, and maintainability requirements have been explicitly defined in the requirements definition phase.

13.1.1.3 Manufacturing phase

The manufacturing phase includes the following major tasks:

- Use *statistical process control* to build quality into the product and to detect possible "assignable causes" for process instability.
- Use acceptance sampling (and testing) and other means to ensure the quality of raw and other input materials.

13.1.1.4 Acceptance phase

The acceptance phase includes the major task to use *acceptance sampling (and testing)* to determine the acceptability of final products to the external customer or to downstream internal customers in the same corporation.

13.1.1.5 Operations phase

The operations phase includes the following major tasks:

- Predict in real-time possible imminent failures or functional deviation (*intelligent maintenance*).
- Collect data on failure modes and life distributions for *continuous improvement*.

13.1.1.6 Some remarks about the modern methodology

The two steps in the requirements definition phase are the first two steps of what the HQA community refers to as *quality function deployment* (QFD). As will become clearer shortly, the SQA community has defined *quality factors* and *quality criteria* (including more detailed *quality subcriteria* and *quality metrics*), and they somewhat correspond to the external and internal quality characteristics defined for HQA. In SQA, however, quality factors contain some quality characteristics that are of little direct concern to the user. In fact, the structure spanning from quality factors through quality metrics is more of a hierarchy of quality characteristics than a process for deriving detailed quality measures from vague customer expectations. Nevertheless, the process of mapping quality factors into the more detailed quality criteria and quality metrics is consistent in spirit with the QFD process. Some commonalities between software and hardware quality assurance is further evidenced by some recent literature on linking software quality directly to measures of customer satisfaction [6].

The HQA community has moved away from the earlier reliance on acceptance testing (of the final products rolling off the mass-production line) to the current emphasis on statistical process control in order to build quality into the product or, at least, to prevent bad products from being made in the first place. Software testing has long been recognized as an important task, but the importance of building quality into software or, at least, preventing software quality problems has also been recognized and emphasized in recent years.

Some software development organizations have a quality assurance department. Although such a department has its own mission, we take the position that SQA, just like HQA (and perhaps even more so than for HQA), is a job for everybody involved in the life cycle of software development, not just the personnel of a quality assurance department. This position will be further justified by the differences between HQA and SQA, to be explained in the next section.

It is well known and accepted that the correctness of software with respect to its requirements cannot be ascertained 100% by testing; there are simply too many tests required. Software quality assurance should follow the trend of hardware quality assurance and avoid relying exclusively or primarily on testing to achieve software quality.

13.1.2 Major differences between hardware and software quality assurance

Some key differences between hardware and software quality assurance result from the fact that software does not suffer from wear and tear and the fact that software duplication is error-free and not subject to any manufacturing imperfections. While a vast majority of HQA is devoted to coping with the innate variability of the manufacturing process or the inevitable product wear and tear, they are of no concern to SQA.

On the other hand, typical HQA personnel lack the domain knowledge required in the design phase, and therefore, proper design for quality assurance has received little attention in the HQA community. Because software involves no wear and tear and enjoys error-free duplication, the design and coding phases must become focal areas for SQA if quality assurance is expected to make a contribution to software quality that is as significant as the contribution made by quality assurance to hardware quality.

The major differences between HQA and SQA are summarized in the next few sections as follows.

13.1.2.1 Design phase

In the design phase:

- Design for reliability (to overcome product aging) is not a concern for SQA.
- Design for manufacturability is not a concern for SQA.
- Unlike HQA, the design phase must be a focal area for SQA.

13.1.2.2 Manufacturing phase substituted with implementation (of design) or coding phase

The manufacturing phase is substituted with implementation (of design) or coding phase. In the implementation phase:

- The discipline for software implementation is not as well defined.
- Statistical process control is not really relevant, although metrics can indeed be used to gauge the quality of the software product and that of the development process.

13.1.2.3 Operations phase

In the operations phase, once a software bug is fixed, it is fixed for good and the quality is improved unless new bugs are introduced during the correction process.

13.1.2.4 Some remarks about the major differences

Some of the differences listed above have long been recognized [7]. It is important, however, to contrast the differences in a systematic way and with respect to the entire product development and quality assurance process. This contrast leads to the following observation regarding whether SQA actually constitutes a distinct discipline from SD or is more like a different job function within software development.

Note that HQA includes statistical process control, statistical acceptance testing, and design for reliability (which involves system reliability computation, statistical analysis of component and system life time data, optimal resource allocation for reliability growth), all of which require much statistical and mathematical expertise and hence cannot be fulfilled by hardware designers. This may explain why HQA has become a discipline that is distinct from hardware development and has been recognized by hardware development and other engineering disciplines. Note also that none of these types of statistical analysis is part of SQA. Therefore, the distinction in terms of the expertise required between the SD and SQA is much weaker than that between HD and HQA. In other words, SQA may not have as much a distinctive identity as its hardware counterpart, and hence it should perhaps be considered as a job function within a software development organization. More explicitly, software developers can perform SQA tasks if they can expand their expertise somewhat and are willing to do the very important work. (As will become clear in Chapter 16, much statistical analysis is needed to establish statistical relationships between software quality and possible explanatory variables. However, such analysis is required of the researchers, not necessarily the practitioners.)

A key to distinguishing SQA from the rest of SD and accentuating the value added of SQA is to conduct SQA in a rigorous manner (while being pragmatic when necessary, for example, not overwhelming the software developers) and with clear rationale. We emphasize such rigor in developing the integrated development and quality assurance process for software components to be detailed in Section 13.4.

Other than system testing, most quality assurance departments deal primarily with process issues and provide guidelines for individual process steps. Since function and performance provided by the code is not easily visible to the SQA personnel and some SQA personnel may actually be unfamiliar with SD methodologies and techniques, SQA should be a job for everyone involved in the life cycle of a software product, just like HQA.

R. Dunn and R. Ullman [7] pointed out another difference between SQA and HQA; they state, "Hardware can be built of standardized components, from devices to complete assemblies, the reliability of which are known. For the most part, software contains few program elements with which there has been prior experience." With the increasing popularity and maturity of software components, this difference may gradually diminish.

13.2 Software quality assurance

We first briefly discuss the essence of SQA, and then describe those tasks that are considered by SQA specialists as part of their job. An integrated SD and SQA process will be the subject of Section 13.4.

13.2.1 The essence of software quality assurance: process and discipline

The essence of software quality assurance is to establish and continually improve a clearly defined process and to exert discipline for each and every step of the process. Although metrics play a pivotal role in HQA and a large number of software metrics have been developed and reported in the literature, software metrics seem to play a limited role in practice. For example, metrics for measuring the complexity of a software module or system have been developed (e.g., McCabe's cyclometric complexity measure [8] and Halstead's measure of conciseness [9]). Although they can be used to measure the complexity of a software module or system, it is probably costly for a software development organization to rewrite a program after the program has been completed but found to be highly complex. However, it is a good practice to learn what may cause high degrees of complexity and how to write programs with simplicity. Therefore, it is always good to understand the complexity measures, develop some guidelines to ensure moderate or low complexity, and follow the guidelines during the design and coding phases. Chapter 16 has more detailed discussion about software metrics.

13.2.2 Major software quality assurance tasks

In addition to system testing, the major tasks considered by many SQA specialists as part of the job of a quality assurance department include the development and implementation of a software quality program for the whole software development organization and the development and implementation of software quality plans for individual software projects. The task of developing a software quality program includes all steps required to start SQA in a software development organization and to complete an SQA system. This big task includes many technical tasks as well as managerial tasks such as developing and implementing organizational structure to support SQA and assigning SQA tasks to individuals. Due to space limitations, we focus here on technical tasks. A big part of the technical tasks is to support the development of SQA plans for individual software projects.

A software quality plan is commonly defined as a document that specifies the details of all procedures, policies, quality factors, quality criteria, quality metrics, reviews, audits, and tests that will be used to track software projects through their life cycles for assuring that quality is built in. Given the essence of the SQA discussed in Section 13.2.1, major categories of generic tasks required for a software quality plan include the following:

- Develop the software development process and quality assurance process (or an integrated process) and ensure the adherence to the processes.
- Develop discipline guidelines for each and every step of the processes (e.g., requirement templates, design methodologies, coding standards).
- Develop checklists for each and every step of the processes.
- Check the results of each and every step against the corresponding guidelines and checklists.

Important specific tasks include the following:

- Develop quality factors, quality criteria, and quality metrics:
    - Identify and define a complete set of quality factors. This set consists of a group of customer-oriented quality characteristics, which is similar to the external quality characteristics defined for HQA, and a group of high-level software quality categories important to the development organization alone (i.e., not necessarily to the customer). A dozen or so quality factors have been widely accepted [10–13], and they will be briefly described in Section 13.4.
    - Based on the quality factors identified and defined in the previous step, identify and define a complete set of more detailed quality criteria and subcriteria. Two dozen or so quality criteria have been widely accepted [10–13]; they will be discussed in Section 13.4.
    - Based on the quality criteria identified and defined in the previous step, identify and define measurable quality metrics and determine a detailed set of requirements and specifications. Many specific quality metrics have been defined. J. Vincent et al. [11] summarize a large number of quality metrics. Development of software metrics has also been an active area of research [14].
- Develop guidelines for every step of the SD and SQA processes based on the quality metrics, requirements, and specifications defined in the previous task.
- Develop review and audit checklists based on the guidelines defined in the previous task. Vincent et al. [11] considered these checklists as the primary tool of the SQA effort.
- Conduct reviews, including management reviews, peer reviews, and walk-throughs, of the results of every major step of the processes (e.g., requirements specification, design, coding and documentation) and assess the quality level achieved by the major steps based on the checklists defined.
- Conduct periodic audits based on the checklists defined.
- Develop and implement improvement plans based on the results of the reviews and audits.

Note the tight interdependence of these tasks. Many SQA specialists consider the following tasks as a job for the quality assurance department. However, most software developers also consider these tasks as part of their job.

- *Conduct testing and evaluation (T&E):* The IEEE [1] adopts a broad definition for testing: "Testing is the process of exercising or evaluating a system or system component by manual or automated means to verify that it satisfies specified requirements or to identify differences between expected and actual results."
- *Conduct verification and validation (V&V):* The IEEE [1] provides the following definitions: "Verification is the process of determining whether or not the products of a given phase of the software development cycle fulfill the requirements established during the previous phase." (This is why we include a verification step in each and every phase of the integrated development and quality assurance process, detailed in Section 13.4.) "Validation is the process of evaluating software at the end of the software development process to ensure compliance with software requirements." (There does not seem to be a consensus within the SD and SQA community regarding the definition of validation. Some definitions of validation do not include testing software through execution. Chapter 16 has a more detailed discussion.)
- *Conduct software configuration management (SCM):* F. P. Ginac [6] defined configuration to be a particular version of a software product. Accordingly, he defined software configuration management as "the set of processes and procedures that an organization follows and tools that it uses to manage the source code from which various configurations of a product are derived." Vincent et al. [11] used a broader definition for configuration to include any significant component of the final product (e.g., some key requirements, specifications, or design documents). SCM, according to them, works to ensure the positive identification, control, auditing, and status accounting of such significant components.

13.3 Main issues involved in QA for software components

This section discusses the main issues involved in quality assurance for software components. We focus on the new issues brought on by the special features of software components. We begin by summarizing the main features of software components when compared to the conventional software, and then summarize the key SQA issues resulting from these main features.

We focus on the differences between a software component and conventional software in the following four closely related aspects:

- *External reuse:* We focus on the case of component reuse by users who are unknown to the developer and have no access to the source code but have documents about the functionality, performance, interoperability, and portability of the software component. Although software components can be developed for internal reuse where the user has full access to the source code, we deal with this case because it is more difficult.
- *Physical reuse procedure:* The reuse requires packaging by the producer, unpacking by the user, configuration by the user (if multiple configurations are provided, and configuration by the user is a feature), and finally assembled by the user into a component-based software system.
- *Potential impact:* Bad component quality in terms of functionality, performance, interoperability, portability, and documentation clarity may wreak havoc because of the possibly widespread use of the component stated in the first bullet item: external reuse.
- *Possible certification requirements:* Certification of quality may be necessary, given the criticality just stated. Note that there exist two possibly different perspectives: the user perspective and the certifier perspective.

Note the relationship between any two consecutive items, and hence the relationship among them all. Some key differences between SQA and component SQA are summarized in Table 13.1.

13.4 Integrated development and QA process for software components

Many software experts think that software engineering, unlike most other engineering professions, has little or no discipline (i.e., has little or no well-defined and well-accepted methodologies and processes to follow) [15]. They believe that few developers understand and practice any such discipline. Because of the high potential impact of the quality of software components on the potentially large number of software systems employing the components, it is critical to have and follow a clear set of rules and guidelines for developing software components and component-based software systems. Depending on the type of software, relevant rules and guidelines vary, ranging from basic good coding practices, to method of entity-relation diagram, to object-oriented design. Some such sets of rules and guidelines pertaining directly to developing software components and component-based software already exist in the market place (e.g., EJB, COM+, and CORBA). These sets

Table 13.1 Additional SD-SQA Complexities Associated with Software Components

SD-QA Task	*Major Additional Complexities Associated with Software Components*
Task category: develop SD/SQA process	Addition of the prerequirements phase, including the tasks of: Domain analysis Selection or development of a component model
Task category: develop guidelines and standards for all phases	Guidelines for the additional prerequirements phase on domain analysis and on selection or development of a component model Additional guidelines related to reuse by unknown users for every phase of the conventional life cycle
Specific QA task: develop quality factors, criteria, and metrics	Tailor the conventional quality factors, criteria, and metrics to capture all important quality characteristics of a software component, including the quality factors of functionality (correctness), interoperability, portability, verifiability (including certifiability), maintainability, security, usability, reusability, and the related quality criteria and metrics The possible need for an additional quality factor of "certifiability," from both certifier and user perspectives; possibly accommodating varying degrees of certification
Specific SD-QA task: develop requirements	Explicit quality requirements regarding portability, interoperability, reusability Significantly more stringent requirements for these and other conventional quality factors Requirements to satisfy a possibly diverse spectrum of unknown users regarding function, performance, interface, application environment and operating environment
Specific SD-QA task: review requirements, design, code, and so forth, and audit the corresponding processes	Verification of satisfaction of all the additional requirements Compliance to the component model and additional phase guidelines
Specific SD-QA task: test and evaluate software components	Additional testing and validation tasks necessitated by the additional explicit requirements regarding interoperability, portability, usability, reusability Additional testing and validation tasks resulting from the criticality of component quality, in terms of functionality, performance, interface, interoperability, and portability, due to the high potential of widespread reuse in diverse application and operational environments
Specific SD-QA task: manage software configuration	Packaging by the producer Unpacking by the user Producer's facilitation of "wrapping" by the user Configuration by the user through interface Assembly by the user into a component-based software system

of rules and guidelines are often referred to as component models. A brief introduction to the discipline introduced and perhaps imposed by these component models can be found in Section 17.3; the purpose of that introduction will be to illustrate the usefulness of rigid rules and guidelines (i.e., the discipline), rather than to attempt to provide complete sets of rules. In this section, we discuss a general integrated process for the development and quality assurance of a software component. An integrated process for the development and quality assurance of component-based software systems will be discussed in Section 14.8.

As mentioned earlier, SQA must go deeply into the design and coding phases and address proper software design and coding if it is to contribute to software quality and the productivity of software development in any significant fashion. It is important for the developers and programmers to recognize that software quality assurance is an integral part of all phases of the life cycle—from prerequirements study through maintenance and upgrade—and is the responsibility of all personnel involved, not just the quality assurance personnel.

In the software literature, the discussion of software life cycle and that of SQA seem to be disconnected. In the literature on SQA, many address quality factors, quality criteria, and quality metrics, but such factors, criteria, and metrics are rarely addressed in the literature on software life cycle. Numerous specific software metrics have been proposed in the literature, and there has been much research on the software metrics [11, 14]. In the SQA literature, although several general steps (reviews, walk-throughs, inspections, and audits) have been commonly discussed and accepted, there is little detail about what the life-cycle process, particularly the SD process, should be so that the life-cycle process will help achieve the desired quality levels and help avoid quality problems. Perhaps it has been tacitly accepted that the best or the most practical thing one can do is to be mindful of the quality metrics while designing and coding. (Another use of the metrics is for measuring the quality of the development personnel and its improvement through time.) On the contrary, discussion of these metrics should be an explicit part of the discussion of software life cycle. See Sections 14.5 through 14.7 for a discussion of very recent attempts to make the connection.

In general, there exist two groups of metrics: product metrics and process metrics. The former are intended to capture the quality of the software product, while the latter are used to gauge the quality of the development process. In the software literature, the discussions of these two groups of metrics also seem to be disconnected in the sense that improvements in the process metrics do not necessarily contribute to improving the metrics of product quality. Actually, it may be too much to expect software metrics, either product metrics or process metrics, to play a role that remotely resembles the role of hardware metrics in (hardware) statistical process control.

These disconnections, in addition to other factors already mentioned in previous sections, motivate the following discussion of an integrated process for development and quality assurance for software components. There exist a number of life-cycle models. For ease of conveying the key differences between SD-SQA for software components from that for conventional computer software, we base our integrated process on the waterfall model. While there exist many books on software development [16] and many others on software quality assurance [7, 10–12], few books [2] have treated these two critical and intertwined activities together as an integrated process. T. Gilb [17] used the approach of *design by objectives* in managing software engineering and advocated the practice of explicitly treating quality objectives, in addition to functional capabilities, at the outset of a software project. He started by exploring how to set objectives for a software development project, especially the critical but elusive quality objectives for the software, and then discussed ways to find and evaluate solutions for reaching the quality objectives, subject to resource constraints. Recently, J. Musa [18] focused on software reliability, which is a key software quality objective, and addressed in detail the concept of software reliability engineering and how to apply it in software development and software testing.

13.4.1 Prerequirements (study) phase

Developing software components may incur considerably higher development costs than their single-use counterparts. A thorough needs assessment is required before developing a software component. Software components can be grouped in two categories: horizontally reusable components and vertically reusable components. The former can be used across a broad range of application areas; examples include graphical user interface, sorting algorithms, and database management systems. The latter are used in similar applications within a particular problem domain. The success of vertical reuse hinges upon domain analysis. Major steps involved in the prerequirements phase are summarized next, followed by a brief explanation of the steps when necessary.

- Determination of functionality in the application environment
  - *Needs assessment:* For vertically reusable software components, perform a domain analysis (to anticipate use as a component by unknown users or to anticipate growth into a component beyond the current project). For horizontally reusable software components, identify "middleware" needs. See Section 17.5 for a discussion of a four-layer component architecture containing three software-component layers proposed by M. L. Griss [19].
  - *Use case development:*

Functionality;

Performance;

Interface.

- Determination of the scope of reuse.
  - *Assembly into to different (component-based) application software systems,* including configurability and customizability if applicable.
  - *Portability to different operating systems and hardware.*
- Determination of the certification process: development, tracking, and satisfaction of certification requirements imposed by the certification agency or by the user.
- Initiation of the development of test plans.

K. C. Kang [20] stated the following about domain analysis:

> Domain analysis is an activity to produce a domain model, a dictionary of terminologies used in a domain, and a software architecture for a family of systems. The output from the domain analysis
>
> - facilitates reuse of domain knowledge in systems development,
> - defines a context in which reusable components can be developed and the reusability of candidate components can be ascertained,
> - provides a method for classifying, storing and retrieving software components,
> - provides a framework for tooling and systems synthesis from the reusable components,
> - allows large-grain reuse across products, and
> - can be used to identify software assets.

A common method for developing user requirements is to identify and develop use cases. Since the user of a software component may not be clearly identifiable or there may exist different groups of users, in terms of differences in the intended use, the use cases may be much diverse than those desired of traditional software. Moreover, the software and hardware environment of the software systems to be built with the software component may vary, and the portability issues become critical. In traditional software development, configuration is a main issue, and system configuration management has been a major part for software quality assurance. Typically, such configuration is performed at the system or product level by the software vendor. Due to the possibly varying application environments and operational environments, including operating systems and hardware, into which a software component may be assembled, configuration may best be performed by the user, and hence configurability and customizability become important design issues as well as an SQA issue for both software components and component-based software. Note that in addition to the issue of portability to different hardware or operating systems, interoperability of a software component in the sense of feasibility and easy assembly into different component-based software systems is also critical.

Varying degrees of certification may exist. A major distinction is product certification versus process certification. Certification for software components will be a focus of Chapter 15.

The development of test plans should begin at the outset of the prerequirement activities. The test plans should be expanded as more requirements are imposed in later phases. This incremental approach introduces the rigor in testing required for ensuring the quality of software components, which may be used widely by many component-based software systems.

13.4.2 External-requirements phase (including user and certifier requirements)

Conventional software development typically begins with the definition of user requirements. Since software-component developers may request certification of quality by a third party and customers of software components may also require such certification, certifier requirements must also be considered together with the user requirements when developing the external quality characteristics. We refer to this phase as the external-requirements phase, rather than the user-requirements phase.

We group external requirements in three categories. As mentioned earlier, we include verification in all phases. We also include test plan expansion in all phases.

- Specification of requirements.
    - User requirements regarding (eight categories):

        Functionality;

        Performance (efficiency, reliability, and so forth);

        Interface;

Assembly into to different (component-based) application software systems after unpacking and configuration by the user (interoperability);

Porting to different operational environments, including operating systems and hardware (portability);

Standards;

Certification (verifiability);

Documentation.

- Certifier requirements;
- Other quality requirements:

  User-imposed quality factors that are among the commonly accepted quality factors;

  Other possible quality factors for the software components being developed;

  The importance measures of these factors and possible trade-off among them.

- Verification: Traceability of these requirements from all the findings accomplished in the prerequirements phase; reflection by the requirements of all the concerns identified in the prerequirements phase.
- Expansion of test plans.

Note again the complexity of the user requirements due to a possibly diverse user population in terms not only of use cases, functionality, performance, and interface but also of the assembly, porting, and certification requirements. Also note that interoperability and portability have long been considered as important but vague quality factors. In the context of software components, however, they should be regarded as explicit user requirements. More precisely, requirements for a software component to be portable to specific operating environments and to be interoperable with specific application environments should be explicitly defined, to the maximum degree possible. For ease of discussion, we will refer to the eight categories of user requirements for a software component [i.e., functionality, performance, interface, assembly into to different (component-based) application software systems (after unpacking and configuration by the user), porting to different operational environments (including operating systems and hardware), standards, certification, and documentation].

The software quality assurance community seems to have accepted a dozen or so quality factors. Some of the accepted factors are of direct concern for the user and tend to be imposed by the user, while some others are of direct concern for the developer and tend to be imposed by the developers themselves. It is appropriate to introduce the development of software quality factor and metrics in a way that is consistent with its hardware counterpart; we distinguish the quality factors typically imposed by the user from those that are typically imposed by the developers themselves. Besides, since the user and the developer in the context of software components are both software developers, such distinction helps clarify the different perspectives of the two. Since certification may be required or desirable for software components, certification perspective is also important. We organize the quality factors in three categories, according to these three different perspectives: user, developer, and certifier.

Certifiability has not been a big issue for conventional software, but it could be a critical one for software components and their reuse. Although certifiability would make a useful quality factor, we have decided not to add such a quality factor to the existing ones for two reasons. First, to the best of the authors' knowledge, there has been virtually no research into the concept of certifiability, and no criteria or metrics have been defined for it. Second, it is closely related to the quality factor of verifiability, and one can expand the definition of verifiability to include certifiability as a special case.

We tailor the definitions for the commonly accepted quality factors for our particular context of software components. Although some factor names may appear in more than one category, the meaning may actually differ due to the difference in perspective. Note that the tailoring may be significant because of the additional complexities associated with software components. For example, the usability factor has been employed to refer to the ease of use and learning of a conventional software product. In the context of software components, usability refers to the ease of unpacking, configuration by the user, and assembly into a possibly wide variety of component-based application software systems; it also refers to the ease of required learning.

In the rest of this section, we discuss the quality factors of the first two categories: user and certifier. The third category of quality factors, from the developer's perspective, will be discussed in Section 13.4.3.

- *Quality factors from the perspective of the user of a software component:* Here, the "user" refers to all the users of a software component, in a collective sense, rather than any particular user. Table 13.2 summarizes the quality factors and their definitions. (Those quality factors tailored for software components are in italic.) Several sets of quality factors have been proposed and used in the literature. We base our selection on the work of J. A. McCall et al. [10]. However, we include several improvements based on ISO/IEC 9126-1—Standard for Software Engineering—Product Quality—Part 1: Quality Model [13]. See Section 15.1 for more details and references about that standard.

Table 13.2 Quality Factors: Perspective of the User of a Software Component

Quality Factor	*Definition*
Functionality	The extent to which a component satisfies its specifications and fulfills the stated or implied needs of the user
Efficiency	The amount of computing resources required by a component to fulfill its functions under stated conditions
Reliability	The extent to which a component can be expected to fulfill its functions for a stated period of time under stated conditions
Security (Integrity)	The extent to which access to a software component, a component-based software using the software component or the companion data by unauthorized persons can be controlled
Usability	The extent of ease to which a software component can be unpacked by possibly a variety of users, configured by these users for selecting the particular configurations that best satisfy the needs of these users (if such configurability is provided), and assembled by these users into the application environments of their component-based application software systems (this also includes understandability and ease of learning)
Portability	The extent to which a software component can be ported to a possibly wide variety of operational environments, including operating systems and hardware, and the amount of effort required for porting
Interoperability	The extent to which a software component can be assembled with a possibly wide variety of component-based software systems employing the component or with other software components
Verifiability	The effort required to verify, with or without access to the source code, architecture, design and the developers, that the software design and implementation satisfies the specifications of the software component (This goes beyond the testability, which refers to the effort required to ensure that it performs its intended function and performance, and, for software components, includes the verification of interface, assembly, porting, and certification requirements in the scope. Testability is a commonly accepted quality factor [10, 11]. Details about verifiability in the conventional SQA context can be found in [12]. This quality factor may include the concept of certifiability, which can be defined to be the extent to which a software component can be certified, the extent to which certification implies the quality of the software component, and the cost of such extents of certification.)
Maintainability	The effort required to replace a software component with a corrected version, to upgrade a current software component (of an operational component-based software system), and to migrate an existing software component from a current component-based software system to a new version of the system

- *Additional quality factors from the perspective of the certifier:* Verifiability (the extent to which a software component can be certified), the extent to which certification implies the quality of the software component, and the effort required for such extents of certification

Verification has been commonly defined as the process of ensuring that the requirements imposed by the previous phase have been satisfied by the current phase. Therefore, verification is a step that is required for all phases except for the very first phase, and it will be a common task for all the following phases. Additional requirements are developed in every phase, and test plans should be augmented in every phase to ensure completeness.

13.4.3 Internal-requirements phase

Because internal quality characteristics may be defined for either the product or the process, we group internal requirements in two separate categories. Specifications of these two categories of requirements together with verification and expansion of test plans constitute the four major tasks of this phase of internal requirements.

- Specification of product requirements:
  - Software component requirements regarding the eight user-requirement categories defined in Section 13.4.2;
  - Other component quality requirements:

    Quality factors to be addressed later;

    Quality criteria to be addressed later;

    Quality metrics to be addressed later;

    The importance measures of the quality factors, criteria, and metrics and possible trade-off among them.
- Specification of process requirements:
  - Component certification requirements:

    Those imposed by the user;

    Those imposed by the certifier.
  - Other requirements for the life cycle of a software component, to be detailed in the rest of the phases;
  - Other possible requirements, including possibly the many software process metrics proposed in the literature, to be addressed below.

- Verification: Traceability from the external requirements developed in the external-requirements phase;
- Expansion of test plans.

We now state the important quality factors with respect to the perspective of the developer. Note that some quality factors of importance to either the user or the certifier are also important to the developer; such quality factors are not repeated here unless their meanings differ under the developer perspective. Table 13.3 summarizes the additional quality factors considered important by the developer, in addition to those already identified for the user and certifier perspectives. Note that the definition of the maintainability factor here differs from that of the same factor defined for the user perspective in that it contains additional developer's concerns. Here, reusability continues to refer to the extent to which a software component can be reused in developing other software products, by the developer or by the user; it actually is not restricted to the extent to which a software component can be reused by a possibly wide variety of unknown users because of the developer perspective. Therefore, reusability actually overlaps with several user-oriented quality factors (e.g., usability, portability, and interoperability). Finally, these quality factors did not result directly from user expectations, but are of direct concern for the software producer. Although they may not be directly related to any customer expectation, higher quality in terms of these factors does eventually benefit the user, at least in the form of lower cost.

Table 13.3 Additional Quality Factors Considered Important by the Developer

Quality Factor	*Definition*
Maintainability	The effort required to locate and fix an error in an operational software component; the effort required to replace a software component with a corrected version, to upgrade a current software component (of an operational component-based software system), and to migrate an existing software component from a current component-based software system to a new version of the system
Expandability	The effort required to increase the capability of a software component
Flexibility	The effort required to modify an operational software component
Reusability	The extent to which a software component can be reused in developing component-based software systems, other software components or other software products in general
Interoperability	The effort required to couple a software component with other programs in general, not necessarily with component-based software systems employing the component or with other software components

These quality factors are important for any generic software components; other factors may be important for the particular software component under consideration.

Part of the modern quality assurance methodology is to refine the internal quality characteristics into specific measures through possibly multiple tiers. The software quality assurance community has accepted a quality hierarchy: quality factors, quality criteria, and quality metrics. The dozen or so commonly accepted quality factors have been refined into two dozen or so independent and more specific quality criteria; many quality metrics have subsequently been developed as measures of quality. These commonly accepted quality criteria include: traceability, completeness, consistency, accuracy, error tolerance, simplicity, modularity, generality, expandability, instrumentation, self-descriptiveness, execution efficiency, storage efficiency, access control, access audit, operability, training, communicativeness, software systems independence, machine independence, communications commonality, data commonality, and conciseness.

The relationships between these criteria and the dozen or so commonly accepted quality factors have also been defined. For software components, critical ones among these commonly accepted quality factors include reusability, usability, portability, interoperability, functionality (correctness), verifiability, and maintainability. The related quality criteria for these quality factors are listed in Table 13.4. Multiple quality metrics have been defined for each and every one of the quality criteria. The reader is referred to [11–14] for details. Quality metrics will be discussed in detail in Chapter 16.

13.4.4 Design phase

Software requirements describe the problem. Pfleeger [16] stated, "Design is the creative process of transforming the problem into a solution; the description of a solution is also called design." Major tasks of the design phase include:

- Design (see Table 13.5 for several important aspects of design).
- Verification (walk-throughs, peer reviews [21], management reviews, inspections, and possible certifier involvement):
    - Traceability from all the internal requirements developed in the internal-requirements phase;
    - Design review to ensure the:

      (Conventional)

Table 13.4 Quality Criteria Related to Critical Component Quality Factors

Quality Factor	*Related Quality Criteria*
Functionality (correctness)	Traceability, consistency, completeness
Reusability	Generality, modularity, software system independence, machine independence, self-descriptiveness (including document availability, accessibility, accuracy, completeness, and clarity)
Interoperability	Modularity, communications commonality (standards compliance), data commonality, machine independence, software system independence, system compatibility
Portability	Modularity, machine independence, software system independence, self-descriptiveness
Usability	Operability, training, communicativeness
Verifiability	Simplicity, modularity, self-descriptiveness
Maintainability	Consistency, simplicity, conciseness, modularity, self-descriptiveness

Table 13.5 Key Aspects of Design of Software Components

Design Aspect	*Definition*
Goal	Fulfillment of all internally set requirements and specifications developed in the internal-requirements phase; maximizing quality with respect to the commonly accepted and other generic or special quality factors, criteria, and metrics
Methodology	Varying with the application domain of the software component (e.g., signal processing, enterprise resource planning, human-computer interface)
Generic software design guidelines	For example, entity-relation diagram, normal forms of a relational database, object-oriented design, UML as a language for defining software components
Guidelines for designing software components	For example, compliance to a commercially available or in-house developed component model and the companion standards
Organizational structure	Ranging from the traditional development to team development [22] to pair programming [23]

Technical correctness of design for achieving functional requirements;

The feasibility for achieving the performance requirements;

Adherence to conventional formal methodologies (e.g., entity-relation diagram, normal forms of a relational database, object-oriented design, and so forth).

(Component-specific)

Adherence to the selected component model and the companion standards (e.g., EJB, COM+, CORBA);

The feasibility of the requirements for assembly into a variety of (component-based) application software systems;

The feasibility of the requirements for portability to a variety of operating systems and hardware;

The feasibility of the requirements for configurability by the user;

The feasibility of the requirements for customization by the user;

Clear and robust interface with the customers' (component-based) application software;

Facilitation of certification of the software component through a certifier;

Ease of understanding, for easy certification, evolution, maintenance, and so forth;

Satisfaction of these and other internal requirements developed in the internal-requirements phase.

- Expansion of test plans.

13.4.5 Implementation phase (or coding phase)

This phase implements the results of the design phase. We group the tasks in two categories: conventional tasks (with difference in degree for software components) and tasks peculiar to software components, followed by verification and expansion of test plans.

- Conventional (with difference in degree for software components):
  - Fulfilling the design specified in the design phase;
  - Adherence to generic guideline for coding;
  - Coding to satisfy quality criteria: maximizing quality with respect to the commonly accepted and other generic or special quality criteria;
  - Organizational structure for coding: ranging from the traditional coding organization to team coding (e.g., "pair programming" for a module).
- Software component–specific:

    - Adherence to rules and guidelines specified by the selected component model, (e.g., COM+);
    - Facilitation of certification;
    - Ease of understanding, for easy component certification, evolution, and maintenance.
- Verification of the fulfillment of the design developed in the design phase through code walk-throughs, peer code reviews, and management code reviews, with possible involvement from the certifier;
- Expansion of test plans.

13.4.6 Test, evaluation, and validation phase

This phase implements the test plans generated so far.

- Test the software component against the component requirements, including:
    - Individual modules of the software component (i.e., unit testing);
    - The software component as a whole regarding the eight user-requirement categories defined in Section 13.4.2.
- Evaluate:
    - The software component with respect to all other quality requirements (e.g., the commonly accepted quality criteria);
    - Software development process with respect to:

      Process quality criteria and metrics;

      The certification requirements.

The software component as a product is validated when all the steps have been verified, all the test plans have been implemented, all the problems identified in the testing process have been corrected, all the quality requirements are satisfied, all the certification requirements are satisfied, and all the process requirements are also met. Test coverage and adequacy has been discussed in Chapters 7 and 9 of this book.

13.4.7 Deployment phase

In the deployment phase, it is critical to provide clear documentation for or about:

- Component description, particularly the eight user-requirement categories defined in Section 13.4.2;
- Component quality:
  - Product quality metrics;
  - Process quality metrics;
  - Results of verification for all steps;
  - Tests planned, tests performed and test results;
  - Certification process and results.
- Adoption procedure;
- Anticipated upgrades of the software component.

Documents on component quality should be part of the deliverables to the user, to be delivered by the SQA personnel. The rest should be delivered by the software developers.

13.4.8 Postdeployment phase

The postdeployment phase of a software component involves at least the following three important tasks.

- *Monitoring.* Data collection and analysis on successes or failures: The vendor of the software component may be required to share the collected data on a real-time basis with the certifier and/or all those customers that have purchased the component. It may be requested and willing to share with prospective customers the data, including:
  - Exception reports on external requirements: functionality, performance, interface, ease of assembly into (component-based) application software systems, portability to operating systems and hardware, and other quality factors, criteria, and metrics;
  - Confirmation of satisfaction of these external requirements.
- *Correction of error.* After receipt of a failure report and upon confirmation of the failure, the vendor may be obligated to correct the failure and

distribute a free corrected version to all the customers with the necessary documentation regarding the problem, the corrected component, and the replacement procedure, possibly with a new set of documents.

- *Upgrade.* Software components may evolve and can be improved to provide better performance, more functionality, or higher quality in other ways without changing the interface, assembly, and portability requirements. Component-based application software systems assembled with the software component may evolve too, as well as the operating systems and the hardware on which they run. The developer of software components should anticipate such evolution and provide appropriate upgrades to accommodate it.

13.5 Summary

While a vast majority of HQA is devoted to coping with the innate variability of the manufacturing process or the inevitable product wear and tear, these issues are of no concern to SQA. On the other hand, typical HQA personnel lacks the domain knowledge required in the design phase, and therefore, proper design for quality assurance has received little attention in the HQA community. We believe that the design and coding phases of software development must become focal areas for SQA if quality assurance is expected to make a contribution to software quality that is as significant as the contribution made by quality assurance to hardware quality.

This chapter has been focused on the classical approach of SQA, which is predominantly process oriented. Motivated by the inadequacy of the process orientation, recent research efforts produced product-oriented approaches to SQA and companion software product quality models. With the help of these approaches and models, SQA has begun to penetrate into the design and coding phases. They will be discussed in the next chapter.

A major reason for the success of HQA is an extensive use of quality measures. Such an extensive use is enabled by the fact that hardware quality can often be clearly defined, precisely measured, and accurately estimated. However, software quality in general is not as amenable to clear definition, precise measurement, and accurate estimation because again of the differences between HQA and SQA just reiterated. We will discuss major differences between hardware quality measurement and software quality measurement in Chapter 16, with a simple example. Because of the quantitative nature of measurement, software measurement requires a rigorous approach. In Chapter 16, we will discuss, among other subjects, a framework for software measurement developed by Fenton [5] and we will also propose an extension to it.

References

[1] IEEE, Standard Glossary of Software Engineering Terminology (IEEE Std 610.12-1990), Software Engineering Technical Committee, IEEE Computer Society, Los Alamitos, CA.

[2] Jalote, P., *An Integrated Approach to Software Engineering*, New Delhi, India: Narosa Publishing House, 1991.

[3] Dromey, R. G., "A Model for Software Product Quality," *IEEE Trans. on Software Engineering*, Vol. 21, No. 2, February 1995, pp. 146–162.

[4] Bansiya, J., and C. Davis, "A Hierarchical Model for Object-Oriented Design Quality Assessment," *IEEE Trans. on Software Engineering*, Vol. 28, No. 1, January 2002, pp. 4–17.

[5] Fenton, N. E., *Software Metrics: A Rigorous Approach*, New York: Chapman and Hall, 1991.

[6] Ginac, F. P., *Customer Oriented Software Quality Assurance*, Upper Saddle River, NJ: Prentice Hall, 1998.

[7] Dunn, R., and R. Ullman, *Quality Assurance for Computer Software*, New York: McGraw-Hill, 1982.

[8] McCabe, T. J., "A Complexity Measure," *IEEE Trans. on Software Engineering*, Vol. 2, No. 4, 1976, pp. 308–320.

[9] Halstead, M. H., *Elements of Software Science*, New York: Elsevier North-Holland, 1977.

[10] McCall, J. A., P. K. Richards, and G. F. Walters, *Factors in Software Quality Assurance, Vol. I*, RADC-TR-77-369 (Rome Air Development Center), U.S. Department of Commerce, November 1977.

[11] Vincent, J., A. Waters, and J. Sinclair, *Software Quality Assurance—Volume 1: Practice and Implementation*, Englewood Cliffs, NJ: Prentice Hall, 1988.

[12] Evans, M., and J. Marciniak, *Software Quality Assurance and Management*, New York: John Wiley, 1987.

[13] International Organization for Standardization (ISO), Standard for Software Engineering—Product Quality—Part 1—Quality Model (ISO/IEC 9126-1: 2001), Geneva, Switzerland, 2001.

[14] IEEE, Proc. of 8th International Software Metrics Symposium, Technical Council on Software Engineering, (sponsored by IEEE Computer Society, Los Alamitos, CA), Ottawa, Canada, June 4–7, 2002.

[15] Heineman, G. T., and W. T. Councill, (eds.), *Component-Based Software Engineering: Putting the Pieces Together*, Reading, MA: Addison-Wesley, 2001.

[16] Pfleeger, S. L., *Software Engineering: Theory and Practice*, 2nd ed., Upper Saddle River, NJ: Prentice Hall, 2001.

[17] Gilb, T., *Principles of Software Engineering Management*, Reading, MA: Addison-Wesley, 1988.

[18] Musa, J., *Software Reliability Engineering: More Reliable Software, Faster Development and Testing*, New York: McGraw-Hill, 1998.

[19] Griss, M. L., "Software Reuse: Objects and Frameworks Are Not Enough," *Object Magazine*, February 1995, pp. 77–87.

[20] Kang, K. C., "Features Analysis: An Approach to Domain Analysis," *Proc. of Reuse in Practice Workshop*, Software Engineering Institute, Pittsburgh, PA, July 1989.

[21] Wiegers, K. E., *Peer Reviews in Software: A Practical Guide*, Reading, MA: Addison-Wesley, 2002.

[22] Humphrey, W. S., *Introduction to the Team Software Process*, Reading, MA: Addison-Wesley, 2000.

[23] Beck, K., *eXtreme Programming Explained: Embrace Change*, Reading, MA: Addison-Wesley, 2000.

CHAPTER

14

Contents

Quality assurance for component-based software systems

This chapter addresses quality assurance for *component-based software systems* (CBSS). This abbreviation has also been used to refer to COTS-based software systems. Since the focus of this part of the book is on COTS components, this dual use of terminology should not cause any confusion. Section 14.1 reviews the salient features of CBSS and identifies the unique issues associated with QA for CBSS. It also justifies the rest of the sections. Near the end of this chapter, we describe an integrated process for the development and quality assurance of CBSS in Section 14.8. The sections in between are devoted to addressing the issues identified in Section 14.1 or to describing the details of the steps of the integrated process to be summarized in Section 14.8. They serve three main purposes: (1) describe the additional tasks of prerequirements evaluation, adaptation/assembly, and validation of software components; (2) describe a general product-oriented approach to SQA proposed recently that complements the conventional process-oriented approach to SQA and, together with the conventional SQA, form a more complete QA methodology for computer software; and (3) describe an extension to the recently proposed product-oriented SQA approach to the specific design and implementation paradigm of object-orientation. Concluding remarks are given in Section 14.9.

It was pointed out in Chapter 13 that in order for SQA to fulfill its potential, it must go beyond the traditional process orientation, penetrate deep into the design and implementation phases of the software development process, and link the desired high-level quality attributes to the detailed rules of design and implementation. It should be clear by now that the conventional SQA process summarized in Chapter 13 does not provide the penetration and linkage. It is therefore imperative to point out the limitations of the conventional SQA approach and to introduce the new product-oriented SQA approach. One of the salient features of CBSS is that object-orientation facilitates component development and reuse, although it is certainly not necessary. The facilitation is evidenced by the fact that all three commercial component models are object oriented. Therefore, CBSS QA must include a detailed discussion of QA for object-oriented software.

14.1 Main issues in quality assurance for component-based software systems

We first discuss the salient features of the development of CBSS, and then use them to discuss the main differences between QA for CBSS and conventional SQA. We in turn use these main differences to discuss the main issues of CBSS QA.

14.1.1 Salient features of the development of component-based software systems

We partition the salient features into two categories: development process and design/implementation paradigm. We first state the features and then explain some of them in more detail.

14.1.1.1 Salient features in the development process

The salient features in the development process include:

- *Prerequirements analysis—selection and implementation of a component model:* This selection process is usually conducted across different specific CBSS products. The three commercial component models are not compatible with one another. Although some argue that they can be "bridged," such bridging may not be easy. Therefore, this selection will likely impose architecture constraint on CBSS development.
- *A "bottom-up" and spiral requirement process converging to selected components and system requirements:* This feature is explained in more detail later.
- *Unpacking, adaptation, and assembly of components.*

- *Postassembly validation of components, including integration tests.*
- *Validation of CBSS.*

A distinct feature of CBSS development is the "bottom-up" and spiral requirement process converging to selected components and system requirements. The development of conventional software systems typically follows a waterfall model or its more complex variations and begins with specification of requirements. However, CBSS development necessitates many additional activities. Rather than the usual top-down approach of first defining functional and nonfunctional (including quality) requirements for the system followed by defining module functionality, a major distinct activity is the identification and evaluation of available software components in the market that may be useful for the CBSS being developed. The accommodation of existing software components, including not just the software code but also the design, architecture, and even business processes, in requirements engineering goes against the sequential or spiral top-down order of the conventional development process and hence gives rise to the term "bottom-up requirements engineering." K. C. Wallnau et al. [1] referred to this phenomenon as requirements engineering under the "market regime." In fact, the requirement process is a spiral one containing converging iterations of balancing the:

- Stakeholder (including user, developer, and so forth) needs;
- The component availability;
- The architecture/design constraints;
- Managerial considerations, including cost, schedule, and risk.

At the end of the requirements engineering process, components will be selected and system requirements specified. Note again that the process may lead to possible modification of the user's business processes.

14.1.1.2 Salient features in the design and implementation paradigm

A salient feature in the design and implementation paradigm is the object orientation of the component models. All three commercial component models are object oriented. This is a critical feature in practice, although object orientation is not required for building CBSS. Therefore, it is particularly relevant in this chapter to discuss building quality into object-oriented software systems.

14.1.2 The main differences between SQA for conventional software systems and SQA for component-based software systems

The differences in the development process will be used to drive the identification of main issues for QA for CBSS.

14.1.2.1 The differences in SQA resulting from the differences in the development process

A bottom-up and spiral SQA process is required. The conventional SQA process no longer works, which is designed for the conventional top-down software development process. This chapter will culminate in an integrated development and QA process for CBSS, to be summarized in Section 14.8. The differences and the corresponding sections addressing them are itemized here:

- Prerequirements evaluation of components (Section 14.2). Particularly relevant is the task of assessing the quality of candidate components with respect to the system being built and the selected component model.
- Ensuring a quality adaptation and assembly process (Section 14.3).
- Validation of the quality of the components, now that the components can be tested as part of the assembled system (Section 14.4).
- A "bottom-up" SQA approach (Sections 14.5 and 14.6). The conventional SQA process is top-down in another sense. It starts with defining high-level quality factors, criteria, and metrics but offers little guidance for detailed software design and implementation. We will discuss and summarize a recent bottom-up SQA approach that complements the conventional top-down SQA process by specifying explicitly the detailed design and implementation rules and linking them explicitly with desired high-level quality attributes.
- Ensuring the quality of the whole system, with components built by a third party and without access to the source code (Section 14.8).

14.1.2.2 The difference in SQA resulting from the predominant object-oriented design and implementation paradigm

The difference in SQA resulting from the predominant object-oriented design and implementation paradigm is QA for object-oriented software (Section 14.7).

14.2 Evaluation of software components

In this section, we first address criteria for evaluating software components based on the discussion in the previous section and Chapter 13. Several evaluation processes for software components have been proposed recently, with different degrees of concreteness and specificity. We discuss some of them in the following sections.

14.2.1 Component evaluation criteria

Although both the developer and the user of a CBSS will be affected by the selected components, the developer should serve as an agent on behalf of the user. Therefore, we focus on only the issues facing the developer—the direct user of components. The direct user cares about the same eight categories of user requirements for software components as those listed in Section 13.4.2. However, the quality perspective of such a particular user is simpler and more concrete than that of the component vendor because the particular user only needs to verify the component against its own possible requirements, instead of a variety of possible requirements desired by various possible users unknown to the vendor. Factors to be considered for component evaluation are summarized in Table 14.1.

14.2.2 Multicriteria selection

It is clear that component selection involves a large number of selection criteria, some of which are conflicting. Selecting from multiple alternatives according to multiple criteria is a very common problem facing decision-makers in practically all professions. Many formal methods are available, but they are not easy to implement. C. Ncube and J. C. Dean [2] considered *multiattribute utility theory* (MAUT), *analytic hierarchy process* (AHP), and *multicriteria decision aid* (MCDA), and argued their shortcomings for evaluating software components. They also proposed an alternative method. This is a very difficult issue; more research is definitely needed to resolve it.

Although MAUT is considered to be the one with the most solid theoretical foundation, AHP seems to be a pragmatic compromise between theoretical rigor and ease of use. See T. Saaty [3] for details about AHP. The most common way of selection with multiple criteria remains to be through simple weight assignment. A main thesis of Saaty's AHP is that assigning weights to a large number of criteria simultaneously is difficult, and as a result, the assigned weights may not be valid. A main feature of AHP is that it calls for only pairwise comparisons and requires, for each possible pair, weight assignment to only two competing alternatives.

Table 14.1 Factors for Consideration in Component Evaluation

Factor Category	*Factors*
User requirements	Functionality
	Performance (efficiency, reliability, and so forth)
	Interface and architecture
	Adaptation or direct assembly into to the application software system being built after unpacking and configuration by the user (interoperability)
	Porting to the operational environment supporting the CBSS, including the operating system and hardware (portability)
	Standards
	Certification (verifiability)
	Documentation
Other quality factors	Usability
	Maintainability
	Security
Third-party code development	Ongoing use of the business processes supported by the COTS components, rather than the organization's own business processes
	Product direction
	Commitment to future software quality
Programmatic constraints	Project schedule
	Available expertise to use, integrate, and maintain the software components
	Cost

14.2.3 A specific process and technique for evaluating software components

K. Ballurio et al. [4] proposed a technique called *Base Application Software Integration System* (BASIS) for selecting COTS software, including a process and a set of explicit criteria and metrics. BASIS consists of three components: component evaluation process, vendor viability evaluation, and assessment of difficulty of integration. We briefly discuss the general approach of BASIS as follows.

The *component evaluation process* (CEP) consists of five high-level activities:

- Determine evaluation scope (set expectations for the level of required effort);
- Search and screen candidates;

- Define evaluation criteria;
- Evaluate component alternatives;
- Analyze evaluation results.

Based on the results of CEP, a CEP value is assessed. The scope of vendor viability evaluation includes:

- Financial viability;
- Market stability;
- Management viability;
- Research and development viability;
- Product support, including responsiveness to requests for change.

Based on the result of vendor viability evaluation, a *vendor viability* (VV) value is assessed for each component candidate.

A component's degree of difficulty of integration with a CBSS hinges upon the interface complexity, the degree of interface mismatch, and the degree of difficulty in resolving the mismatches. They proposed a measure for *difficulty of integration* (DOI).

Based on the CEP, VV value, and DOI values, and three weights assigned for the relative importance of the three values, an overall BASIS Factor is calculated to measure integration risk. Higher values of the BASIS Factor indicate higher risk levels. The overall BASIS Factor can be used to compare candidate components for the same function; choose the candidates with lower values. It can also be used to sequence the integration of different components selected for integration with a CBSS to reduce the risk; components with higher BASIS Factor values should be integrated sooner. The definitions of CEP, VV, and DOI measures and the BASIS Factor involve a number of metrics (proposed by themselves or other researchers), formulae, and constants; more research may be needed to justify or modify the formulae and constants. The reader is referred to [4] for details.

14.2.4 A general process for evaluating software components

S. Comella-Dorda et al. [5] proposed a general process for evaluating software components, based on ISO 14598 [6]. The process consists of four major steps: planning the evaluation, establishing the criteria, collecting the data, and analyzing the data. The focus is on the methodology (e.g., how to construct evaluation criteria), rather than on suggesting concrete requirements or criteria for evaluating a software component.

14.3 Ensuring a quality adaptation and integration process

This section first discusses major strategies for adapting or integrating a software component with a software system. It then addresses the issue of sequencing multiple software components for sequential integration into a software system.

K. Bergner et al. [7] summarized five adaptation strategies and used the implementation of a product model browser to illustrate the five strategies. Note that it is possible that a software component may be designed with "out-of-the-box" readiness for assembly. We deal with the situation where some adaptation effort is required. The five strategies are:

- *Adaptation interface:* Many generic components are equipped with functionalities and interfaces for use in multiple specific and popular contexts, in addition to the generic functionality and interface. Such an arrangement enables configurability-by-the-user. This involves only nominal adaptation effort.
- *Wrapper:* This entails embedding an acquired software component in a customized component through the custom code serving only as a conduit for delivering the required input or support services to the acquired component and delivering the output or functionality from the acquired component. The acquired component requires certain services from its environment in order to provide the functionality. The customized component arranges for those services and delivers the functionality through its own interface with the environment. Because of the wrapping, the acquired component is not "visible" to the environment in the sense that there is no direct interface between the environment and the acquired component.
- *Composition with adaptor component:* This strategy differs from the wrapper strategy in that the environment can interact directly with the acquired component, but an adaptor component is added to facilitate particular uses of the acquired component.
- *Inheritance:* All the strategies discussed so far deal mainly with interface issues, not with functionality. In some cases, however, the functionality provided by the acquired component is not adequate, and new functionality is needed. Inheritance allows the addition of new functionalities and interfaces to the environment in order to implement an adequate set of functionalities for the environment.

- *Reimplementation:* If the source code is available, this strategy calls for drastic change of the original implementation. Some parts of the original code can be reused for the new implementation.

The degree of complexity of their implementation varies, and so does the complexity of testing. Quality assurance for the first three strategies is accomplished primarily by testing. The quality of the interface portion of the last two can be assured by testing, while that of the additional design and coding can be assured by following the process-oriented SQA methodology already described in Chapter 13 and the product-oriented SQA approach to be described later in this chapter.

When multiple components are to be integrated with a CBSS, the BASIS Factor developed by Ballurio et al. [4] and summarized in Section 14.2 can be used to determine the integration sequence. As discussed there, the BASIS Factor is a measure of integration risk, and therefore, the components with higher BASIS Factors should be integrated earlier.

T. G. Baker [8] classified component-system coupling at four levels and offered some lessons learned at the Boeing company about integrating software components into software systems. The adaptation strategies discussed above involve different levels of coupling. In general, the lower level of coupling is better. It is cheaper, less risky, and more flexible for future changes [8].

14.4 Validation of the quality of software components

Testing should be performed to validate the eight categories of user requirements and other quality factors like usability, maintainability, and security, as explicitly stated as part of the component selection criteria discussed in Section 14.2. The scope of testing is similar to that associated with testing components by the vendor, except that in this context the testing has much more specific and concrete use cases, requirements, application software environments, operating systems, and hardware platforms. Also, the likely absence of the source code is another difference.

In addition to the testing conducted by the developer of the system, the assessment of component quality can be supplemented by (1) the background information provided by the vendor, including the quality plan and its implementation, test plans, and test results, (2) information provided by possible certifiers, and (3) information provided by other customers of the same software component. The importance of information regarding how other customers of the same software component think of the quality of the software component is asserted repeatedly in the literature and should not be underestimated.

14.5 Several major limitations of the conventional SQA process

As already argued in Chapter 13, unlike hardware quality assurance, software quality assurance must penetrate deep into the design process. However, very much like HQA, the conventional SQA approaches address little, if at all, design and implementation. Recently, the effectiveness of the conventional SQA approaches has been examined by many researchers (e.g., R. Howe et al. [9] and B. Kitchenham and S. L. Pfleeger [10]). Chapter 13 discussed the conventional approaches to software quality assurance. This section discusses the major limitations of the conventional SQA approaches. Some promising approaches to SQA in general have been proposed recently. One of the new approaches has been extended for developing quality object-oriented software. These will be the subjects of the next two sections.

Chapter 13 summarized the classical work on SQA proposed by J. McCall et al. [11]. Their approach evolved into an international standard: ISO 9126. Although the two are significantly different in several aspects, they share a common approach. Despite the tremendous contribution made by McCall et al.'s seminal work and by ISO 9126, this common approach suffers from some shortcomings.

The fundamental problem is the approach's inability to provide explicit guidance to software developers in their design, implementation, and inspection, and hence, there is no explicit linkage from the high-level quality attributes to design, implementation, and inspection rules. The approach works well at the higher-level steps (e.g., the requirement process). It points out the importance of quality factors, criteria, and metrics (in the language of McCall et al.) and some quality characteristics or subcharacteristics (in the language of ISO 9126), but it offers only some vague guidelines for achieving these quality attributes through good design and implementation. The recent approaches have been motivated by this inability, and some are even able to help with the inadequacy. The emphasis of the recent approaches on providing clear guidance on design and implementation, particularly when contrasted with the conventional SQA approaches' inability to reach these lower levels of the development activities, perhaps motivated the reputation of their being "bottom-up" approaches. (This is different from the bottom-up requirements engineering discussed in Section 14.1.)

More precisely, major limitations of the conventional SQA approaches include:

- Absence of clear guidelines for measuring product quality characteristics;
- No clear relationship between the existing metrics and quality characteristics;
- Over-reliance on process control.

Kitchenham and Pfleeger [10] stated, "The standard recommends measuring quality characteristics directly, but does not indicate clearly how to do it. Rather, the standard suggests that if the characteristic cannot be measured directly (particularly during development), some other related attribute should be measured as a surrogate to predict the required characteristic. However, no guidelines for establishing a good prediction system are provided." This lack of guidance has led to a small number of useful metrics that can actually help the designer or the programmer build quality into the software at the corresponding stages.

Kitchenham and Pfleeger [10] also stated, "No means for verifying that the chosen metrics affect the observed behavior of a [quality] factor." This lack of relationship is evidenced by the fact that few quality metrics have actually been in use. (See S. H. Kan [12]; this will be discussed in more detail in Chapter 16.) Again, even the available metrics cannot provide explicit guidance for design or coding.

Perhaps because of the difficulties in linking the high-level quality attributes to the design and implementation rules, much attention has turned to the development process and the use of process control to prevent poor quality. Again, this route did not provide concrete guidance to design and implementation, and it provided no direct link between quality and design and implementation. N. Fenton [13] argued that software standards overemphasize process, and many standards are not actually standards. Kitchenham and Pfleeger [10] stated, "There is little evidence that conformance to process standards guarantees good products."

14.6 A complementary bottom-up approach for software quality assurance

Few software professionals question the value of SQA processes, but it is time that the community understands the limitations of the conventional SQA approach, particularly for software components, and move forward. Several researchers have heeded the continued inability of conventional SQA to penetrate into software design and implementation, and have developed new approaches to SQA [14–17]. Most notable is the work of R. G. Dromey [14, 15] and an important extension of his work to object-orientation by J. Bansiya and C. Davis [16, 18, 19]. Before describing Dromey's work, we first briefly describe some additional background against which these new approaches were developed. After describing Dromey's work, we briefly discuss the work of I. Tervonen [17], which also attempts to link high-level quality attributes with design and implementation rules.

Many researchers argue that software engineering lacks concrete discipline, especially when compared to other engineering disciplines; some even question if it should be regarded as an engineering discipline at all [20]. There is certainly some truth to these observations. However, over the past two decades, more and more software development methods and paradigms have emerged. On the data side, the method of the entity-relationship diagram, the relational database, and the norm forms have contributed to a body of rules for data modeling. Some even credited the advent of the entity-relationship diagram as the key step toward forming the discipline of information engineering. The object orientation bridged the gap between data and procedures and incorporated a number of desirable design and implementation techniques in the form of clear and rigid constructs. Some of these constructs were actually motivated by quality issues associated with earlier paradigms; quality metrics developed for object orientation finally began to link the top-level quality attributes to the bottom-level design and implementation rules [21–23].

The object orientation actually facilitated the development of software components and component-based software systems. Recently, with the apparent successes of the seven-layer *open system interconnection* (OSI) reference model for communication and the six-layer VLSI architecture for microelectronics, the need for a software architecture becomes clearer and more urgent [15, 24]. The point is that through the advances of methods and standards, more and more discipline has been accumulated for software engineering. Perhaps more importantly, the growing body of discipline spanning the life cycle of software development has provided a timely opportunity for developing new approaches to SQA that can begin to penetrate from the top-level quality factors into the bottom-level design and implementation rules. This may have sparked the recent movement toward a new bottom-up SQA paradigm.

14.6.1 Dromey's method of quality-carrying properties

In Dromey's view [14, 15], a product is composed of elements, and product quality is largely determined by:

- The choice of elements that make up the product and how they are implemented;
- The tangible properties of the individual elements;
- The tangible properties associated with element composition.

Note that Dromey [14, 15] uses the term "component" rather than "element." To avoid possible confusion with the principal subject of "software component" of this book, we use the term "element" instead to substitute for Dromey's "component."

Dromey [14, 15] also developed a generic product quality model consisting of three principal components:

- A set of high-level quality attributes;
- Product properties (including properties of individual elements and properties of element composition) that influence quality;
- A means of linking the high-level quality attributes and the product properties.

The linkage is made by the concept of "quality-carrying property." This generic model can be applied to the implementation process, the design process, and the requirement process of the software development life cycle.

For each such application, Dromey provided a five-step procedure for constructing the product quality model:

1. Identify a set of high-level quality attributes for the product.
2. Identify the product elements.
3. Identify and classify the most significant, tangible, quality-carrying properties for each product element.
4. Propose a set of axioms for linking product properties to quality attributes.
5. Evaluate the model, identify its weaknesses, and either refine it or scrap it and start again.

Dromey [15] cited N. Wirth's statement, "In programming, the devil hides in the detail," and dealt first with implementation. He developed an implementation quality model, followed by a requirement quality model and a design quality model. We briefly describe the implementation quality model to provide a flavor of his approach, and we refer the reader to his papers for the details of the implementation quality model [14] and the other two quality models [15]. The first four of the five steps for developing the implementation quality model are briefly described next. Note again that in this description, the term "product" refers to implementation (i.e., coding or programming).

1. *Identify quality attributes:* ISO 9126 [25] provides six high-level quality attributes: functionality, reliability, efficiency, usability, maintainability, and portability. For software components, reusability should be added; Dromey used a set of rules to define code reusability [14]. He stated, "A structural form is reusable if it uses standard language features, it contains no machine dependencies and it implements a single, well-defined, encapsulated and precisely specified function whose computations are all fully adjustable and use no global variables or side-effects.

   All ranges associated with computations and data structures in a reusable module should have both their lower and upper bounds parameterized. Also no variable is assigned to a number or any other fixed constant and all constants used should be declared."

   (In ISO 9126, understandability is a subattribute of maintainability. This is appropriate for a software system. However, understandability is critical for software components and may need to be elevated to a top-level attribute. Also, understandability may also be needed as a top-level quality attribute for the requirements quality model because the requirements must be made clearly understandable for their readers.)

2. *Identify product elements:* It is relatively easy to identify product elements (i.e., implementation or coding elements in this implementation quality model) because the programming language's syntax identifies them. Dromey addressed generic programming languages. (Tervonen's GRCM [17] was developed for design, not for implementation, i.e., coding. Bansiya and Davis [16, 18, 19] dealt with object-oriented design, not implementation.) Dromey [14] grouped implementation elements into two categories: computation elements

(e.g., loops, if-statements, guards, assignments) and expression and data elements (e.g., variables, constants, and types).

3. *Identify quality-carrying product properties:* Developing a set of quality-carrying properties for each implementation element (i.e., coding element) and for element composition is an empirical process. Dromey classified them into four categories: correctness properties (minimal generic requirements for correctness), structural properties (low-level, intramodule design issues), modularity properties (high-level, inter-module design issues), and descriptive properties (various forms of specification/documentation). Like many other classification schemes, this one is not immune from ambiguity among the categories. To avoid classification difficulties, these four categories are given and to be used in order of precedence.

4. *Link quality-carrying product properties to quality attributes:* Dromey summarized a large number of quality-carrying properties for implementation elements, including eight correctness properties, nine structural properties, six modularity properties, and three descriptive properties. He also identified the applicable implementation elements and the quality attributes impacted and provided sample defects. For example, in the category of correctness properties, the property "Assigned" is defined, and "A variable is assigned if it receives a value either by assignment, input or parameter assignment prior to its use." It is applicable to the implementation element "Variable" and has quality impact on "Functionality" and "Reliability." A sample defect would be "Unassigned." In the category of modularity properties, the property of "Loosely Coupled" is defined, and "A module or a program is loosely coupled if all module calls are data-coupled to the calling program/module." This property is applicable to "module calls" and has impact on quality attributes "Maintainability, Reusability, Portability and Reliability." Sample defects include "control-coupled, stamp-coupled, content-coupled, common-coupled, externally coupled." (See G. Myers [26] for details.) Note that loose coupling is particularly important for developing software components and component-based software systems.

The quality attributes, implementation elements, and quality-carrying properties of the implementation elements are summarized in Table 14.2.

There exists a many-to-many relationship between the implementation elements (i.e., the structural forms) and the quality-carrying properties of implementation elements. There also exists a many-to-many relationship

Table 14.2 Quality Attributes, Product Elements, and Quality-Carrying Properties for Dromey's Quality Model for Implementation

Implementation Elements (Structural Forms)	*Quality-Carrying Properties of Implementation Elements*	*Quality Attributes*
PROCESS	CORRECTNESS PROPERTIES	Functionality
System (set of programs)	Computable	Reliability
Library (set of reusable ADTs, function and procedures)	Complete	Usability
Meta-program (e.g., shell script using program I/O)	Assigned	Efficiency
Program	Precise	Maintainability
User interface	Initialized	Portability
Objects (ADT)	Progressive	Reusability
Module (encompassing functions and procedures)	Variant	
Sequence	Consistent	
Statement	STRUCTRUAL PROPERTIES	
Loop	Structured	
Selection	Resolved	
Function/procedure call	Homogeneous	
Assignment	Effective	
Guard	Nonredundant	
Expression	Direct	
DATA	Adjustable	
Records	Range-independent	
Variables	Utilized	
Constants	MODULARITY PROPERTIES	
Types	Parameterized	
	Loosely coupled	
	Encapsulated	
	Cohesive	
	Generic	
	Abstract	
	DESCRIPTIVE PROPERTIES	
	Specified	
	Documented	
	Self-descriptive	

between the quality-carrying properties of implementation and the quality attributes. (See [14] for details about the two many-to-many relationships.) Combining these two many-to-many relationships produces another many-to-many relationships relating the implementation elements directly to the quality attributes and vice versa. This third many-to-many relationship provides guidance to programmers during implementation for higher software quality.

In addition to this implementation quality model, Dromey also applied his product quality model to requirements engineering and software design and developed a requirements quality model and a design quality model. These two models were also discussed in [15]. Finally, SQA began to penetrate into software design and implementation in a systematic way.

This approach helps achieve two very important goals for SQA:

- *Building quality into software:* This reduces to systematically ensuring that all quality-carrying properties are satisfied.
- *Systematic detection and classification of software quality defects:* This reduces to systematically checking whether any of the quality-carrying properties that imply high-level quality attributes is violated.

Dromey [14, 15] provided much more detail on the new approach, and the reader is encouraged to read the two papers. Dromey's design quality model addresses generic design methodology; in the next section, we will discuss an extension tailored for object-oriented software design.

Software is often classified with respect to some of their key features (e.g., safety critical systems, commercial-off-the-shelf software components). The relative and absolute importance of the high-level quality attributes may vary accordingly. Such importance can be translated into appropriate requirements for the high-level quality attributes. Through Dromey's approach, one can trace these quality attributes deeply into the product elements that possess the corresponding quality-carrying properties and hence ensure the product quality by strictly following the rules during development and by inspecting the product for possible violation of the rules.

Some of these quality-carrying properties are well known prior to Dromey's work, but Dromey was able to *link* these properties with high-level quality attributes in a *systematic* fashion. Note that Dromey's methodology is equally applicable to developing software components. When applied to software components or component-based software systems, additional steps must be taken. (See the integrated development and QA process for CBSS to be discussed in Section 14.8.) Apparently, much more research in this direction is needed.

14.6.2 Tervonen's goal-rule-checklist-metric model

I. Tervonen [17] proposed a *goal-rule-checklist-metric* (GRCM) model to link top-level quality factors to the bottom-level design rules. The rules not only are used as design discipline but also constitute parts of inspection checklists to be used during inspection and other quality assurance activities. Perhaps the methodology is best illustrated with a simple example of the GRCM thread provided by Tervonen [17]. In his terminology, *goal* is synonymous with the "quality factor" of McCall's approach and the "quality characteristic" of ISE 9126; *rule* is synonymous with the "quality criteria" of McCall's approach and the "quality subcharacteristic" of ISO 9126. One of the top-level goals is expandability, which is translated into several rules at the next level (e.g., readability, upgradability, and so forth). Let us focus on the readability rule, which is defined as "aiming at understandable class structure, method interfaces, and methods." Multiple checklists may be defined for the readability rule [e.g., "Checklist Rea. 1: Check that the number of methods in a class is less than 20" and "Checklist Rea. 2: Check that the class hierarchy nesting level is less than 6 (counted in frameworks from the bottom)"]. The metrics used in these two cases are *number of instance methods* (NIM) and *hierarchy nesting level* (HNL), respectively [23]. Note that a checklist may involve multiple metrics. This model provides concrete guidance for software design and implementation as well as for the corresponding inspections as part of SQA.

14.7 Building quality into software: the case of object-oriented design

Based on Dromey's software product quality model for SQA, Bansiya and Davis [16] developed a hierarchical model for assessing the quality of the design of object-oriented software. Before discussing the work of Bansiya and Davis, we compare at a high level their work with the methods we already covered in this chapter and in Chapter 13, including McCall's work [11], ISO 9126 [25], and the GRCM model [17]. All previous methods accommodate software quality attributes in two or three layers. McCall's method organizes them as quality factors in the first layer and quality criteria in the second layer; ISO 9126 organizes them as quality characteristics in the first layer and quality subcharacteristics in the second or the third layer; and the GRCM model organizes them as the goals in the first layer and the rules in the second layer. Bansiya and Davis used only one layer for software quality attributes. However, because of their explicit focus on object orientation, they are able to identify a set of desirable object-oriented design properties and link them directly to the single layer of software quality attributes. Bansiya and

Davis stated, "Design properties are tangible concepts that can be directly assessed by examining the internal and external structure, relationship, and functionality of the design components, attributes, methods and classes." Moreover, a primary goal of their method is that each of the design properties represents an attribute or characteristic of a design that is sufficiently well defined to be objectively assessed by using one or more well-defined design metrics during the design phase. The design properties can be viewed as a second layer of quality attributes, but only for object-oriented software design. Together with a third layer of design metrics, the three layers of their method at least look very similar to the factors/criteria/metrics structure of McCall's method and the quality attributes-quality/subattributes/metrics structure of ISE 9126. But, the big differences are Bansiya and Davis' focus on design, the identification of design elements, and the explicit linkage from them to the design metrics (and subsequently to design properties and to high-level quality attributes). Also, the framework of their method is actually the same as GRCM, except that GRCM includes checklists for inspection purposes. Note again that the scope of Bansiya and Davis' method can be easily extended to cover checklists.

Dromey's product quality model applies to design, implementation, requirements, and any other intermediate software product; Bansiya and Davis' work focuses on design. GRCM addresses design as well as inspection. Although the examples used to illustrate GRCM are all object oriented, the methodology is actually applicable to other design paradigms.

We now briefly describe the Bansiya and Davis method [16]. Like Dromey [14, 15], they use the term "component" to describe in general an element of a software product; to avoid confusion, we use the term "element." For ease of comparison to the other methods, our description deviates from the original one slightly. Like Dromey's design quality model, Bansiya and Davis linked quality attributes with design elements through quality-carrying properties. With the special focus on object orientation and with the availability of some metrics for object orientation [21, 22], Bansiya and Davis were able to identify one or more metrics to define each of the OO design properties they identified, using existing OO metrics or new OO metrics they developed to supplement the existing ones. This results in a primary difference from Dromey's work in that Bansiya and Davis provided not only full details about the quality attributes, design elements, and quality-carrying properties of design elements but also a prototypical example involving numerical evaluation of actual software products with respect to the selected (top-layer) quality attributes. For each quality attribute, such evaluation requires amalgamation of multiple metrics assessed for all the relevant design elements present in a software design.

The selection of quality attributes depends on the software being built. Bansiya and Davis selected reusability, flexibility, understandability, functionality, extendibility, and effectiveness. Note that their emphasis on reusability, understandability, and extendibility makes their results particularly useful for developing software components and component-based software systems.

The 11 object-oriented design properties included are design size, hierarchies, abstraction, encapsulation, coupling, cohesion, composition, inheritance, polymorphism, messaging, and complexity. Bansiya and Davis chose to use one metric for each of the 11 design properties. The corresponding 11 design metrics are design size in classes (total number of classes in the design), average number of ancestors, data access metric (the ratio of the number of private attributes to the total number of attributes declared in the class; ranging from 0 to 1, with higher values as more desirable), direct access coupling, cohesion among methods of class, measure of aggregation, measure of functional abstraction, number of polymorphic methods, class interaction size, and number of methods. At the highest level, design elements can be grouped into classes and the relationships among them. Many design elements exist. A class consists of data and methods. Relationships among classes may take the form of generalization-specialization structures and class hierarchies.

Design elements and quality-carrying properties for OO design are detailed in [18]. All such quality-carrying properties can be mapped into 11 design properties. To obtain a numeric measure for each of the quality attributes, the following procedure is used. First, map a given OO software design, including the design elements and element composition, into the 11 design metrics and obtain the corresponding values. These values represent the degree to which the 11 design properties are achieved. The procedure then requires a set of weights to map the values of the 11 metrics to quantitative measures of the six quality attributes. Bansiya and Davis proposed a set of weights. The set was chosen because of its straightforward implementation. As mentioned earlier in this chapter, making selection in the presence of multiple criteria is a common but difficult issue. Any research into this general issue should benefit the OO SQA process.

Although the weights chosen by Bansiya and Davis may not be theoretically justified, the methodology is more complete than Dromey's. Moreover, Bansiya and Davis used a software tool QMOOD++ (implemented for C++) to obtain the overall measures of the six quality attributes for different versions of two products (and to refine their models): *MicroSoft Foundation Classes* (MFC) and Borland *Object Windows Library* (OWL). For both products, all six quality measures increase with the version number. A comparison between the quality measures predicted by QMOOD++ and the corresponding evaluations by human experts was also conducted. For all the human evaluations that were conducted without identifiable flaws, the hypothesis that there exists no correlation between the quality measures predicted by QMOOD++ and those assessed by human experts was rejected at the significance level of 0.05. The Spearman's rank correlation coefficient, which is a nonparametric statistic with a range between –1 and +1, was used in the test.

14.8 An integrated process for development and QA of CBSS

Several software component evaluation processes have been proposed in the literature, with much similarity [e.g., the *Information Technology Solutions Evolution Process* (ITSEP) proposed by Albert and Brownsword [27]]. ITSEP is to be used with the *Rational Unified Process* (RUP) [28], and together (i.e., RUP/ITSEP) they provide a way to develop, deploy, and support a software system containing one or more software components, custom code, and possibly implementation of required changes to business processes. Based on the earlier discussion in this chapter and ITSEP, we propose an integrated development and quality assurance process for CBSS. We first briefly describe ITSEP and RUP and then propose the process.

14.8.1 ITSEP

The development process for conventional software can be characterized as top-down. Although there exist several development models for it (including spiral models [29]), it is typified by the (sequential) waterfall model summarized in Chapter 13. A waterfall model contains three sequential steps, among others: requirements, architecture/design, and implementation. However, this is not adequate for CBSS because of the need to identify and, more importantly, understand the available components in the marketplace and the companion need to negotiate stakeholder buy-in.

The fundamental approach of ITSEP is the need to simultaneously define and trade-off among four *spheres of influence*: marketplace, stakeholder needs/business processes, architecture/design, and "programmatics"/risk. The rationale for considering the first two has been explained earlier. Architecture/design has long been a significant influence on conventional software development, and the rationale is well known. In developing CBSS, it becomes more important because the three commercial component models have their own architectures and design rules, and they are not compatible. The additional tasks and the risks introduced by the use of components add much to the project management for CBSS development and quality assurance.

The objective of the ITSEP is to induce iteratively converging decisions through accumulating knowledge and increasing stakeholder buy-in. The knowledge includes increasingly detailed understanding of:

- Capabilities and limitations of the candidate components;
- Implications of the candidates on the requirements for the CBSS and even the end-user business processes, and the planning necessary to implement the needed changes;
- Architectural alternatives and integration mechanisms that bind the candidates with the CBSS;
- Risk, cost, and schedule associated with implementing and deploying the CBSS.

Stakeholders include the end-user, developer, the vendor, and perhaps also the component certifier. Note again that the end-user business processes may need to change in order to accommodate the software components. The ITSEP process is summarized in Table 14.3.

14.8.2 Rational unified process

In the integrated SD and SQA process, each product/project increment is managed with the *rational unified process* (RUP) [28]. The RUP consists of four phases and three anchor points separating the four phases; it is summarized in Table 14.4. The four phases are inception, elaboration, construction, and transition; the three anchor points are life-cycle objectives, life-cycle architecture, and initial operational capacity. Each of the four phases consists of multiple ITSEP iterations. Iterations in each phase build on the knowledge and stakeholder buy-in achieved during the previous iterations.

The inception phase achieves a common understanding among the stakeholders of the objectives of the product/project increment and ends with the *life-cycle objectives* (LCO) anchor point, which marks an increase in project intensity. The information to be gathered includes a high-level understanding of user needs and user business processes. The scope of activity for the elaboration phase is the same as that of the inception phase, but the level of detail is deeper and the resource commitment is higher. A candidate component is selected and acquired, with risk, cost, and schedule fully considered. This phase ends with the *life-cycle architecture* (LCA) anchor point. The deliverable of the construction phase is a production-quality release of the CBSS with the

Table 14.3 ITSEP: Iteratively Accumulating Knowledge in Four Spheres of Influence and Iteratively Increasing Stakeholder Buy-In

Sphere of Influence	*Influence Via Understanding and Buy-In*
Marketplace	Capabilities and limitations of the candidate components
Stakeholder needs/business processes	Implications of the candidates on the requirements for the CBSS and even the end-user business processes, and the planning necessary to implement the needed changes
Architecture/design	Architectural alternatives and integration mechanisms that bind the candidates with the CBSS
"Programmatic"/risk	Risk, cost, and schedule associated with implementing and deploying the CBSS

Table 14.4 The Rational Unified Process

Phase or Anchor Point	*Objective*
Inception phase	A common understanding among the stakeholders of the objectives of the product/project increment
Anchor: LCO	Stakeholder agreement to the *life-cycle objectives* (LCOs)
Elaboration phase	Selection and acquisition of a component, with cost, schedule, and risks fully considered
Anchor: LCA	Stakeholder agreement to the *life-cycle architecture* (LCA)
Construction phase	Production-quality release of the system with selected and acquired components
Anchor: IOC	Stakeholder agreement to the *initial operational capacity* (IOC)
Transition phase	Initial deployment (i.e., beta test) through retirement

selected component(s). This phase ends with the *initial operational capacity* (IOC) anchor point. The IOC anchor point allows the "beta test." In other words, it allows the stakeholders to verify that a production-quality release is ready to be deployed to at least a subset of the users. The transition phase begins with an initial deployment or beta test. After bugs are fixed and features are adjusted, the transition phase ends when the CBSS increment is retired and replaced by a new increment. At the anchor points, the development team makes sure that the objectives of the corresponding phases are met, and seeks approval to move on to the next phase.

14.8.3 RUP/ITSEP with SQA

This RUP/ITSEP development process can be augmented with SQA activities. In the inception phase, a preliminary evaluation of the quality of the candidate components should be conducted according to the discussion of Section

14.2, and a screening process eliminates unqualified candidates and reaches a set of most viable candidates. In the elaboration phase, in-depth hands-on experiments with the most promising candidates are conducted in an experimentation facility that mimics as closely as possible the operational environment. The quality of the process of adaptation and assembly should be conducted according to the discussion of Section 14.3. Validation of the quality of the most promising components should be conducted according to the discussion of Section 14.4. In the construction phase, the limitations of the SQA process currently practiced by the development team should be examined with respect to the discussion of Section 14.5, and the SQA process should be augmented with the bottom-up SQA approach according to the discussion of Section 14.6. For object-oriented CBSS, the design process should be augmented with the bottom-up SQA approach developed for object-oriented software systems by Bansiya and Davis [16], which was briefly summarized in Section 14.7.

14.9 Summary

We discussed many techniques for the quality assurance of component-based software systems. The techniques provide many useful rules for requirements engineering, design, implementation, and inspection, and significantly contribute to quality assurance for component-based software systems. Due to space limitations, we are only able to discuss the overarching principles of the techniques. For details of these techniques, the reader is encouraged to read the papers and books cited. Some of these techniques were developed very recently, and may mature in due course. The reader is encouraged to follow the literature. More importantly, these techniques should be tailored to suit the reader's specific development context.

Many of the subjects covered in this chapter are equally applicable to developing software components. The user of such components should pay attention to the software development and QA practices of the component vendor, and may actually require the vendor to demonstrate its use of the type of bottom-up SQA approaches discussed in this chapter.

References

[1] Wallnau, K. C., S. A. Hissam, and R. C. Seacord, *Building Systems from Commercial Components*, Reading, MA: Addison-Wesley, 2002.

[2] Ncube, C., and J. C. Dean, "The Limitations of Current Decision-Making Techniques in the Procurement of COTS software Components," *Proc. of 1st International Conference on COTS-Based Software Systems (ICCBSS 2002)*, February 2002, pp. 176–187.

[3] Saaty, T., *The Analytic Hierarch Process*, New York: McGraw-Hill, 1990.

[4] Ballurio, K., B. Scalzo, and L. Rose, "Risk Reduction in COTS Software Selection with BASIS," *Proc. of 1st International Conference on COTS-Based Software Systems (ICCBSS 2002)*, February 2002, pp. 31–43.

[5] Comella-Dorda, S., et al., "A Process for COTS Software Product Evaluation," *Proc. of 1st International Conference on COTS-Based Software Systems (ICCBSS 2002)*, February 2002, pp. 86–96.

[6] International Organization for Standardization, ISO/IEC 14598-1:1999—Information Technology—Software Product Evaluation, Geneva, Switzerland, 1999.

[7] Bergner, K., et al., "Adaptation Strategies in Componentware," *Proc. of Software Engineering Conference 2000*, Australia, 2000, pp. 87–95.

[8] Baker, T. G., "Lessons Learned Integrating COTS into Systems," *Proc. of 1st International Conference on COTS-Based Software Systems (ICCBSS 2002)*, February 2002, pp. 21–30.

[9] Howe, R., D. Gaeddert, and M. Howe, *Quality on Trial*, Maidenhead, England: McGraw-Hill, 1992.

[10] Kitchenham, B., and S. L. Pfleeger, " Software Quality: The Elusive Target," *IEEE Software*, January 1996, pp. 12–21.

[11] McCall, J. A., P. K. Richards, and G. F. Walters, *Factors in Software Quality Assurance, Vol. I*, RADC-TR-77-369, U.S. Department of Commerce, November 1977.

[12] Kan, S. H., *Metrics and Models in Software Quality Engineering*, Reading, MA: Addison-Wesley, 1995.

[13] Schneidewind, N., and N. Fenton, "Do Standards Improve Quality? (Point vs. Counterpoint)," *IEEE Software*, January 1996, pp. 22–24.

[14] Dromey, R. G., "A Model for Software Product Quality," *IEEE Trans. on Software Engineering*, Vol. 21, No. 2, February 1995, pp. 146–162.

[15] Dromey, R. G., "Cornering the Chimera," *IEEE Software*, January 1996, pp. 33–43.

[16] Bansiya, J., and C. Davis, "A Hierarchical Model for Object-Oriented Design Quality Assessment," *IEEE Trans. on Software Engineering*, Vol. 28, No. 1, January 2002, pp. 4–17.

[17] Tervonen, I., "Support for Quality-Based Design and Inspection," *IEEE Software*, January 1996, pp. 44–54.

[18] Bansiya, J., "A Hierarchical Model for Quality Assessment of Object-Oriented Designs," Ph.D. dissertation, University of Alabama in Huntsville, 1997.

[19] Bansiya, J., and C. Davis, "Automated Metrics for Object-Oriented Development," *Dr. Dobb's Journal*, Vol. 272, December 1997, pp. 42–48.

[20] Heineman, G. T., and W. T. Councill, *Component-Based Software Engineering: Putting the Pieces Together*, Reading, MA: Addison-Wesley, 2001.

[21] Chidamber, S. R., and C. F. Kemerer, "A Metrics Suite for Object-Oriented Design," *IEEE Trans. on Software Engineering*, Vol. 22, No. 1, June 1994, pp. 476–493.

[22] Li, W., and S. Henry, "Object-Oriented Metrics That Predict Maintainability," *Journal of Systems and Software*, Vol. 23, 1993, pp. 111–122.

[23] Lorenz, M., and J. Kidd, *Object-Oriented Software Metrics*, Englewood Cliffs, NJ: Prentice Hall, 1994.

[24] Garlan, D., and M. Shaw, *An Introduction to Software Architecture*, Technical Report CMU/SEI-94-TR-21, Software Engineering Institute, Carnegie Mellon University, Pittsburgh, PA, 1994.

[25] International Organization for Standardization, ISO/IC Standard 9126—Information Technology—Software Product Evaluation—Quality Characteristics and Guidelines for Their Use, Geneva, Switzerland, 2001.

[26] Myers, G., *Software Reliability: Principles and Practices*, New York: John Wiley, 1976.

[27] Albert, C., and L. Brownsword, "Meeting the Challenge of Commercial-Off-The-Shelf (COTS) Products: The Information Technology Solutions Evolution Process (ITSEP)," *Proc. of 1st International Conference on COTS-Based Software Systems (ICCBSS 2002)*, February 2002, pp. 10–29.

[28] Kruchten, P., *The Rational Unified Process: An Introduction*, 2nd ed., Reading, MA: Addison-Wesley, 2000.

[29] Boehm, B., "A Spiral Model of Software Development and Enhancement," *Computer*, May 1998, pp. 61–72.

CHAPTER

15

Contents

Standards and certification for components and component-based software

Widespread development and use of software components has the potential of drastically improving both the quality of those software systems using such components and the productivity of the system developers. Previous chapters in Part IV addressed various software development and quality assurance methods that can be used to help achieve this potential. Some of those methods have been in existence for a long time, while others are relatively new and may still be in the research stage. In the midst of a plethora of published results, software development practitioners need clear guidance in their day-to-day development activities. Responding to this need, a number of international and national organizations have developed standards, in the sense of best practices, to provide the guidance. This chapter addresses this same need for not only the producer but also the user of a software component and discusses many standards responding to this need. Note that since a standard may take several years to develop and software components are a relatively recent phenomenon, current published standards may not reflect the state of the art. Our focus will be on existing standards. We address standards related to developing software components as well as those related to developing component-based software systems. Although there exist many software standards, few specialize in software components or component-based software. The few developed specifically for software components do not seem to cover the whole spectrum of the activities involved, and therefore, despite the complaint made by many about the proliferation of software standards, there seems to be a need for more standards to guide the development and the use of software components. A goal of this chapter is to provide a "big picture" about existing software standards and a more detailed discussion on standards specifically developed for software components.

Widespread use of software components requires widespread confidence in them, their development processes, and their developers. An effective way to help build such confidence is third-party certification. This chapter addresses certification of software components at three different levels: product, development process, and development personnel. In addition to building confidence by improving the quality of software components, certification also helps reduce user testing of the components.

Current standards related to software, including components or component-based software as special cases, are predominantly process oriented, and existing certification programs focus predominantly on the related processes (e.g., the software engineering processes or simply the test processes). Certification of a software process is typically performed against a published standard, although what is really needed is satisfaction of the requirements set by the certifying organization. The fact that few current standards pertain exclusively to software components may hinder widespread certification of software components in the near future.

Product producers have been facing liability issues; software component producers are no exception. It has been argued that software producers may be able to use the so-called "state-of-the-art argument" to fend off possible liability lawsuits resulting from software defects, and such an argument is stronger if the producer can demonstrate that it has followed pertinent standards (i.e., best practices). It can be even stronger if its product, the software engineering process, and even the development personnel have been certified by a credible third party. A companion issue is the liability on the part of the certifying agency. This issue seems to have received much less attention in the literature. However, if a standard exists for a process and the certifying agency indeed verified the process against the standard, the liability risk associated with certifying that process may be significantly reduced.

This chapter is organized as follows. Section 15.1 addresses existing general standards for software processes and products (i.e., those standards developed for computer software in general). Section 15.2 expands on the standards for software testing. Section 15.3 deals with standards pertaining explicitly to the development of software components and component-based software systems. It also discusses briefly how general software standards can be specialized to meet some of the current needs of the software component producers. As the collection of standards has grown in recent years, their proper application becomes more and more difficult. Section 15.4 summarizes the result of two recent efforts to organize software engineering standards. Section 15.5 discusses certification of software processes, software products, and software developers. Due to the existence of few standards developed explicitly for software components, we summarize in Section 15.6 some recent research activities and the companion results regarding certification of software components. The liability aspects of producing software components and their certification will be the subjects of Section 15.7. Concluding remarks are given in Section 15.8.

15.1 Standards for software processes and products

This section addresses existing standards for software processes and products. Some particular industrial sectors have their own standards. For example, the aviation industry has a standard on avionics software [1, 2]. Our focus is on standards applicable across all different domains. S. Magee and L. L. Tripp [3] identified 315 standards produced and maintained by 55 organizations. These organizations include *International Organization for Standardization* (ISO), *International Electrotechnical Commission* (IEC), *Institute of Electrical and Electronics Engineers* (IEEE), *American National Standards Institute* (ANSI), *British Standards Institution* (BSI), RCTA, Inc., and *Electronic Industries Association* (EIA). ISO and IEC agreed in 1987 to form the *Joint Technical Committee 1* (JTC1) to deal with the area of *information technology* (IT). These standards and related documents can be obtained from several organizations [e.g., Global Engineering Documents; Information Handling Services; and Rapidoc, Technical Indexes, Ltd. (TIUK)]. The reader is referred to [3, 4] for a guide through the maze of these standards. Due to space limitations, we will focus on key standards developed by IEEE, ISO, IEC and BSI. Managing standards compliance is beyond the scope of this chapter; the reader is referred to [5].

The value of some published software standards and the validity of some standard development processes have been a subject of debate. While software standards in general seem to receive much support, some have criticized the predominant process orientation and lack of empirical evidence in their value [6]. We partition our discussion into two sections, one devoted to process standards and the other devoted to product standards. The most current versions as of this writing are used; we attach the year of publication or revision for all the standards cited in this chapter.

15.1.1 Standards for software processes

This section first briefly reviews the standards pertaining to software development and quality assurance. Although some standards carry a title involving only quality assurance, they actually are standards also for software development. This is because quality assurance cannot be achieved in separation from the software development process. For example, as mentioned in Chapter 13, software configuration management has been regarded as a software development activity as well as a quality assurance activity. ISO standards related to software configuration management include ISO 10007, which is put explicitly under the umbrella of quality management standards. A common criticism of ISE 9000 is that it aggregates all possible desirable aspects of software under the umbrella of "quality." In Tables 15.1 through 15.4, respectively, we list key standards under four categories: (1) quality management, (2) software engineering, (3) quality assurance and project management, and (4) risk analysis.

Of those quality management stndards listed in Table 15.1, IEEE 730 will be used in Section 16.1.

In addition to the standards on quality management and planning, IEEE, ISO, and other major standard-setting organizations have published a number of standards pertaining to software life cycle or specific phases of the cycle. We list key standards in this regard in Table 15.2. Some standards pertain to both software development and systems engineering; they are also within the scope of Table 15.2. Other systems engineering standards can be found in [4].

Of those software engineering standards listed in Table 15.2, IEEE 610.12 and IEEE 830 will be used in Section 16.1.

Table 15.3 lists key standards for quality assurance and project management. Note again that software development, quality assurance, and project management are not separate functions and overlap significantly. To avoid duplication, no standards are listed in more than one table. Other tables may contain related standards.

Software safety is a critical aspect for many industrial sectors (e.g., avionics and nuclear power generation), and many software safety standards exist [1, 2, 7, 8]. Major elements of existing approaches to achieving software safety include hazard analysis and risk analysis. Related techniques include fault tree analysis, which traces the factors that contribute to the occurrence of an undesirable event affecting the performance of the system, and *failure mode and effects analysis* (FMEA), which can be used to identify failures that have a significant impact on system performance. The latter has been extended to *failure mode, effects and criticality analysis* (FMECA) to consider the severity of the consequences of a failure. Although not all software components will be deployed in safety-critical systems, producers of software components should

Table 15.1 Standards for Management of Software Quality

Standard	*Title*	*Remarks*
ISO 9001:2000	*Quality Management Systems—Requirements*	Replacing ISO 9001:1994 (listed next nevertheless because ISO 9000-3, developed to provide guidance for application of ISO 9001:1994, has not been updated yet)
ISO 9001:1994	*Quality Systems - Model for Quality Assurance in Design, Development, Production, Installation, and Servicing*	The only relevant one for software among ISO 9001, ISO 9002, and ISO 9003 because only it covers a life cycle sufficiently broad to include software development processes
ISO 9000-3:1997	*Quality Management and Quality Assurance Standards—Part 3: Guidelines for the Application of ISE 9001 to the Development, Supply, Installation, and Maintenance of Software*	Specific guidance, in software development terminology, for applying the manufacturing-oriented ISO 9001 to design-centered software development; source of mandatory quality assurance requirements
ISO 10005-1995	*Quality Management—Guidelines for Quality Plans*	
ISO/TR 10013: 2001	*Quality Management—Guidelines for Quality Management System Documentation*	
ISO 10007	*Quality Management—Guidelines for Configuration Management*	
IEEE 730-2002	*Standard for Software Quality Assurance Plans*	
IEEE 730.1-1995	*Guide for Software Quality Assurance Planning*	
IEEE 1298-1992	*Software Quality Management System—Part 1: Requirements*	

proactively perform these analyses to anticipate failure modes and consequences, due to the components' intended widespread use and the potential disastrous effects associated with undetected defects. (It should not be surprising that the UL 1998 (1998)—*Standard for Software in Programmable Components*—evolved from UL 1998 (1994)—*Standard for Safety-Related Software*.) Standards related to these analysis techniques are listed in Table 15.4.

The *Capability Maturity Model*® (CMM®) developed by the Software Engineering Institute has been considered by many as a standard for software engineering and has been widely used as a tool for software process improvement. The CMM® was first introduced by the Software Engineering Institute in 1986 as a maturity framework, after modeling and benchmarking many top software development organizations. Through the experience accumulated in

Table 15.2 Software Engineering Standards

Focus	*Standard*	*Title*
Terminology	IEEE 610.12-1990	*Standard Glossary of Software Engineering Terminology*
Life cycle	IEEE 1074-1997	*Standard for Developing Software Life Cycle Processes*
Life cycle	IEEE 1074.1-1995	*Guide for Developing Software Life Cycle Processes*
Life cycle	ISO/IEC 12207:1995/Amd 1:2002	*Information Technology—Software Life Cycle Processes*
Life cycle	ISO/IEC TR 15271:1998	*Information Technology—Guide for ISO/IEC 12207 (Software Life Cycle Processes)*
Life cycle	ISO/IEC TR 14759: 1999	*Software Engineering—Mock up and Prototype—A Categorization of Software Mock up and Prototype Models and Their Use*
Life cycle	IEEE/EIA 12207.0-1996	*Information Technology—Software Life Cycle Processes*
Life cycle	IEEE Handbook	*Handbook to IEEE/EIA Std 12207.0*
Life cycle	IEEE/EIA 12207.1-1997	*Guide for Information Technology—Software Life Cycle Processes—Life Cycle Data*
Life cycle	IEEE 1362a-1998	*Guide for Information Technology—System Definition—Concept of Operations Document: Content Map for IEEE 12207.1*
Life cycle	IEEE/EIA 12207.2-1997	*Guide for Information Technology—Software Life Cycle Processes—Implementation Considerations*
Operations concept	IEEE 1362-1998	*Guide for Concept of Operations Document*
Operations concept	AIAA G-043-1992	*Guide for the Preparation of Operations Concept Documents*
System requirements	IEEE 1233-1998 IEEE 1233a-1998	*Guide for Developing System Requirements Specifications*
System design	IEEE 1471-2000	*Recommended Practice for Architectural Description for Software-Intensive Systems*
Requirements	IEEE 830-1998	*Recommended Practice for Software Requirements Specifications*
Design	IEEE 1016-1998	*Recommended Practice for Software Design Descriptions*
Design	IEEE 1016.1-1993	*Guide to Software Design Descriptions*
Testing	IEEE 1008-1987	*Standard for Software Unit Testing*
Testing	IEEE 829-1998	*Standard for Software Test documentation*
Verification and validation	IEEE 1012-1998	*Standard for Software Verification and Validation*
Verification and validation	IEEE 1012a-1998	*IEEE Standard for Software Verification and Validation—Content Map to IEEE 12207.1*
Verification and validation	IEEE 1059-1993	*Guide for Software Verification and Validation Plans*
Review and audit	IEEE 1028-1998	*Guide for Software Reviews*
Documentation	ISO/IEC TR 9294:1990	*Information Technology—Guidelines for the Management of Software Documentation*
Documentation	IEEE 1063-2001	*Standards for Software User Documentation*
Maintenance	IEEE 1219-1998	*Standard for Software Maintenance*

Table 15.3 Standards for Quality Assurance and Project Management

Focus	*Standard*	*Title*
Anomalies	IEEE 1044-1993	*Standard Classification for Software Anomalies*
Anomalies	IEEE 1044.1-1995	*Guide to Classification for Software Anomalies*
Quality metrics	IEEE 1061-1998	*Standard for a Software Quality Metrics Methodology*
Measurement: functional size	ISO/IEC 14143-1:1998	*Information Technology—Software Measurement—Functional Size Measurement—Part 1: Definition of Concepts Software Measurement—Functional Size Measurement—Part 1: Definition of Concepts*
Measurement: functional size	ISO/IEC 14143-2:2002	*Information technology—Software Measurement—Functional Size Measurement—Part 2: Conformity Evaluation of Software Size Measurement Methods to ISO/IEC 14143-1:1998*
Measurement: functional size	ISO/IEC TR 14143-4:2002	*Information technology—Software Measurement—Functional Size Measurement—Part 4: Reference Model*
Measurement: performance	ISO/IEC 14756:1999	*Information technology—Measurement and Rating of Performance of Computer-Based Software Systems*
Dependability and integrity	ISO/IEC 15026:1998	*System and Software Integrity Levels*
Reliability	IEEE 982.1-1988	*Standard Dictionary of Measures to Produce Reliable Software*
Safety	IEEE 1228-1994	*Standards for Software Safety Plans*
Project management	IEEE 1058-1998 IEEE 1058a-1998	*Standard for Software Project Management Plans*
Productivity metrics	IEEE 1045-1992	*Standard for Software Productivity Metrics*
Configuration management	IEEE 1042-1987	*Guide to Software Configuration Management*
Configuration management	EIA 649 (1998)	*National Consensus Standard for Configuration Management*
Acquisition	IEEE 1062-1998 IEEE 1062a-1998	*Recommended Practice for Software Requisition*

the several years after its introduction, a new CMM® was developed [9, 10]. In addition, several studies [11, 12] have been conducted to compare CMM® to some popular software standards, particularly the ISO 9000 family of software engineering standards.

15.1.2 Standards for software products

While much similarity exists among the development and quality assurance processes for individual software products, the products themselves may differ drastically from one another. This may explain the fact that there exist few

Table 15.4 Standards for Techniques of Risk Analysis

Standard	*Title*
IEC 61025 (1990)	*Fault Tree Analysis*
IEC 60812 (1985)	*Analysis Techniques for System Reliability—Procedure for Failure Mode and Effects Analysis (FMEA)*

standards for software products. However, although software products differ, the relationships between properly defined product elements (e.g., program statements) and product quality may be similar. Dromey's software product quality model discussed in Section 14.6, which links software product elements to high-level product quality attributes through "quality-carrying properties," and other software product quality models developed in a similar approach may eventually be useful for developing software product standards. In Table 15.5, we summarize key existing standards for evaluating software products.

Note that a major difference between the ISO/IEC 14598 family of standards and ISO/IEC 12119 is that the former is to be used during software development and in conjunction with software process standards, while the latter is to be used for evaluating software "packages" (a bundle of programs, data, and documentation) (e.g., spreadsheets, word processors, and utilities) in isolation from the development process. ISO/IEC 14598 is useful for developers of software components and certification agencies, while ISO/IEC 12119 is particularly useful for the user. Note that both are particularly relevant because COTS software components can be viewed as a special type of software packages.

15.2 On standards for software testing

This section expands on the testing phase of software life cycle. As mentioned earlier, the CMM® developed by the Software Engineering Institute has been considered by many software professionals in the United States as a standard for software process improvement; it specifies, among others, test requirements. H. K. N. Leung [13] compared ISO 9001 (and the companion standard ISO 9000-3), IEEE, and CMM® test requirements in an attempt to improve the software testing process of a bank group in Hong Kong and reported several key differences. He identified 19 test requirements, each of which is required by at least one of the three sets of test requirements. The source of the IEEE test requirements that Leung analyzed consists of five IEEE standards: 829-1983, 1008-1987, 1012-1986, 730-1984, and 983-1984. Note that IEEE 983 has been withdrawn since his study and is no longer endorsed by the

Table 15.5 Standards for Software Product Evaluation

Standard	*Title*
ISO/IEC 9126-1:2001	*Software Engineering—Product Quality—Part 1: Quality Model*
ISO/IEC 14598-1:1999	*Information Technology—Software Product Evaluation—Part 1: General Overview*
ISO/IEC 14598-2:2000	*Software Engineering—Product Evaluation—Part 2: Planning and Management*
ISO/IEC 14598-3:2000	*Software Engineering—Product Evaluation—Part 3: Process for Developers*
ISO/IEC 14598-4:1999	*Software Engineering—Product Evaluation—Part 4: Process for Acquirers*
ISO/IEC 14598-5:1998	*Information Technology—Software Product Evaluation—Part 5: Process for Evaluators*
ISO/IEC 14598-6:2001	*Software Engineering—Product Evaluation—Part 6: Documentation of Evaluation Modules*
ISO/IEC 12119:1994	*Information Technology—Software Packages—Quality Requirements and Testing*
ISO 9127:1988	*Information Processing Systems—User Documentation and Cover Information for Consumer Software Packages*

IEEE. The four other IEEE standards, along with the years of their most recent versions, have been listed earlier. The 19 requirements and the key differences are listed in Table 15.6.

Leung concluded that neither ISO 9001 nor CMM® is very useful in providing the specifics of test elements. However, they are useful in pointing out the need to improve the test process, to set up an independent test team and to test all supporting software components, including customer-supplied or purchased software products. On the other hand, IEEE standards provide detailed descriptions of unit tests and documentation format.

Since CMM® does not adequately address testing issues, the concept of a mature testing process was developed. I. Burnstein et al. [14] reported the effort on the development of their *Testing Maturity Model*℠ (TMM℠) and discussed the relationships between TMM℠ and CMM®. Like CMM®, their TMM℠ contains a set of maturity levels for software testing, a set of recommended practices at each of the maturity levels, and a maturity assessment model.

15.3 Standards for software components and component-based software systems

This section deals with standards pertaining directly to the development of software components and component-based software systems. Since development and quality assurance of software components should be integrated with its counterpart of component-based software systems, existing standards related to software components address both. We also address them in one section. We also discuss briefly how general software standards can be

Table 15.6 19 Test Requirements and Comparison Among ISO, IEEE, and CMM®

Category of Test Requirements	*Test Requirements*	*Comparison*
Test phases	Unit/module test Integration test System test Acceptance test	Only ISO 9001 does not specify different test phases, although acceptance testing is implied. IEEE standards provide a very detailed description of unit testing.
Test activities	Test planning Test case design Test execution Result evaluation/reporting Regression test	Both IEEE and CMM® standards identify all five test activities. ISO 9001 does not include regression testing.
Test management	Test plan review Test staff training Use of test standard Test strategy/test technique Testing by independent group Test documentation format	All three require the first three elements. Only CMM® requires an independent test group and gives examples of test strategies and techniques. Only IEEE standards give a detailed format description for test documentation.
Testing supporting software	Purchased software Customer-supplied software Software tools	IEEE and CMM® standards specify only the testing of purchased software components. ISO 9001 requires testing of all three.
Improving test process	Continuous improvement in test process	Both ISO 9001 and CMM® advocate continuous process improvement. IEEE standards do not address this issue.

specialized to meet some of the current needs of software component producers. Existing standards for software components and component-based software systems are listed in Table 15.7.

IEEE 1517 describes software reuse processes and how they relate to traditional software life-cycle processes. This standard identifies the required processes and specifies minimum requirements for the processes, but does not prescribe an overall process or particular techniques. The scope of IEEE 1517 is broader than that of IEEE 12207.0 because reuse activities transcend the life cycle of any particular software system. Most software reuse processes are not distinct from the normal life cycle, but all software reuse processes must be integrated into other life-cycle processes. IEEE 1517 specifies software reuse processes and how they should be incorporated into the life cycle process framework already provided by 12207. For a more detailed discussion of IEEE

Table 15.7 Process Standards for Software Components and Component-Based Software System

Standard	*Title*
IEEE 1517-1999	*Standard for Information Technology—Software Life Cycle Processes—Reuse Processes*
IEEE/EIA 12207.2-1997	*Information Technology—Software Life Cycle Processes—Implementation Considerations (including Software Reuse)*
UL 1998-1998 (ANSI/UL 1998)	*Standard for Software in Programmable Components*

1517 against the background of IEEE 12207, the reader is referred to Chapter 17 and [15].

Based on IEEE 12207.0, IEEE 12207.2 provides guidance on software reuse; software process management indicators; categories for problem reporting; and guidance on software/system architecture, development strategies, tailoring and build planning, software product evaluations, alternate means of compliance for joint reviews, configuration management, and acquirer-supplier interaction.

For obvious reasons, safety-critical systems have been subjected to certification and standard compliance [1, 8]. Since many such systems contain software components and hence can be viewed as component-based systems, the corresponding standards and certification programs explicitly address issues associated with software components and component-based software systems. As mentioned earlier, the 1998 version of UL 1998 is entitled Standard for Software in Programmable Components, and it evolved from the previous (1994) version called UL 1998 Standard for Safety-Related Software. Although this and other safety-related standards explicitly address the safety issues and the safety issues provided the rationale for high software quality, the resulting process requirements serve the quality needs of component-based software systems and particularly the quality needs of software components. This is because both safety-critical software systems and software components may have a tremendous impact on their users, despite the difference in the nature of the impact. The scope of ANSI/UL 1988 covers software whose failure could result in a risk of personal injury or loss of property. (This risk may result in tort liability, which will be discussed in Section 15.7.) The requirements of UL 1988 are intended for evaluating products that rely on software and microelectronics to perform safety-related functions. The requirements are not intended to be used as the sole basis for evaluating safety-related functions. Instead, they are applicable when used in conjunction with an application-specific standard, directive, regulation or purchasing specification. ANSI/UL 1998 emphasizes, among other things, risk-based analysis and design, consideration of provisions for hardware malfunction and software failure, qualifications for off-the-shelf software, labeling that clearly and uniquely identifies the specifics of the product interface, the hardware platform, the software configuration, and the process for handling software changes. Note that these issues are of major concerns for the developers and users of software components and for those of component-based software systems as well. The 1998 version was recognized by ANSI in February 1999, and has also been referred to as ANSI/UL 1998. For more information about ANSI/UL 1998, the reader is referred to [16]. Other standards related to software safety include those listed in Table 15.8.

Several standards have been developed for libraries of reusable components. They are summarized in Table 15.9.

The *British Standards Institution* (BSI) has developed BS 7925-2, which is a standard for testing software components where a software component is defined to be "a minimal program for which a separate specification is available." BS 7925-2 will be addressed in more detail in Chapter 16. For completeness reasons, we list it in Table 15.10, but defer the discussion to Chapter 16.

Table 15.8 Some Other Safety-Related Software Standards

Standard	*Title*
IEC 60601-1-4 (2000)	*Medical Electrical Equipment—Part 1-4: General Requirements for Safety—Collateral Standard: Programmable Electrical Medical Systems*
IEC 61508 (7 Parts)	*Functional Safety of Electrical/Electronic/Programmable Electronic Safety-Related Systems*
ANSI/ISA-S84.01-1996	*Application of Safety Instrumented Systems for the Process Industries*
RCTA DO-178B	*Software Considerations in Airborne Systems and Equipment Certification*

Table 15.9 Standards for Reuse Libraries

Standard	*Title*
IEEE 1420.1-1995	*Standard for Information Technology—Software Reuse—Data Model for Reuse Library Interoperability: Basic Interoperability Data Model (BIDM) Standard*
IEEE 1420.1a-1996	*Supplement to IEEE Standard for Information Technology—Software Reuse— Data Model for Reuse Library Interoperability: Asset Certification Framework [developed in cooperation with the Reuse Library Interoperability Group (RIG)]*
IEEE 1420.1b-1999	*IEEE Standard for Information Technology—Software Reuse—Data Model for Reuse Library Interoperability: Intellectual Property Rights Framework*
AIAA G-010-1993	*Guide for Reusable Software: Assessment Criteria for Aerospace Applications*

Table 15.10 Standards for Testing Software Components

Standard	*Title*
BS 7925-2:1998	*Standard for Software Component Testing*

However, we note again that this definition of software component is less stringent than ours. We also note that the test techniques specified in BS 7925-2 pertain mostly to white-box testing, and therefore, the standard is useful as a guide to testing and assuring quality for software components during their development and certification.

Although there are few standards explicitly addressing the development or quality assurance processes for software components, the general software standards discussed in the previous sections of this chapter can be tailored to suit the special case of software components. The tailoring can begin with the explicit treatment and refinement of the following categories of user and certifier requirements.

- User requirements regarding (eight categories):
  - Functionality;
  - Performance (efficiency, reliability, and so forth);
  - Interface;
  - Assembly into to different (component-based) application software systems after unpacking and configuration by the user (interoperability);
  - Porting to different operational environments, including operating systems and hardware (portability);
  - Standards;
  - Certification (verifiability);
  - Documentation.
- Certifier requirements.

Note that all these requirement categories should capture all unique features of software components. These categories should be refined into a set of clearly stated specifications, which should be subsequently implemented, tested, and clearly documented.

15.4 Organizing software standards

Standards benefit software producers in that they can be used to help not only build quality into products but also build customer confidence in the products and in the producers themselves. As will become clear later in Section 15.7, standards may also help alleviate producers' liability risks. J. W. Moore [17] pointed out, "With such benefits, you would expect a nearly universal application of software engineering standards. Unfortunately, this is difficult due to the vast amount of available and occasionally inconsistent information."

There have been at least two major efforts to organize existing software standards and to anticipate possible addition of standards in the future. We discuss briefly in this section two such efforts. This discussion is consistent with a goal of this chapter, which is to provide a "big picture" about existing software standards.

15.4.1 ISO/IEC 12207 or IEEE/EIA 12207

IEEE/EIA 12207 is the U.S. version of ISO/IEC 12207. Both standards are accompanied by two additional guides. ISO/IEC 12207 establishes a common framework for software life cycle processes, with well-defined terminology, that can be referenced by the software industry. Its encompassing scope of the whole life cycle makes it a natural candidate for an umbrella standard. In addition, it has been widely regarded as a framework within which other standards should be unified. In fact, the *Software Engineering Standards Committee* (SESC) of the IEEE Computer Society has adopted a policy that its other standards will be harmonized with 12207, in an attempt to reduce the proliferation of standards, procedures, methods, tools, and environments for developing and managing software. The 12207 standards specify three categories of processes: primary, organizational, and supporting. These three categories, their relationships, and the constituent processes are summarized in Table 15.11.

The scope of 12207 is broad and covers the entire life of software from conception through retirement. Each of the processes is hierarchically decomposed into a set of activities and then tasks, which in some cases can be viewed as requirements on the activities. Although there are separate processes for verification, validation, and quality assurance, evaluation is actually built into each of the processes of all three categories, and the three separate processes (of verification, validation, and QA) are included in addition to the fundamental requirements for built-in evaluation. The 12207 does not prescribe particular development methodologies or technologies and is useful for any life-cycle model (e.g., waterfall, spiral, and so forth).

Table 15.11 ISO/IEC 12207 Standard: Processes and Their Relationships

Process Category	*Process*	*Relationships*
Primary	Acquisition Supply Development Maintenance Operation	These correspond to five major roles played by an organization (in the sense of an enterprise). An acquirer and a supplier enter into an agreement; the supplier then executes one or more of the other three primary processes according to the agreement.
Organizational	Management Infrastructure Improvement Training	These are regarded as inherent to an enterprise. Initiating any of the primary processes causes instantiation of the management process and other appropriate organizational processes.
Supporting	Documentation Configuration management Quality assurance Verification Validation Joint review Audit Problem resolution	Any of the three primary processes can invoke one or more of these supporting processes to accomplish appropriate objectives.

15.4.2 IEEE SESC Vision 2000 Architecture

Major standards-making bodies in the United States include the IEEE and the EIA. EIA has played an important role in converting to civilian use standards for complex software systems originally developed for use in the U.S. defense industry. The SESC of the IEEE Computer Society manages approximately 50 software engineering standards developed or adopted, in some cases with modifications, since 1979. Since 1991, SESC has endeavored to integrate its collection of standards. In 1995, SECS decided to develop an integrating architecture for all the standards managed by SESC and coined the term Vision 2000 Architecture. The Vision 2000 Architecture has three organizing criteria, which are summarized in Table 15.12.

In 1999, IEEE published a four-volume set of all the standards managed by SESC [18]. That set reflects the Vision 2000 Architecture; the four volumes correspond in principle to the four objects of software engineering.

Table 15.12 IEEE SESC Vision 2000 Architecture

Organizing Criterion	*Organization of Criterion*
Objects of software engineering	Four objects:
	Customer
	Process
	Product
	Resources
Normative levels	Six layers:
	Terminology
	Overall guide
	Principles
	Element standards
	Application guides
	"Toolbox" of detailed techniques
Relationships to other disciplines	Many other disciplines:
	Principles from more general disciplines (e.g., systems engineering, quality management, project management)
	Crosscutting disciplines of safety, security, dependability, and so forth
	Principles of mathematics, engineering, computer science, and so forth

15.5 Certification of software processes, products, and developers

Certification of some hardware products against safety hazards has been a common practice for a long time (e.g., the certification of electrical appliances by the Underwriters Laboratories, Inc. against fire hazards). Certification for systems consisting of hardware and software against safety hazards has also been performed for decades (e.g., certification of on-board avionics [1]). More examples can be found in [8]. However, perhaps all certification programs in existence for software involve certification of software engineering processes, not the products produced by the processes. The subject of software certification has received some research attention in recent years; the reader is referred to [19, 20].

A common software process certification is performed against ISO 9000 family of software engineering/quality assurance standards. (ISO does not provide any certification services, but provides standards against which software development organizations can be certified as well as guidance for how certification should be conducted by a certifying organization.)

Certification is usually performed against a standard. Section 15.1 summarized standards for software engineering processes and software products.

Tables 15.1 through 15.5 listed, among many others, some key standards for software processes and products.

Because of the special discipline required for development and quality assurance of software components, developers of software components may need to be certified too. Developer certification for some other special computer skill-sets has been a common practice in the software profession. Currently, *American Society for Quality* (ASQ) also certifies software quality engineers.

15.6 Certification of a software component

The few current standards developed explicitly for software components have been discussed in Section 15.3 and listed in Tables 15.7 through 15.10. The inadequacy of the current standards on software components may make certification of a software component difficult. However, as mentioned earlier, the current general standards for software engineering and quality assurance can be tailored to accommodate some of the special needs. As pointed out earlier, the vast majority of existing software engineering standards are about the software engineering processes, and few standards have been developed to evaluate, not to mention certify, a software product itself. Since a good software engineering process does not guarantee a high-quality software product and quality plays a critical role in the development and use of software components, evaluation of a software product itself is a critical issue. In the rest of this section, we briefly discuss the benefits of software-component certification and then provide references to some recent research efforts on such certification.

15.6.1 The benefits of certification of software components

This issue of component evaluation is also a difficult one because the user of a software component usually does not have access to the source code. Moreover, a user of a software component does not know whether or how well the producer has followed a good software engineering process. It is common knowledge that no testing, particularly black-box testing, can achieve 100% correctness or 100% of any other quality goals. Nevertheless, black-box testing should be a basic driver in component testing to ensure that a component conforms to its specifications. If black-box testing is the only way to test a software component and the specifications are complete and clear, then such testing alone can offer good software quality provided that the component is used strictly according to the specifications. However, it can be anticipated that every user may have to spend much resources on testing the component. This amount of effort on the part of one individual user can be quickly multiplied to constitute a huge amount of overall effort, and more importantly, much of this effort may be overlapped a multitude of times. This causes an inordinate amount of waste. These reasons point to the tremendous possible benefit that can be offered by a certification program. A common understanding of the benefit of software component certification is that it can improve the quality of a software component and can help build customer confidence. In addition to the economy of reuse expected of software component reuse, this economy of testing through certification can further reduce the cost of using software components. We will address in Section 15.7 the liability issue involved in selling defective software products and will explain some further benefits that software component certification can offer.

15.6.2 Research into certification of software components

In the end, what is required to obtain a certificate hinges upon what the certifier requires. Therefore, absence of standards for engineering software components does not really prevent certification from being performed. However, existence of standards that the producer can follow and the certifier can use to verify and evaluate software quality certainly helps reduce the liability risk and facilitates the widespread demand and supply of software component certification.

The whole issue of software component certification is still in the research stage. There are at least three approaches to software component certification: certification of process, certification of product, and a combination of the two. Many researchers believe that certification of software components requires the certifier's monitoring and active participation in the development, quality assurance, and even quality management processes. Note that the integrated software development and quality assurance process provided in Chapter 13 involves the certifier from the very beginning of the process. This approach is consistent with the approach adopted in the ISO/IEC 14598 family of standards. However, J. Voas [21] argued for a product-oriented certification scheme and proposed a "high-assurance certification pipeline" consisting of three pipes: (1) requirements and specifications, (2) software development, and (3) certification through stress. He argued that certification must be independent of the software development process. His approach can be supported by ISO/IEC 12119, which was developed to evaluate software packages, without the benefit of any knowledge of the actual development processes. He also suggested, among other things, that in the requirements and specifications process, we should also define what we do not want the software to do so that we can develop appropriate error handling. He commented, "Since defect-free software is, in general, an oxymoron, it is prudent to acknowledge certain problems while concentrating on thwarting the problems that are totally unacceptable." Certification of product, rather than process, will certainly require extensive testing by the certifier. All the techniques introduced in earlier parts of this book and those included in BS 7925-2 are also useful for the certifier.

Certification of a software product itself is a general concept and is used to ascertain the quality of the software. Since many software quality attributes exist, certification can be performed to measure the extent to which the software satisfies some particular attributes. For example, J. Voas and J. Payne [22] proposed a measure of the dependability of a software component and, based on it, proposed a method to certify and compare the dependability of software components.

It should be apparent by now that developing or assuring the quality of software components involves a very strict discipline. Therefore, certification of software component developers should be a natural consequence. Voas proposed a "software quality certification triangle" consisting of process, product, and personnel; the rationale should be quite clear. Finally, since component-based software development involves domain engineering, personnel certification may require some domain knowledge and hence may also be conducted for individual industrial sectors.

15.7 Liability aspects of producing software components and their certification

As pointed out by J. W. Moore [17], the liability benefit of certification has been underappreciated in the software engineering community. This section briefly discusses the liability issues associated with selling software and certifying software. Many argue that without standards and certification, extensive reuse of software components will probably not occur [23]. Therefore, liability is a critical issue for the success of component-based software engineering. There has been a significant amount of literature on software liability in general, but little on the liability associated with software components in particular. We summarize key findings reported by I. Lloyd [24] and C. Kaner et al. [25]. The former discusses certification of software from the perspective of British law while the latter from the U.S. perspective. The two judicial systems are similar, but they have different laws and distinct legal precedents. Legal situations may have changed significantly since their writing so that some findings may no longer apply. The authors of this book are not legal experts, and the discussion in this section is intended to provide a flavor of the legal aspects of software certification that pertain to software components. In Section 15.7.1, we provide a "big picture" of several key concepts in liability litigation. We address possible liability lawsuits first and then possible defenses. The following sections provide more details.

15.7.1 A "big picture"

Complaints about a product can be grouped into two categories: contractual and noncontractual. A sales contract usually includes specifications of what the product does, but not what the product does not or should not do. If a buyer believes that the product does not do what the seller promised that the product would do, the buyer can sue the seller for breach of contract and the resulting contractual liability. Kaner et al. [25] stated, "Judges usually award limited damages in contract lawsuits. For example, within contract law, the natural award for a product that does not work as advertised is a refund."

But because a contract usually does not specify what the product should not do, when loss of property, personal injury, death, or other damages not explicitly addressed in the contract result from a product defect, the buyer cannot sue the seller for breach of contract. However, the buyer can pursue noncontractual liability. In the case of injury, death, or loss of property, the buyer can sue for breach of tort duty and the resulting tort liability. Kaner et al. [25] stated, "The requirements to win a tort lawsuit are quite strict." If won, tort lawsuits may result in large damage awards, much larger than what contract lawsuits may.

The core of tort liability is fault or negligence. Due to the difficulty for a buyer to prove the fault of a seller in some cases, the concept of product liability or "no-fault" liability emerged where proof by the buyer of the fault or negligence on the part of the seller is no longer required.

A response to this reduction of burden of proof on the buyer is the legal technique of the state-of-the-art defense, with which the seller can argue that given the state of art at the time of the product being put on the market, the existence of the defect could not have been discovered. This technique is also useful in a tort lawsuit.

Another response is the use of liability disclaimer. Such disclaimers can be used to limit not only noncontractual liability, including tort liability and product or no-fault liability, but also contractual liability, although the effectiveness differs significantly.

15.7.2 Software component: a product, not a service

Liability for services and that for goods (i.e., products) differ significantly in both U.S. and British laws. In the United States, custom-designed software (i.e., software written exclusively for a customer according to the customer's requirements—by the supplier through exercising skills and labor on the customer's behalf) is considered as a service [24]. An extreme case would be that a contractor is hired to write a piece of software from scratch and is paid by the hour [25]. Software that has been developed first according to some generic requirements and then put on the open market for sale is considered as a product. Lloyd [24] reported that this issue had been extensively litigated in U.S. courts, but not in British courts; he suggested that this distinction is the "best view" for the British market in the absence of decisive legal authority. Therefore, it is safe to assume that a software component should be considered as a product and is subject to product liability laws.

Liability laws regarding product defects are more stringent than their service counterparts because, in addition to the contract agreement if any, liability for product defects is also based on buyer expectations. The "big picture" provided in Section 15.7.1 and the discussion to follow in the rest of this section all pertain only to product defects, as opposed to service defects.

15.7.3 Liability lawsuits against product defects: contract, tort, and product ("no-fault") liability

Lloyd [24] categorized liability associated with product defects into contractual liability and noncontractual liability. The British Sale of Goods Act of 1979 states that goods (i.e., a product) supplied under a contract of sale must correspond with any description that may have been applied to them and must also be of merchantable quality and reasonably fit for any specific purpose or purposes for which they are supplied [24]. (The criteria for "merchantable quality" and "reasonably fit for any specific purpose" involve buyer expectations, and therefore, liability for product defects goes beyond contract agreements.) Lloyd [24] stated, "Any statement, however, that the goods conform with specified standards or have been accredited by specified agencies or have undergone particular forms of testing will be regarded as descriptive." Such descriptions become part of the contract in United Kingdom, according to the British Sale of Goods Act of 1979.

U.S. laws on this are similar. Anything used by the seller in the bargaining negotiation to sell the product constitutes an "express" warranty. (Warranties do not have to be written, and do not even have to be called "warranties." There are also implied warranties.) For example, under the U.S. Uniform

Commerce Code (UCC), product brochures and product specifications are considered to be express warranties.

Contractual remedies are available in the event of a supplier's breach of contract, including rescission of contract (with a full refund) or claim of compensation for the diminished value of the product. The appearance of the word "reasonably" in pertinent laws serves to limit the extent of the supplier's liability. Often an aggrieved customer is satisfied with a corrective "software patch." As mentioned earlier, judges usually award limited damages in contract lawsuits.

The category of noncontractual liability includes tort liability and product liability. The foundation of law of tort is tort duty, which means that "everyone, whether individual or company, owes a duty of care to avoid actions which are likely to cause harm" [24]. Harm here refers to injury, death, and property damage. A primary requirement for tort liability is negligence [24] or fault [25], among others. If a standard exists for the development and quality assurance of software components but the seller of a component neglected to practice it and the component caused harm, then the seller may be held liable due to its negligence. (If the seller claims, during the bargaining process, that it followed the standard but actually did not follow it, this would also constitute breach of contract. Recall that such a claim constitutes an express warranty under the U.S. law.)

But breach of tort duty requires proof of fault or negligence committed by the seller, which may be difficult in some cases. Such difficulties gave rise to the concept of "no-fault" product liability. The resulting laws relieve the buyer from the burden of proof of seller's fault. Such tort or "no-fault" product liability lawsuits sometimes result in the award of large compensatory damages and even punitive damages to the buyer.

15.7.4 Disclaimer as a liability defense

A common legal technique to limit tort and no-fault product liability is liability disclaimer. Such a technique has also been used to limit contractual liability, and such a use is often referred to in the United States as a warranty disclaimer. However, the effectiveness of such disclaimers depends on the contractual or noncontractual nature of the breach. Regarding *contractual* liability under relevant British laws, Lloyd stated, "In business contracts, the Unfair Contract Terms Act provides that any clause attempting to restrict or exclude liability will be effective only in so far as it is fair and reasonable. The question of whether an exclusion clause is fair and reasonable or not will have to be determined by reference to the facts of an individual transaction." After all, a seller cannot expect to be able to agree to a contract and negate it immediately afterwards with a disclaimer. In fact, the liability associated with some particular types of breach of contract not only cannot be excluded with a disclaimer but may even be considered as criminal. Lloyd [24] pointed out, "... the British Trade Descriptions Act of 1968, which makes the issuing of a false description a criminal offence, specifically refers to claims that goods have been tested or approved by any person or that they conform with any standards." In the United States, warranty disclaimers are not effective against express warranties [25].

Regarding noncontractual liability under related British laws, Lloyd [24] stated, "In relation to noncontractual liability the [British Sale of Goods] Act provides that no attempt may be made to restrict or to exclude liability for death or personal injury resulting from negligence. Any other form of liability may, however, be excluded subject to the provision of a suitable notice to this effect. This may take the form of a disclaimer notice which appears whenever the software is loaded and run."

Some researchers pointed out the fact that most software producers use liability disclaimers to attempt to avoid or minimize their liability in case of software defects, and cited this fact as a reason why software certification would not work. The argument is that even the software sellers themselves do not want to be liable for their own products, why would any certification agency be willing to assume that risk? They suggested that a way to deal with this situation is to let potential buyers use the software component for a given amount of time so that the buyers can test the components and make a purchase decision accordingly. Note that such testing is typically performed without the source code (i.e., black-box testing) and may require a large amount of resources.

15.7.5 The roles of standards and certification in liability defense

Another defense against such tort or no-fault product liability lawsuits is the "state-of-the-art defense." Software producers may be able to use this defense if they adhere to existing standards and have their products, processes, and/or personnel certified by a credible third party. This brings out a key value of standards and certification as a means to limit liability risks. Again, why would a third-party certifier be willing to assume the liability risks? In addressing the inspections required for software certification, Lloyd [24] pointed out, "Should this inspection be carried out negligently the certifying agency may be held liable to any third party who suffers loss as a result of a software defect that should have been identified by the certification process." As in the case of liability exclusion or limitation for software sellers, such a certifier's risk of tort liability can be reduced if there exists credible state-of-the-art guidance for how certification should be performed and the certifier adheres to that guidance, preferably with adequate documentation to prove the adherence. In addition to providing the ISO 9000 family of quality standards, ISO also provides guidance for certifying compliance with the ISO 9000 standards, although it does not offer any such certification services.

15.8 Summary

Although there exist many software standards, few specialize in software components or component-based software. We provided a "big picture" about existing software standards and a more detailed discussion on standards specifically developed for software components or their reuse. The few standards developed specifically for software components do not seem to cover the whole spectrum of the activities involved, and therefore, despite the complaint made by many about the proliferation of software standards, there seems to be a need for more standards to guide the development and the use of software components. Some of the software standards mentioned in this chapter will be further discussed in Chapters 16 and 17, particularly those developed specifically for software components and their reuse. For example, IEEE 610.12, IEEE 730, IEEE 830, and BS 7925-2 will be discussed in more detail in Section 16.1. Also, more details about IEEE 1517 will be provided in Chapter 17, with IEEE 12207 as the backdrop. Needed areas of standardization will be the subject of Section 17.3.

It is likely that black-box testing is the only way for a user to test a software component. Such testing can help achieve good software quality provided that the specifications are complete and clear and that the component is used strictly according to the specifications. However, it can be anticipated that every user may have to spend much resource on testing the component. In addition to the economy of reuse expected of reuse of software components, the economy of testing through certification can further reduce the cost of using software components. Note that the certifier of a software component may not be limited to black-box testing, and may even have full access to the code, related documents, the developers, and much more.

Selling software components and component-based software systems as well as certifying such software products involve liability risks. We provided a "big picture" about these risks and discussed the possible roles of software standards and third-party certification in minimizing these risks. So far, neither contractual liability nor tort liability seems to have inordinately burdened the software industry, but this may change with the possible widespread reuse of defective software components. In any event, adherence to relevant software standards may enable a software vendor to use the "state-of-the-art" defense to minimize the often-dreaded tort liability.

References

[1] RCTA, Inc., RTCA/DO-178B: Software Considerations in Airborne Systems and Equipment Certification, Washington, D.C., December 1992.

[2] Hesselink, H. H., "A Comparison of Standards for Software Engineering Based on DO-178B for Certification of Avionics Systems," *Microprocessors and Microsystems,* Vol. 19, No. 10, December 1995, pp. 559–563.

[3] Magee, S., and L. L. Tripp, *Guide to Software Engineering Standards and Specifications,* Norwood, MA: Artech House, 1997.

[4] Moore, J. W., *Software Engineering Standards: A User's Road Map,* Los Alamitos, CA: IEEE Computer Society, 1998.

[5] Emmerich, W., et al., "Managing Standards Compliance," *IEEE Trans. on Software Engineering,* Vol. 25, No. 6, 1999, pp. 836–851.

[6] Schneidewind, N. F., and N. Fenton, "Do Standards Improve Quality? (Point vs. Counterpoint)," *IEEE Software,* January 1996, pp. 22–24.

[7] Geary, K., "Military Standards and Software Certification," in *Software Certification,* B. de Neumann (ed.), London: Elsevier Applied Science, 1989, pp. 60-72.

[8] Herrmann, D. S., *Software Safety and Reliability: Techniques, Approaches, and Standards of Key Industrial Sectors,* Los Alamitos, CA: IEEE Computer Society, 1999.

[9] Paulk, M. C., et al., *Capability Maturity Model® for Software, Version 1.1,* Technical Report CMU/SEI-93-TR-024, Software Engineering Institute, Carnegie Mellon University, Pittsburgh, PA, February 1993.

[10] Paulk, M. C., C. V. Weber, and B. Curtis, *The Capability Maturity Model®: Guidelines for Improving the Software Process,* SEI Series in Software Engineering, Reading, MA: Addison-Wesley, 1995.

[11] Paulk, M. C., "How ISO 9001 Compares with the CMM®," *IEEE Software,* January 1995, pp. 74–83.

[12] van der Pijl, G. J., G. J. P. Swinkels, and J. G. Verrijdt, "ISO 9000 Versus CMM®: Standardization and Certification of IS Development," *Information & Management,* Vol. 32, 1997, pp. 267–274.

[13] Leung, H. K. N., "Improving the Testing Process Based upon Standards," *Software Testing, Verification, and Reliability,* Vol. 7, 1997, pp. 3–18.

[14] Burnstein, I., T. Suanassart, and R. Carlson, "Developing a Testing Maturity Model for Software Test Process Evaluation and Improvement," *Proc. of 1996 IEEE International Test Conference,* October 1996.

[15] McClure, C., *Software Reuse: A Standards-Based Guide,* Los Alamitos, CA: IEEE Computer Society, 2001.

[16] Desai, M., "UL 1998 — Software in Programming Components," *Proc. of Embedded Systems Conference (ESC),* Spring 2000.

[17] Moore, J. W., "An Integrated Collection of Software Engineering Standards," *IEEE Software,* November/December 1999, pp. 51–57.

[18] Institute of Electrical and Electronics Engineers, *Software Engineering, 1999, Vols. 1–4,* Piscataway, NJ: IEEE Press, 1999.

[19] Rae, A. K., H. L. Hausen, and P. Robert, (eds.), *Software Evaluation for Certification: Principles, Practice, and Legal Liability,* New York: McGraw-Hill, 1995.

[20] de Neumann, B., (ed.), *Software Certification,* London: Elsevier Applied Science, 1989.

[21] Voas, J., "Certifying Software for High-Assurance Environments," *IEEE Software,* July/August 1999, pp. 48–54.

[22] Voas, J., and J. Payne, "Dependability Certification of Software Components," *Journal of Systems and Software,* Vol. 52, 2000, pp. 165–172.

[23] Flynt, J., and M. Desai, "The Future of Software Components: Standards and Certification," in *Component-Based Software Engineering: Putting the Pieces Together,* G. T. Heineman and W. T. Councill, (eds.), Reading: MA: Addison-Wesley, 2001.

[24] Lloyd, I., "Certification of Computer Software: The Legal Aspects," in *Software Certification,* B. de Neumann, (ed.), London: Elsevier Applied Science, 1989, pp. 1–22.

[25] Kaner, C., J. Falk, and H. Q. Nguyen, *Testing Computer Software,* 2nd ed., New York: John Wiley, 1999.

CHAPTER

16

Component quality verification and measurement

Contents

Major potential benefits of developing and using software components include higher productivity and better quality. Achieving the quality potential requires stringent discipline in both software development and quality assurance, including the development of and adherence to an integrated process for software development and quality assurance of software components. Chapter 13 provided an example of such a process. A critical part of such an integrated process is verification of component quality. Quality measurement and metrics play an important role in this critical part. This chapter addresses verification of component quality from the perspective of a component developer, with an emphasis on quality measurement. Verification of the quality of a software component from the perspective of a component user (i.e., the developer of a component-based software system using the software component) is within the scope of Chapter 17.

A major part of an integrated SD & QA process is to (1) define standards for each step of the process, (2) verify if the standards for each step are followed, and (3) validate the whole product, including the software itself and various related documents. Validation of software includes (3a) checking if all requirements can be traced and are built into the software and (3b) testing, through executing the software, if the software does satisfy the requirements. In this chapter, the scope of verification of software quality includes (1), (2), (3a), and (3b).

Verification of software quality has been an integral part of commercial software development processes. We briefly discuss major classical techniques for software quality verification in

Section 16.1; references will be provided for interested readers. Section 16.2 addresses some recent approaches and proposes a way to integrate them with the classical approach discussed in Section 16.1. Throughout this chapter, concepts and methods particularly important for software components will be discussed after a brief introduction to the general context of verification of software quality.

Much of the performance and quality of a hardware product can be quantified, and quality measures have played a pivotal role in hardware quality assurance. The software community has also known the importance of quality measures for decades. Frameworks have been developed for software measurement to guide the development and justification of software quality measures. We briefly discuss the framework proposed by N. E. Fenton [1] and expand it to include a discussion of the purposes of developing software quality measures in Section 16.3. That section also provides a brief contrast between hardware quality measurement and software quality measurement. Although our emphasis is on quality measurement, it goes without saying that not all aspects of quality can be easily quantified and measured. Section 16.4 summarizes practical software quality metrics that have been used successfully by some major software development organizations and then discusses software metrics that are particularly important for measuring the quality of software components. Of all the important quality metrics for software components, those related to defects are perhaps the most important ones. They will be the focus of Sections 16.5. Although many software metrics have been defined, the true relevance of some of them to software defects is still being studied and remains inconclusive. Some recent empirical studies will be summarized, and difficulties involved in such empirical studies will also be discussed in Section 16.5. Since object-oriented programming is of particular importance to the development of both software components and component-based software systems, some of our discussions in Section 16.5 pertain to object orientation exclusively. Section 16.6 discusses other practical measures that are important for the quality of a software component. Concluding remarks are given in Section 16.7.

16.1 The classical approach to verification, validation, and testing

Verification of software quality involves the classical tasks of *verification and validation* (V&V) and *testing and evaluation* (T&E). We clarify their definitions, and then briefly discuss related standards and methods.

16.1.1 Terminology and tasks

The software development and quality assurance community does not seem to have reached a consensus on the definitions of validation and verification. We use the definitions given in IEEE 610.12—*Standard Glossary of Software Engineering Terminology* [2]: "Verification is the process of determining whether or not the products of a given phase of the software development cycle fulfill the requirements established during the previous phase.... Validation is the process of evaluating software at the end of the software development process to ensure compliance with software requirements.... Testing is the process of exercising or evaluating a system or system component by manual or automated means to verify that it satisfies specified requirements or to identify differences between expected and actual results."

The disagreement seems most obvious for the definition of verification. S. L. Pfleeger [3] stated, "...validation ensures that the system has implemented all of the requirements, so that each system function can be traced back to a particular requirement in the specification. System testing also verifies the requirements; verification ensures that each function works correctly. That is, validation makes sure that the developer is building the right product (according to the specification), and verification checks the quality of the implementation." In her definition, verification is performed to ascertain the correctness of individual functions implemented and other quality attributes of the implementation, not to check the fulfillment of the requirements generated by a generic phase of the development cycle by the next. Validation as defined in the IEEE Glossary includes testing in the scope. Although some authors do include testing as an integral part of validation (e.g., [4]), others do not. For example, G. G. Schulmeyer and J. I. McManus [5] stated, "The difference between V&V and T&E is that V&V can be accomplished without testing the software (but usually is not). The essence of V&V rests in the traceability of requirements.... The essence of T&E is in the actual exercising of the computer programs during test. As each test is made and evaluated, the validation assurance which traces requirements to software under test can be checked off for each test until all tests have been passed." Despite the differences, traceability of requirements is a critical component of validation across different definitions.

These tasks of verification, validation, and testing have been discussed in Chapter 13 in the context of an integrated software development and quality assurance process for software components. This chapter focuses on these three aspects and expands on the discussion of Chapter 13 related to these aspects.

16.1.2 Standards for specification of requirements, design, implementation, and testing

Given the standards for specification of requirements, design, implementation, and testing, a major task of verification of component quality is to check each of these steps for the adherence to the standards. We briefly discuss these standards next. Note that in addition to the standards approved by professional standard-setting organizations, any software development organization can and should develop its own sets of standards, perhaps by expanding or tailoring the public standards.

As discussed in Chapter 13, principal topics to address in a *software requirements specification* (SRS) document for a software component include:

- User requirements regarding (eight categories):
    - Functionality;
    - Performance (efficiency, reliability, and so forth);
    - Interface;
    - Assembly into to different (component-based) application software systems after unpacking and configuration by the user (interoperability);
    - Porting to different operational environments, including operating systems and hardware (portability);
    - Standards;
    - Certification (verifiability);
    - Documentation.
- Certifier requirements;
- Other quality requirements:

- User-imposed quality factors that are among the commonly accepted quality factors;
- Other possible quality factors for the software components being developed;
- The importance measures of these factors and possible trade-off among them.

Standards for SRS documents include IEEE STD 830. For a detailed discussion of good practices in requirements engineering, the reader is referred to

[6]. Requirement traceability is a critical part of validation. Although many commercially available CASE tools support some degrees of requirement tracing, more research is needed. B. Ramesh and M. Jarke [7] recently proposed a reference model for requirement tracing based on an empirical study of user requirements (through focus groups across software development organizations of different nature), and they developed a prototype tool accordingly. They also provided a good literature review.

Design standards in general address the following aspects of software development [8]:

- Naming convention;
- Interface format;
- System and error messages;
- Defect classification;
- Software size;
- Design representation.

Detailed discussion of software design quality and related metrics can be found in [9].

Implementation standards address the following issues [8]:

- Standards review;
- Naming, interface, and message standards;
- Coding standards;
- Size standards;
- Defect standards;
- Defect prevention.

16.1.3 Reviews, audits, walk-throughs, and inspections for verification of adherence to standards

IEEE Standards for Software Quality Assurance Plan (IEEE Std 730-2002) recommends the following reviews and audits as a minimum: *system requirements review* (SRR), *preliminary design review* (PDR), *critical design review* (CDR), *software verification and validation review* (SVVR), functional audit, physical audit, four in-progress audits (code versus design documentation; interface specifications, including hardware and software; design requirements versus functional capabilities; functional capabilities versus test descriptions), and periodic managerial reviews.

Vincent et al. [4] recommended more reviews and audits, including structured walk-throughs and code structured walk-throughs. E. Yourdon [10] discussed the mechanics and psychology of walk-throughs and management's role in walk-throughs, and provided guidelines for analysis, design, and code walk-throughs important for software development. For these reviews, audits and walk-throughs, Vincent et al. [4] provided a listing of 178 itemized quality requirements (attributes) extracted from [11] as well as 66 metrics. The itemized quality requirements are grouped under quality subcriteria, which are in turn grouped under quality criteria. (Some criteria have only one subcriterion, in which case the subcriterion bears the same name as the criterion.) Some of these itemized quality requirements and metrics are more practical and useful than others. An example itemized quality requirement is: Free from Input/Output References under the Sub-criterion and Criterion of Machine Independence, and the corresponding metric is

$$1 - \frac{\text{number of modules with I/O references}}{\text{total number of modules}}$$

Vincent et al. [4] suggested "...that the review and audit checklists are an even more powerful bug preventer than software testing, and that these should be the primary tool of the SQA effort."

Going beyond these inspection activities, M. E. Fagan [12] proposed a set of rigorous rules for software inspections; some authors refer to inspections conducted according to those rules as Fagan-type inspections. Some empirical evidence confirmed the value of these reviews, audits, walk-throughs, and inspections. For example, Fagan [12] reported in his much-cited survey that errors in code detected by peer reviews during system development account for 87% of the total number of coding errors detected during system development.

16.1.4 Testing standards for software components

Recently, British Standards Institution has defined a testing standard for software components: BS 7925-2:1998 [13, 14]. In BS 7925-2, a component is defined as a minimal program for which a separate specification is available. Although this definition is not exactly the same as and is actually less restrictive than our definition for a software component, the standard is very important. It needs to be augmented in order to capture all the concerns involved in developing software components; the additional concerns stem from the need for assembly of a software component into a variety of application software systems and the need for portability to a variety of operating systems and hardware. The most important clauses of BS 7925-2 describe a generic test process and define test case design techniques and test measurement techniques.

The standard mandates that at the project level there must be a project component test strategy and a project component test plan. It also mandates that each component must have a specification from which it is possible to derive the expected outcome for a given set of inputs. The project component test strategy requires the specification of:

- Test case design techniques, which must be selected from the techniques to be specified later;
- The criterion for test completion (including rationale for its selection), which must be measured using techniques to be specified later;
- The degree of independence of the testers;
- The approach to component testing (e.g., isolation, top-down, bottom-up, and so forth);
- The test environment, including both hardware and software requirements;
- The test process to be used for software component testing.

BS7925-2 also suggests a generic component test process.

Thirteen specific test case design techniques and 11 specific test measurement techniques are specified in the standard. The 11 test measurement techniques are directly related to 11 of the 13 test case design techniques. We list all test case design techniques, with all the test measurement techniques attached to the corresponding test case design techniques in parentheses:

- Equivalence partitioning (equivalence partitioning coverage);
- Boundary value analysis (boundary value analysis coverage);
- State transition testing (state transition coverage);
- Cause-effect graphing (cause-effect coverage);
- Syntax testing;
- Statement testing (statement coverage);
- Branch/decision testing (branch/decision coverage);
- Data flow testing (data flow coverage);
- Branch condition testing (branch condition coverage);
- Branch condition combination testing (branch condition combination coverage);

- Modified condition decision testing (modified condition decision coverage);
- *Linear code sequence and jump* (LCSAJ) testing (linear code sequence and jump coverage);
- Random testing;
- Other testing techniques (other test measurement techniques).

Testing is the primary subject of this book, and it has been discussed in detail in Parts I, II, and III. Test coverage for the purpose of verifying the quality of a software component by its developer during unit testing was discussed in Chapter 7. (Test coverage for the purpose of verifying the quality of a component-based software system was discussed in Chapter 9, in the context of integration testing by the developer of the system.)

16.2 Recent approaches and their integration with the classical approach

16.2.1 Technical approaches

As discussed in Chapter 14, R. G. Dromey [15, 16] defined the concept of an element of a software product and the concept of "quality-carrying properties" of such a software element. Software products include requirement documents, software design, and its implementation. These software elements are linked to high-level quality attributes of a software product through the quality-carrying properties. Given high-level quality attributes desired by the user, verification of software quality becomes a task of (1) identifying the software elements related to the desired quality attributes and (2) checking if the software elements do possess the quality-carrying properties. For example, to check a program (i.e., the implementation of a software design, as a software product) for the high-level quality attributes *functionality* and *reliability*, first find those program elements that influence functionality and reliability through the elements' (quality-carrying) properties. One of such product elements is a "variable," and one of such quality-carrying properties of this product element is "assigned." (As discussed in Chapter 14, "A variable is assigned if it receives a value either by assignment, input or parameter assignment prior to its use.") To verify functionality and reliability, one would then check, among other things, if all variables are "assigned." Recall that requirement documents, software design, and programs (i.e., implementation of the design) are each considered as a software product in Dromey's approach.

Therefore, his approach can be used for verifying the quality of design, coding, and requirement documents. (Note that test plan can be viewed as a product, and Dromey's approach can be used to develop concrete rules and guidelines for developing test plans too.)

As discussed in Chapter 14, J. Bansiya and C. Davis [17] extended Dromey's approach to object-oriented software design, and I. Tervonen [18] proposed an approach to linking top-level quality factors to bottom-level design rules. Their methods can be used in a similar way as Dromey's approach for verification of design quality.

Recently, many design metrics and implementation metrics have been developed in an attempt to improve software quality, particularly the quality of object-oriented software [17, 19–25]. Many design and implementation guidelines have also been proposed. However, the relationships between the guidelines and software quality are far from being empirically validated. Recently, many such empirical studies have been reported [26, 27]; some of these studies relate to software defects and will be summarized in later sections of this chapter.

16.2.2 Organizational approaches

The concept of team software development perhaps was conceived and also practiced during early days of scientific programming. As software systems became more and more complex, the concept grew and evolved. The concept of "pair programming" is relatively new and has been practiced and often promoted as part of "extreme programming" [28]. Pair programming can be characterized as two programmers working side by side at one computer on the same problem. The quality of software design and implementation is relatively invisible when compared to its hardware counterpart. Although reviews, audits, walk-throughs, and other types of inspection can help uncover design and coding defects, they usually do not get into as much detail as the programmers performing the design and implementation. In addition, building software right according to the standards is much better than inspecting the software to uncover possible defects after it has been built. However, whether a developer knows the standards clearly and whether the developer adheres to the standards may not be clear or "visible" to the peers and managers. Pair programming provides an opportunity for the two programmers to check their programs against the standards and, perhaps more importantly, to check each other. It therefore provides more visibility into software design and its implementation. In addition, many software developers believe and have reported that two developers often produce better designs and implementations. Pair programming, if practiced appropriately, has the potential of achieving higher quality, but with the possibility of lower productivity. The possibly higher development cost may not be a critical factor for developing software components because of the intended widespread reuse; the higher quality is exactly what software components require. Therefore, component development may benefit much from pair programming. In fact, it may be exactly what is needed for development and quality assurance of software components. This concept seems to carry the concept of team development to an extreme, and this may explain why it has been promoted as an integral part of *extreme programming* (XP) [28]. Some recent literature reports success stories with empirical evidence [29].

16.2.3 Integration of the classical and recent approaches

The classical approach tends to emphasize the process but does not provide enough detailed guidelines for the individual steps. The recent approaches complement the classical approach well in that they enable software quality assurance to penetrate more deeply into the individual steps, particularly design and implementation, in a systematic way. We now sketch a "big picture" of seven steps for verifying the quality of software components.

1. Determine the products to be produced during the component development process, including requirements, designs, programs, test plans, and the corresponding documents. Test plans may be produced and accumulated across different phases.
2. Determine the desired high-level quality attributes of the software with respect to the classical approach (e.g., McCall's factors/criteria/metrics hierarchy, ISO 9126, and so forth); determine, for each software product (including the software itself, documents, test plans), the desired high-level quality attributes with respect to the more recent approaches (e.g., those of Dromey and Bensiya and Davis).
3. For each individual software product to be produced at the end of a phase (e.g., the software itself, requirements, design, test plans, and so forth):
    - *Product metrics and rules:* For any generic software product, identify product elements and their "quality-carrying properties" linking the elements to the high-level quality attributes identified in the previous step, and develop product development rules using Dromey's approach [15, 16]. For OO software design, identify pertinent metrics and determine target values and the corresponding rules using extensions to Dromey's approach (e.g., Bansiya and Davis' method

[17]), or similar methods like Tervonen's GRCM method [18]. (These methods can be extended beyond OO design to cover other design paradigms or even to cover any generic software product.) For the software itself, itemized quality requirements and metrics of McCall et al. [11] or Vincent et al. [4] can also be used. For software design in general, Card and Glass' metrics [9] can also be used.

- *Process standards:* Develop or select standards for the process of producing the product.
- Build quality into the products according to (1) the product metrics and rules and (2) the process standards determined above.
- Verify, via reviews, audits, walk-throughs, and other inspections, the quality of the software product against (1) the product metrics and rules and (2) the process standards determined above.

4. Verification across different phases—requirements and test plans:
   - Verify a product against the requirements set in the previous phase;
   - Build up and validate test plans, using BS 7925-2 (including a component test process, test case design techniques and test coverage measurement techniques) and supplement the plans with techniques verifying satisfaction of the need for assembly of the software component into a variety of application software systems and the need for portability to a variety of operating systems and hardware.
5. Testing of code, according to the test plans, and improving the software.
6. User acceptance testing for the software products.
7. Field use of the software products.

Verification of quality is involved in every one of the last five steps; in these steps, the software products are checked for the desired quality attributes. Key success factors of a software component include correctness and reusability. Reusability is a desirable quality attribute for any generic software product. However, it is a requirement for a software component, and the requirement must be specified explicitly and in detail (e.g., the ability and even ease of assembly into specified environments of component-based application systems, portability to specified operating systems and hardware configurations). These reusability requirements, like the more traditional categories of requirements (e.g., functional requirements), must be verified from one phase to the next and validated as a whole. Test plans must be developed and accumulated through all development phases leading to system integration testing; the requirements must be tested according to the plans via execution of software.

16.3 A framework for software quality measurement

Fenton [1] developed a framework for software measurement. We adopt his framework in addressing measurement of software quality, and expand his framework with a discussion of fundamental purposes of quality measurement. We will do this with a brief contrast between hardware quality measurement and software quality measurement.

Fenton [1] grouped entities of interest for measurement into three classes:

- *Processes:* any activities related to software development (regardless of the phases of the life cycle);
- *Products:* any deliverables, documents, or artifacts that arise out of the processes;
- *Resources:* any items that are inputs to the processes.

Fenton distinguishes internal and external attributes in the following way:

- Internal attributes of a product, process, or resource are those that can be measured purely in terms of the product, process, or resource itself.
- External attributes of a product, process, or resource are those that can only be measured with respect to how the product, process, or resource relates to its environment.

Quality is customer satisfaction, and the customer of a software component is interested in external product attributes like functionality (correctness), reliability, verifiability (including testability and certifiability), usability, efficiency, integrity (security), reusability, maintainability, portability, and interoperability. Therefore, quantifying quality attributes requires external measures of a product. However, what a development organization of a software component usually has access to and control over are internal measures about the software development processes, the products, and the resources themselves. Therefore, estimation or prediction of component quality (by a software development organization before releasing a software component) often requires the establishment of statistical and even causal relationships between an external quality measure and some internal measures of the product or the process. Internal product attributes include length (size), functionality, modularity, reuse, redundancy, and syntactic correctness. Internal process attributes include time, effort, number of requirement changes (before software release), and defects found (during different phases).

There is a general relationship between the classical quality models (e.g., the hierarchical factors/criteria/metrics quality model of McCall et al.) and Fenton's framework. Fenton [1] observed "... that the 'quality factors' generally correspond to external product attributes. The 'criteria' generally correspond to internal product or process attributes. The 'metrics' generally correspond to proposed measures of the internal attributes. In each case, we say 'generally' because there is considerable fuzziness in the quality modeling work." Recall that, in Chapter 13, we used the concepts of *external quality characteristics* and *internal quality characteristics* commonly accepted in the hardware quality assurance community to categorize software quality factors and quality criteria. There, the qualifiers external and internal are used in a similar fashion.

Use of internal measures for the purpose of estimating or predicting the external product quality measures requires theoretical as well as empirical justification. Empirical justification typically requires controlled experiments and careful statistical data analysis. Some internal measures commonly used as predictors for software defect number will be discussed in Section 16.5, together with a brief discussion of the pitfalls of their use.

The necessity of justification for their use stands whether or not one makes the distinction between internal and external measures. For example, Fenton [1] stated, "... attributes like quality are so general that they are almost meaningless. Thus we are forced to make contrived definitions of the attributes in terms of some other attributes which are measurable." Such measurable attributes can be referred to as surrogate measures. The relationships between software quality and such surrogate measures must be justified theoretically and empirically.

We now briefly address possible fundamental purposes of quality measurement and expand Fenton's framework for software measurement [1]. The possible purposes include:

- To define quality;
- To measure quality;
- To estimate quality.

We use acceptance testing (or, more precisely, acceptance sampling) taken from hardware quality assurance to illustrate these three purposes and the concomitant actions, and to contrast some major differences between hardware and software quality measurement. A measure of the quality of a batch of hardware product items (i.e., a "lot") is often *defined* to be the percentage of the items in the batch that meet the specifications (i.e., percentage conforming). (Actually, percentage nonconforming is the measure that is commonly used in this context.) The specifications of a hardware product are typically clearly defined, and therefore, whether a particular item is defective or not can typically be defined accurately and precisely. Therefore, this measure of percentage conforming as a quality measure is well defined. With this measure clearly defined, one can actually *measure* the quality of a batch by testing each and every item in the batch and calculating the percentage conforming. In this case, the quality of a batch not only is well defined but also can be precisely measured (by 100% sampling). Because of cost and several other considerations, statistical sampling is usually conducted to *estimate* the quality of the batch. With the help of some basic statistical theory and under some usually justifiable assumptions, the quality of the batch can usually be estimated with arbitrarily high degrees of accuracy. In this case, all three purposes can be achieved. These can be achieved because of mass production and the repeatability and reproducibility usually enjoyed by hardware manufacturing processes that are in statistical control. More complex hardware quality measures include reliability in terms of life distribution, time till first failure, and time between failures, but the three fundamental purposes can often be fulfilled nevertheless because of similar reasons.

Unfortunately, software quality measurement is much more complex and difficult than this. For example, in the quality hierarchy proposed by McCall et al. [11], a key quality factor is *correctness*, and the three quality criteria contributing to correctness are traceability, completeness, and consistency. A list of 24 checks is provided to help ensure satisfaction of these three criteria. While checks like

- "Design agrees with requirements"
- "Code agrees with design"
- "Error analysis performed and budgeted to module"

can certainly help ensure quality, there is no actual definition of the level of correctness of a program. (Some primitive metrics have also been defined in this context. For example, the extent to which "Design agrees with requirements" can be assessed and a "score" between 0 and 1 assigned to represent the degree.) No measurement of correctness level is possible because of the lack of a precise definition; no estimation of correctness level is possible in the absence of a well-defined measure. One closely related concept is the defect number (i.e., the number of defects or bugs contained in a software program);

a normalized version of this concept is the defect density (i.e., the number of defects per unit size of code). Note that this concept does not take into consideration the severity of a defect in terms of the consequences. More importantly, it is really not something that can be accurately measured, except for very small programs. In addition, software defect may not even be a well-defined concept. Of all different kinds of defect numbers or defect densities, the most interesting one is the postrelease defect number (i.e., the number of defects present in a software product at the time of general release).

Let us assume that software defect is a well-defined concept, and there is no ambiguity in identifying defects and counting the number. Although postrelease defect number cannot be measured, it may be related statistically to some other (internal) product or process measures. Whatever the relationship might be, establishing any such relationship will require an estimate of the postrelease defect number. Although the true postrelease defect number is not estimable or even not knowable, one can use the total number of defects uncovered (1) during a long period of field execution after release and (2) under many different field-use scenarios as a surrogate. This surrogate has also been referred to as fault-related customer change requests. However, a much more common name for it is simply the postrelease defect number, although it is only a surrogate for the *true* postrelease defect number. In what follows, we use the term "postrelease defect number" and "postrelease defect density" in the sense of the surrogate unless otherwise specified.

The traditional focus of the software community in this regard has been on identifying and validating measures that are correlated statistically with the postrelease defect number or postrelease defect density (e.g., software size, software complexity). Recently, the statistical relationships between the quality of object-oriented software and some metrics of object-oriented design and implementation have received much research attention. However, due to the fact that some potentially critical factors have not been controlled or the fact that some of these factors are difficult to control or quantify, statistical evidence has not been conclusive.

Without well-defined metrics as direct measures of software quality or as statistically validated predictors for software quality, it is difficult to specify quantitative quality requirements. So far, quality requirements regarding correctness seem to continue to remain vague. Although this does not mean that little can be done to improve software quality, we may not be able to quantify the extent to which some actions may improve or may have improved software correctness. Fenton [30] stated, "... project managers make decisions about software quality using best guesses; it seems to us that will always be the case and the best that researchers can do is 1) recognize this fact and 2) improve the 'guessing' process."

In a book review published in *IEEE Software*, S. R. Rakitin [31] quoted Bill Hetzel, "As an industry we collect lots of data about practices that is poorly described or flawed to start with. This data then gets disseminated in a manner that makes it nearly impossible to confirm or validate its significance." Recently, B. A. Kitchenham et al. [32] developed a set of preliminary guidelines for empirical research in software engineering.

Other software characteristics may be more easily defined than the vague concept of software correctness or "defect-proneness." For example, many measures of software complexity exist (e.g., lines of code, T. J. McCabe's cyclomatic measure [33], M. H. Halstead's volume measure [34], and so forth); criteria have been developed to evaluate and compare the effectiveness of different complexity measures. Object-oriented design and implementation has a more rigid structure, and it is more amenable to precise definition of specific software metrics. The relationships between such metrics and the high-level quality attributes (e.g., defect-proneness) have been a subject of study and debate; they will be discussed briefly in Section 16.5. The theoretical relationships recently proposed by Dromey [15, 16] or Bensiya and Davis [17] between product elements' quality-carrying properties and high-level software quality attributes (e.g., defect-proneness) should also be studied empirically.

Before closing this section, we briefly discuss the difference between measure and metric. While most authors use them interchangeably, some define a measure as a quantity that is directly observable while a metric is defined to be a quantity that is derived from (directly observable) measures. [Some authors (e.g., [35]) use the terms Primitive Metrics and Computed or Calculated Metrics to refer to the two, respectively.] For example, postrelease defect number, prerelease defect number, and the number of source lines of code are considered as measures, and postrelease and prerelease defect densities are considered as metrics.

16.4 Practical quality measures for software components

Many software metrics have been proposed in the literature; some of them are more practical and useful than others. Some of the proposed software metrics are related to software quality. A common advice made by practitioners is to limit the number of metrics to a small number when initiating a software metrics program, and to increase the number gradually after the initial metrics have been proven useful and accepted by the organization [35–37]. Another advice is that if a metric has not been used in making managerial decisions, stop tracking it and stop collecting data for it [37]. This section first discusses practical measures for software quality in general, and then focuses on practical quality measures for software component.

16.4.1 Practical measures for software quality

Practical software quality metrics reported in the literature include those reported by K. H. Moller and D. J. Paulish in [35] and those reported by R. B. Grady and D. L. Caswell in [36]. Throughout this chapter, a defect is defined to be "an error in software that causes the product to produce an incorrect result for valid input." This is the definition of a "fault" adopted in Moller and Paulish [35]. Moller and Paulish [35] distinguished global metrics from phase metrics. The latter refer to those obtained to reflect the quality of the individual phases of the software development process, while the former refer to those obtained either to reflect the quality of the product or the quality of key groups of phases of the software development process. We now provide a list of practical measures of software quality. We will use the commonly accepted names of the measures; the parenthesized terms following the names of the metrics are the names used in Moller and Paulish [35].

16.4.1.1 Global metrics

Global metrics include:

- *Lines of code:* This metric measures the size of a program; it is important because it has been used as the normalizing factor (i.e., the denominator) for other measures (e.g., defect density). (Other metrics for program size include the number of "function points" for procedure-oriented programming and the total number of classes or the total number of methods for object-oriented programming. T. A. Hastings and A. S. M. Sajeev [38] proposed the use of a vector, incorporating both functionality and problem complexity, to measure software size.) There are many different ways to count the number of lines of code. They can be counted with or without comments, including or excluding declarations, before or after incorporation of "includes," and with or without expansion of "macros." J. Rosenberg [39] conducted an empirical study for a dozen software applications written in C, demonstrated that different variations of counting lines of code are highly statistically correlated, and concluded with empirical evidence that any particular variation of counting lines of code can be used as long as it is used consistently across software programs. (Rosenberg [39] compared four versions of the same program files: "as-is"; "maximally formatted" by running the files through the *indent* program with several options to produce code with a maximum number of lines and white space; "minimally formatted" by running the files through the *indent* program with a different set of options to produce code with a minimum number of lines and white space; and "partially processed" by running the files through the C preprocessor *cpp* with a set of options to invoke minimal processing.) J. S. Poulin [37] suggested the use of noncommented source lines of code as the measure. (This measure is an internal product measure.)

- *Prerelease defects (system test faults):* This is the total number of defects uncovered during the system test. (This is an internal process measure.)
- *Postrelease defects (fault-related customer change requests):* This is the total number of defects uncovered during a specified period of time after the release of a program (e.g., 1 year). If the length of the specified time period is "sufficiently" long and the program has been used in sufficiently large number of field-use scenarios, this measure may be a reasonably good surrogate measure for the total number of defects contained in the released program. Statistical models have been proposed to help predict the postrelease defect number based on prerelease defect number. These models and related critiques will be addressed in Section 16.5.
- *Prerelease defect density:* This is the number of prerelease defects per unit size of code.
- *Postrelease defect density:* This is the number of postrelease defects per unit size of code.

16.4.1.2 Phase metrics

Phase metrics include:

- *Requirement specification change requests:* This is the number of change requests that are made to the requirement specification from the time of the first release of the requirements specification document to the time of the beginning of product field use. Some practitioners have observed [35] that a major cause for software project failure is frequent changes in requirements. This measure can be used to track requirements stability.
- *Design defects:* This is defined to be the number of defects uncovered during all design reviews.

- *Code defects:* This is defined to be the number of defects uncovered during all code reviews.
- *Defect discovery rate during testing (test defect rate):* This is the number of defects discovered per unit time (per unit testing staff) within each testing phase. This measure can be used to determine when to stop a testing phase by predicting time of diminishing return on test effort. (Note that a decreasing defect discovery rate may indicate that no or few more defects can be found using the test methods employed in the testing process, and hence may not be a proof for diminishing number or absence of defects in the software.)
- *Customer complaint rate:* This is the number of customer-identified defects per unit time from the time of first field use of the software product through its lifetime.

Although some researchers have argued that software complexity is better than software size for predicting the number of code defects [40], complexity measures do not seem to be as popular with the practitioners for initial introduction of metrics into software development organizations. Such complexity measures include McCabe's cyclomatic complexity measure [33] and Halstead's software science measures [34]. The former is simpler and seems to be more popular than the latter. In addition, the former seems to enjoy a higher degree of theoretical and empirical validation. These measures have been discussed at length in many software engineering books, and hence are omitted from further discussion. For more detailed discussion of practical software quality metrics, the reader is referred to [35], which summarizes several sets of quality measures used by the ESPRIT PYRAMID Consortium (consisting of nine companies from six countries that develop, test and provide consulting services for software systems) and Siemens companies and also discusses ways to design questionnaires for solicitation of customer satisfaction level.

16.4.2 Practical measures for software components

Poulin [37] addresses measurement and metrics for the development and quality assurance of software components. The principles behind the development of the metrics include:

- *Goal:* The goal is to develop a software component on time, on budget, and with quality. Use metrics to reinforce the processes and steps considered important. This reflects the common belief that "what gets measured gets done." In particular, measure schedule, effort, and quality to achieve the goal.

- *Strategy:* Begin with a pilot project with few metrics and then grow.
- *Metrics criteria:*

  - Data availability: do not let measurement become a resource drain;
  - Simple yet meaningful;
  - Consistent and repeatable (regardless of who calculates them);
  - Validated theoretically and empirically;
  - Useful for on-time and on-budget delivery of a quality product.

16.4.2.1 Basic measures with which to start a metrics program

Quality improvement is one of the strongest reasons for development and use of software components. Poulin [37] pointed out that many software projects claimed one order of magnitude higher quality in reused components, in terms of defect density. This resulted from more thorough prerelease testing of components and the enormous amount of field-testing in many different contexts. The basic measures he suggested are all used for defining productivity and quality metrics. We list those basic measures related to quality first, and then those related to productivity.

- Classification of the component: new code, changed code, code built for reuse, and reused code;
- Lines of code (instructions) by component (to measure the size of software);
- Requirement specification change requests by component (to track stability or volatility of requirements);
- Defects by phase of development, defect type, and component (to identify sources of development problems);
- Schedule (planned versus actual);
- Labor hours;
- Cost.

16.4.2.2 Key metrics

The key metrics suggested by Poulin include:

- *Quality:* One can use defect discovery rate (i.e., test defect rate) to determine when to stop a testing phase.
- *Productivity:* Poulin [37] reported that industry standards for productivity ranged from 200 to 500 source lines of code per labor month and that high levels of reuse and product-line development could lead to 40% to 400% higher levels of productivity.
- *Requirement stability:* Poulin [37] suggested that the total number of requirement specification change requests should not exceed 5% to 10% of the baseline number of requirements.

Poulin [37] also suggested other metrics, including response time, throughput of software, and some design metrics related to object orientation. Concrete design metrics related to software correctness, suggested by Poulin and other researchers, will be addressed in Section 16.5, among other things. Software availability, speed, throughput, scalability, and several other key performance measures were discussed in Chapter 11.

16.5 Predictive models regarding defect numbers and guidelines

This section discusses (1) software metrics developed to help predict the postrelease defect number and (2) quality guidelines suggested to help minimize the postrelease defect number. A popular software metric used to help predict postrelease defect number is prerelease defect number. We discuss this metric in Section 16.5.1. Other such metrics include program size and program complexity in general and a variety of design and implementation metrics for the special case of object-oriented programming. These metrics, guidelines on threshold values for these metrics, and their empirical validation are the subjects of Section 16.5.2. Section 16.5.3 discusses some important issues regarding statistical validation of software metrics. Other important metrics will be discussed in Section 16.6.

16.5.1 Prerelease defect number as a predictor for postrelease defect number: a need for caution

A common belief is that the larger the prerelease defect number (or defect density) of a module is, the larger the postrelease defect number (or defect density) would be. This hypothesis has been suggested as a guideline by some [41] while others use the two metrics only to track the numbers of defects detected during the two different phases without suggesting use of the former to predict the latter [35, 36]. A literature review conducted by Fenton and Kitchenham [42] confirmed widespread acceptance of this hypothesis despite little empirical evidence to support it. At least three published empirical studies reported just the opposite of the hypothesis [43–45]. Fenton and Ohlsson [43] found in their empirical study that the modules with highest prerelease defect densities have lowest postrelease defect densities. In fact, many of the modules with a high prerelease defect density have no postrelease defects. Their study produced no evidence to support the hypothesis. In addition, there was evidence to support the converse. The conjecture put forward by them is that the modules with a high prerelease defect density may have been so well tested that all or most defects have been tested out and that those modules with a low prerelease defect density may not have been tested properly. Data on the testing effort involved in their case study was not available, and therefore, this conjecture could not be validated [43]. Fenton et al. [45] extended that study. They classified test modules into five different degrees of complexity. Since no compensating increase in test effort for the more complex modules was provided, they assumed that simpler modules were better tested while more complex modules were more poorly tested. With this assumption and with this additional factor of complexity explicitly considered, they were able to establish "... the better-tested modules yield more defects during prerelease testing and deliver fewer defects. For the more poorly tested modules, the converse is true." As a result, caution must be exercised when using prerelease defect number as a predictor for the postrelease defect number. In fact, the practice of treating the prerelease defect density as the de facto measure of customer-perceived quality should also be reexamined.

As will be made clearer later, a major problem with some existing guidelines and their empirical validation is that only one predictor variable is used to predict the postrelease defect number, while there clearly exist other variables that are not controlled or are difficult to control.

The implications of these three empirical studies [43–45] actually go beyond the validity of the hypothesis about the relationship between the prerelease and postrelease defect numbers because most validation studies for software complexity metrics presuppose the validity of this hypothesis. More precisely, in those studies, a complexity metric is considered valid if it correlates with the prerelease defect density (with a statistically significant positive correlation coefficient). The two empirical studies conducted by Fenton and his colleagues [43, 45] point out the need to reexamine some of those validation studies.

16.5.2 Quality guidelines for software design and implementation

Although measuring the quality of software design and implementation has been a subject of research for a long time [9, 12], the pace of research seems to be much faster with the widespread adoption of object-oriented programming. In this section, we first discuss the current status of the pursuit of good measures of software complexity. This is important because such measures have been used to predict defect-proneness and maintainability of software. We then briefly discuss a number of object-oriented software metrics that are intended for use as predictors for defect-proneness of object-oriented software. Their validity will also be briefly discussed.

16.5.2.1 Complexity measures

Many complexity measures have been proposed. E. J. Weyuker [46] proposed nine desirable properties of a complexity measure for a computer program, and compared four popular such measures. These properties include, for example:

1. A component of a program is no more complex than the program itself.
2. Program complexity is sensitive to the order of the statements and hence the potential interaction among statements.
3. At least in some cases the complexity of a program formed by concatenating two programs is greater than the sum of their individual complexities.

The four complexity measures compared are statement count (i.e., lines of code), McCabe's cyclomatic measure [33], Halstead's programming effort [34], and the data flow complexity proposed by Oviedo [47]. The first two are the most popular measures of complexity. While the first two measures satisfy property (1), the latter two do not. Since Halstead's programming effort measure is designed to directly predict the amount of effort needed to implement the program, the violation of property (1) by Halstead's measure renders the effectiveness of the measure questionable, as Weyuker pointed out [46]. However, the first two measures violate both properties (2) and (3).

16.5.2.2 Predictive models for the defect numbers or densities based on a complexity measure alone

Defect density of a software component, either a module in the conventional procedural paradigm or a class in the object-oriented paradigm, as a function of the size of the component has been a subject of much research interest in recent years. A number of authors suggested that there is an optimal size for software components. For example, T. Compton and C. Withrow [48] suggested that the optimal size for an Ada component is 83 lines of source code; R. Lind and K. Vairavan [49] suggested that the optimal size for Pascal and Fortran code is between 100 and 150 lines of source code. Card and Glass [9] noted that military standards for module size range from 50 to 200 executable statements. Typical empirical evidence supporting such optimal sizes is a U-shape curve of defect density versus component size (or complexity). Compton and Withrow [48] called the concept of optimal component size "Goldilocks Principle," borrowing from the children's story the ideas of "not too big, not to small, and just right." It has been referred to in general as the Goldilocks conjecture. These claims of optimal component size have generated much debate in the software engineering community. We briefly discuss the main points of debate. The first point of debate is about the piece of the U-shape curve to the left of the optimal component size, and that is that a smaller component tends to have a higher defect density. The current general consensus seems to be that this is nothing but a mathematical artifact. Note that defect density is defined as the ratio Y/X of two random variables X and Y, where Y denotes the number of defects in a component and X denotes the size of the component, and that the 1/X factor in the definition of defect density is by mathematical necessity negatively correlated with X. Note that decomposition has been a standard way of managing complexity, and that, if smaller components indeed have a tendency to have higher defect densities, then much of the conventional wisdom about computer programming may be in doubt and must be reexamined.

The second point of debate is about the piece of the U-shape curve that is to the right of the optimal component size. If the defect density indeed suddenly increases at a certain component size, then this size can be viewed as a threshold not to be exceeded. Contrary to this observation of a threshold size, other researchers observed, based on different case studies, a phenomenon that larger modules could actually have lower densities. Fenton and Neil [30] critiqued the Goldilocks conjecture as well as studies intended to validate the conjecture. K. El Eman et al. [50] conducted a case study and provided some empirical evidence against this conjecture. All of the 174 components considered in their case study are reused components, and it is not clear whether the reuse has any bearing on their conclusion. Regardless of the answer to this question, their conclusion is particularly relevant to our focus on software components, which are developed for reuse to begin with. Fenton and Neil's [30] conclusion on all these attempts to use a regression curve to characterize the relationship between defect density and component size is that "the relationship between defects and module size is too complex, in general, to admit to straightforward curve fitting models." As mentioned earlier, many other possible factors may contribute to postrelease defect density, but an overwhelming majority of such studies use component size as the sole predictor of the defect density or defect count. Since the number of possible factors is large, and many of them cannot be controlled or are difficult to control, proper experimental design is a rather difficult task.

16.5.2.3 Object-oriented software metrics and the validation of their effect on software defect-proneness

Recently, many software metrics have been proposed for the object-oriented paradigm, and some of them have been implemented in automated computer tools developed to automate the measurement process. Different tools implement different metric sets. Most notable ones include the sets proposed by S. R. Chidamber and C. F. Kemerer [20], M. Lorenz and J. Kidd [19], F. B. Abreu and R. Carapuca [21], J. Bansiya and C. Davis [17, 24], and W. Li and S. Henry [25]. We focus on those related to defect-proneness.

It has been hypothesized that the structural properties of a software component influence its cognitive complexity, which is defined to be the mental burden on those individuals who deal with the component (including developers, testers, inspectors, maintainers), and that a high cognitive complexity leads to undesirable external quality attributes (e.g., increased defect-proneness and reduced maintainability). In particular, the distribution of functionality across classes in an object-oriented system and the exacerbation of this through inheritance may make object-oriented programs more difficult to develop and/or to understand. It has been suggested by a number of authors that highly cohesive, sparsely coupled, and low inheritance programs are likely to contain less defects. For a recent review of different proposed sets of OO metrics related to defect-proneness and the current status of their validation, the reader is referred to [27]. We provide a brief summary of the results. The 10 OO metrics for a class treated in [27] are related to size, cohesion, coupling, and inheritance, all of which may contribute to complexity and hence defect-proneness, and they are summarized below. (Six of the 10 were proposed by Chidamber and Kemerer [20] while the other four were proposed by Lorenz and Kidd [19].)

1. *Weighted methods per class (WMC) [20]:* This is the count of the methods in a class, and can be classified as a size measure.
2. *Lack of cohesion in methods (LCOM) [20]:* This is a cohesion metric and is defined to be the number of pairs of methods in a class using no common attributes minus the number of pairs of methods using some common attributes. (If the result is negative, reset it to zero.)
3. *Coupling between object (CBO) [20]:* This is a coupling metric. A class is considered as coupled with another class if the methods of one class use the methods or attributes of the other, where "use" can mean a member type, parameter type, or method local variable type or cast. It is the number of other classes to which a class a coupled.
4. *Number of parameters on average (NPAVG) [19]:* This is a coupling metric and is defined to be the average number of parameters per method (excluding the inherited methods). Methods with a large number of parameters may in general require more testing and may lead to more complex and less maintainable programs.
5. *Response for a class (RFC) [20]:* This is a coupling metric. It is the total number of methods in the response set of the class, where the response set of a class consists of the set M of methods of the class and the set of methods invoked directly by the methods in M.
6. *Depth of inheritance tree (DIT) [20]:* This is an inheritance metric and is defined as the length of the longest path from the class to the root in the inheritance hierarchy. It can also be classified as a complexity measure because of the conjecture that the lower a class is on a class hierarchy, the more complex it becomes.
7. *Number of children (NOC) [20]:* This is an inheritance metric and is defined to be the number of classes that inherit from the class under consideration.
8. *Number of methods overridden (NMO) [19]:* This is an inheritance metric and is defined to be the number of inherited methods overridden by a subclass. A large number of overridden methods may indicate a problem.
9. *Number of methods added (NMA) [19]:* This is an inheritance metric and is defined to be the number of methods added by a subclass (with inherited methods not counted). (As this number increases for a class, the functionality of that class becomes increasingly distinct from the functionality of the parent classes.)

10. *Specialization index (SIX) [19]:* This is an inheritance metric and consists of a combination of other inheritance metrics. It is defined to be the product of the number of overridden methods (i.e., NMO) and the class hierarchy nesting level (e.g., DIT) normalized by the total number of methods in the class. Higher values of this metric may indicate a higher likelihood that a class does not conform to the abstraction of its parent classes.

Many empirical studies have been conducted, but some support the hypothesized influence while others do not. A vast majority of existing empirical studies examining the influence of these metrics on the defect-proneness of a class employed univariate statistical analysis where a regression function relating the number of defects to a single metric is fitted to observed data. Based on their own data set, El Emam et al. [27] conducted a multivariate statistical analysis for each of the 10 OO metrics by including one additional metric of class size in the regression, and found that, with the class size included in the regression, none of the 10 OO metrics exhibit statistically significant influence on defect-proneness. More precisely, for each of these OO metrics, they tested the null hypothesis that the metric has no effect on software defect-proneness, but their empirical evidence is not strong enough to reject that hypothesis. El Emam et al.'s findings should not be interpreted as a final conclusion to this matter. Rather, El Emam et al. suggested a reexamination of previous validation studies regarding this matter. Since many other factors are ignored by El Emam et al., the search for truth regarding the impact of some OO metrics on defect-proneness should continue.

Other OO metrics include the *metrics for object-oriented design* (MOOD) proposed by Abreu and Carapuca [21]. Abreu and Melo [22] partitioned a graduate/senior-level *management information system* (MIS) class into eight independent development teams. Given identical requirements for a medium-size MIS project and an identical sequential development process derived from the waterfall model, the eight teams developed eight management information systems. Abreu and Melo [22] showed that individual MOOD metrics were in general statistically correlated with the defect-proneness of the eight systems.

16.5.2.4 Some practical guidelines about object-oriented design and implementation

While the search for truth about the effect of some software metrics on software defect-proneness continues, researchers and practitioners have provided some guidelines. We briefly summarize a few suggested by J. S. Poulin [37] and M. Lorenz [51]:

Poulin suggested the following upper limits for a general program:

- Size in noncommented source lines of code [37]: 100:
- McCabe's cyclomatic measure [33]: 10;
- Halstead's volume measure [34]: 30.

Lorenz [51] suggested the following upper limits for OO programming:

- Average method size: 8 for Smalltalk; 24 for C++ and Java (too big otherwise);
- Average number of methods per class: 20 (too much responsibility otherwise);
- Average number of instance variables: 6 (too many attributes assigned to the class otherwise);
- Depth of class hierarchy: 6 (testing and maintenance difficulties otherwise);
- Number of subsystem-to-subsystem relationships: the number of class-to-class relationships (for low coupling; coupling too high otherwise);
- Number of class-to-class relationships within a subsystem: high (for high cohesion; cohesion too low otherwise);
- Number of comment lines: in-line documentation on function, author, change history, and so forth, according to project standards.

16.5.3 Some problems with predictive models for defects

The validity of validation studies for the effectiveness of software metrics hinges upon proper design of experiment. *Design of experiment* (DOE) for understanding the relationships between hardware quality and its contributing factors is routinely performed, but usually with tight control of major factors (or with paired comparison to cancel out the possible effect of factors that are difficult to control). With such control, repeatability (with respect to different experimenters) and reproducibility (through time) can be more easily achieved, and such relationships can be established with higher degrees of certainty than their software counterparts.

In the hardware context, such tight control is often achieved with the technique of *statistical process control* (SPC). It has also been suggested that statistical control for the software development process may be a required step toward a true understanding of the intricate relationships between software defect-proneness and its possible contributing factors. Fenton and Neil [30] identified other major issues regarding the existing validation studies, and they include:

- Imprecise terminology about defect: In different empirical studies, defects refer to postrelease defects, the total of known defects, and the set of defects discovered after some arbitrary fixed point in the software life cycle (e.g., after unit testing);
- Lack of defect classification and lack of assessment of the seriousness of a defect;
- Problems using size or complexity as a sole "predictor" of defect count;
- Problems in statistical methodology and data quality;
- False claims about software decomposition and the Goldilocks conjecture.

Many metrics can be used to track if the corresponding processes are in control (e.g., prerelease and postrelease defect densities). Recent reports of application of SPC for software development process control include [52, 53]. Both reported a successful application of SPC for tracking code inspection rate. E. F. Weller [52] also applied SPC to tracking a metric closely related to defect discovery rate. (He tracked the problem arrival rate, where problems are "potential defects," and the problem arrival rate is defined to be the number of problems per week.) Weller used SPC "runs rules" to determine if a downward trend of problem arrival rate indicates, with a sufficiently small risk, a diminishing number of remaining defects, and to estimate the end date for testing.

16.6 Other practical software measures and metrics

Some of the metrics that have been discussed in this chapter have been referred to in the literature by other names. In general, the two words "fault" and "defect" have been used interchangeably. A specific example is that, instead of defect discovery rate, Ginac [41] used the term defect arrival rate.

Before closing this chapter, we briefly mention several additional important practical metrics as follows.

- *Defect severity:* Poulin [37] suggested the classification of defect severity as shown in Table 16.1.

- *Actual fix response time, by severity:* This can be used to gauge customer satisfaction or dissatisfaction.
- *Actual fix response effort, by severity:* This can be used to measure rework effort.
- *Delinquent fixes:* This is the number of defects whose fixes took longer than the agree-upon fix response time.
- *Defective fixes:* This is the number of fixes that did not repair the defect reported by the customer.
- *Defect removal effectiveness:* This is the percentage of defects that were discovered during system test (but beyond unit and integration test) (i.e., the prerelease defect number divided by the total number of defects present at the beginning of the system test). This total number is actually unknowable (except for very small programs) but can be approximated by the number of defects discovered during a sufficiently long period of time after the release (i.e., the postrelease defect number) plus the prerelease defect number.

16.7 Summary

In this chapter we briefly discussed major classical life cycle techniques for software quality verification and provided references for interested readers. We also addressed BS 7925-2, a British standard for testing software components. Test coverage is an important concept, and was discussed in detail in Chapters 7 and 9 in the contexts of unit-testing a software component and integration-testing a software system containing a software component,

Table 16.1 Poulin's Classification of Defect Severity

Severity Level	*Definition*	*Typical Agreed-Upon Fix Response Time After Complaint*
1	Business-critical features are absent or do not function; program crashing is possible	Between 4 to 24 hours
2	Business-critical features function most of the time, with no work-around possibilities	1 week
3	Nonbusiness-critical features are absent or do not work, or work-around possibilities exist for those business-critical features that function most of the time	2 weeks
4	Inconsequential functions that may not work as expected; typos in the documentation, and so forth	Next release

respectively. We also discussed ways to integrate some modern SQA approaches with these classical techniques. The modern approaches include the technical method of product quality modeling and the organizational method of pair programming. Product quality models were discussed in Sections 14.6 and 14.7.

Software quality metrics have not played a significant role in SQA; at least, their role is not as significant as the role played by their hardware counterparts in HQA. This is because it is usually not as easy to clearly define quality characteristics of software. Without a clear definition of a software quality characteristic, it is difficult, if not impossible, to measure the characteristic. Without a measure, it is difficult, if not impossible, to estimate the characteristic. Apparently, much more research is needed for rigorous software measurement. In the meantime, we discussed a small number of practical software measures as well as some cautionary notes about their usage.

References

[1] Fenton, N. E., *Software Metrics: A Rigorous Approach*, London, England: Chapman and Hall, 1991.

[2] Software Engineering Technical Committee, *Standard Glossary of Software Engineering Terminology (IEEE Std 610.12-1990)*, Los Alamitos, CA: IEEE Computer Society, 1990.

[3] Pfleeger, S. L., *Software Engineering: Theory and Practice*, 2nd ed., Upper Saddle River, NJ: Prentice Hall, 2001.

[4] Vincent, J., A. Waters, and J. Sinclair, *Software Quality Assurance—Volume 1: Practice and Implementation*, Englewood Cliffs, NJ: Prentice Hall, 1988.

[5] Schulmeyer, G. G., and J. I. McManus, *Handbook of Software Quality Assurance*, 3rd ed., Upper Saddle River, NJ: Prentice Hall, 1999.

[6] Sommerville, I., and P. Sawyer, *Requirements Engineering: A Good Practice Guide*, New York: John Wiley, 1997.

[7] Ramesh, B., and M. Jarke, "Toward Reference Models for Requirements Traceability," *IEEE Trans. on Software Engineering*, Vol. 27, No. 1, 2001, pp. 58–93.

[8] Humphrey, W. S., *Introduction to the Team Software Process*, Reading, MA: Addison-Wesley, 2000.

[9] Card, D. N., and R. L. Glass, *Measuring Software Design Quality*, Englewood Cliffs, NJ: Prentice Hall, 1990.

[10] Yourdon, E., *Structured Walk-Throughs*, 4th ed., Englewood Cliffs, NJ: Prentice Hall, 1989.

[11] McCall, J. A., P. K. Richards, and G. F. Walters, *Factors in Software Quality Assurance, Vol. I*, RADC-TR-77-369, U.S. Department of Commerce, November 1977.

[12] Fagan, M. E., "Design and Code Inspections to Reduce Errors in Program Development," *IBM Systems Journal*, Vol. 15, No. 3, 1976, pp. 182–210.

[13] British Standards Institution, BS 7925-2:1998, British Standards Headquarters, London, England, 1998.

[14] Reid, S. C., "BS 7925-2: The Software Component Testing Standard," *Proc. of 1st Asia-Pacific Conference on Quality Software*, 2000.

[15] Dromey, R. G., "A Model for Software Product Quality," *IEEE Trans. on Software Engineering*, Vol. 21, No. 2, February 1995, pp. 146–162.

[16] Dromey, R. G., "Cornering the Chimera," *IEEE Software*, January 1996, pp. 33–43.

[17] Bansiya, J., and C. Davis, "A Hierarchical Model for Object-Oriented Design Quality Assessment," *IEEE Trans. on Software Engineering*, Vol. 28, No. 1, January 2002, pp. 4–17.

[18] Tervonen, I., "Support for Quality-Based Design and Inspection," *IEEE Software*, January 1996, pp. 44–54.

[19] Lorenz, M., and J. Kidd, *Object-Oriented Software Metrics*, Englewood Cliffs, NJ: Prentice Hall, 1994.

[20] Chidamber, S. R., and C. F. Kemerer, "A Metrics Suite for Object-Oriented Design," *IEEE Trans. on Software Engineering*, Vol. 22, No. 1, June 1994, pp. 476–493.

[21] Abreu, F. B., and R. Carapuca, "Object-Oriented Software Engineering: Measuring and Controlling the Development Process," *Proc. of 4th International Conference on Software Quality*, McLean, VA, October 1994.

[22] Abreu, F. B., and W. Melo, "Evaluating the Impact of Object-Oriented Design on Software Quality," *Proc. of 3rd International Software Metrics Symposium*, Berlin, Germany, 1996, pp. 90–99.

[23] Bansiya, J., "A Hierarchical Model for Quality Assessment of Object-Oriented Designs," Ph.D. Dissertation, University of Alabama in Huntsville, Huntsville, AL, 1997.

[24] Bansiya, J., and C. Davis, "Automated Metrics for Object-Oriented Development," *Dr. Dobb's Journal*, Vol. 272, December 1997, pp. 42–48.

[25] Li, W., and S. Henry, "Object-Oriented Metrics That Predict Maintainability," *Journal of Systems and Software*, Vol. 23, 1993, pp. 111–122.

[26] Briand, L., C. Bunse, and J. Daly, "A Controlled Experiment for Evaluating Quality Guidelines on Maintainability of Object-Oriented Designs," *IEEE Trans. on Software Engineering*, Vol. 27, No. 6, 2001, pp. 513–530.

[27] El Emam, K., et al., "The Confounding Effect of Class Size on the Validity of Object-Oriented Metrics," *IEEE Trans. on Software Engineering*, Vol. 27, No. 7, 2001, pp. 630–650.

[28] Beck, K., *eXtreme Programming Explained: Embrace Change*, Reading, MA: Addison Wesley, 2000.

[29] Williams, L., et al., "Strengthening the Case for Pair Programming," *IEEE Software*, July/August 2000, pp. 19–25.

[30] Fenton, N. E., and M. Neil, "A Critique of Software Defect Prediction Models," *IEEE Trans. on Software Engineering*, Vol. 25, No. 5, 1999, pp. 675–689.

[31] Rakitin, S. R., "Book View: Software Quality Assurance and Measurement: A Worldwide Perspective," *IEEE Software*, May/June 1998, pp. 116–117.

[32] Kitchenham, B. A., et al., "Preliminary Guidelines for Empirical Research in Software Engineering," *IEEE Trans. on Software Engineering*, Vol. 28, No. 8, August 2002.

[33] McCabe, T. J., "A Complexity Measure," *IEEE Trans. on Software Engineering*, Vol. 2, No. 4, 1976, pp. 308–320.

[34] Halstead, M. H., *Elements of Software Science*, New York: Elsevier North-Holland, 1977.

[35] Moller, K. H., and D. J. Plaulish, *Software Metrics: A Practitioner's Guide to Improved Product Development*, London, England: Chapman and Hall, 1993.

[36] Grady, R. B., and D. L. Caswell, *Software Metrics: Establishing a Company-Wide Program*, Englewood Cliffs, NJ: Prentice Hall, 1987.

[37] Poulin, J. S., "Measurement and Metrics for Software Components," in *Component-Based Software Engineering: Putting the Pieces Together*, G. T. Heineman and W. T. Council, (eds.), Reading, MA: Addison-Wesley, 2001.

[38] Hastings, T. E., and A. S. M. Sajeev, "A Vector-Based Approach to Software Size Measurement and Effort Estimation," *IEEE Trans. on Software Engineering*, Vol. 27, No. 4, 2001, pp. 337–350.

[39] Rosenberg, J., "Some Misconceptions About Lines of Code," *Proc. of 4th International Software Metrics Symposium*, Albuquerque, NM, 1997, pp. 137–142.

[40] Schneidewind, N. F., and H. M. Hoffmann, "An Experiment in Software Error Data Collection and Analysis," *IEEE Trans. on Software Engineering*, Vol. SE-5, No. 3, May 1979, pp. 276–286.

[41] Ginac, F. P., *Customer Oriented Software Quality Assurance*, Upper Saddle River, NJ: Prentice Hall, 1998.

[42] Fenton, N. E., and B. A. Kitchenham, "Validating Software Measures," *Journal of Software Testing, Verification, and Reliability*, Vol. 1, No. 2, 1991, pp. 27–42.

[43] Fenton, N. E., and N. Ohlsson, "Quantitative Analysis of Faults and Failures in a Complex Software System," *IEEE Trans. on Software Engineering*, Vol. 26, No. 8, 2000, pp. 797–814.

[44] Adams, E., "Optimizing Preventive Service of Software Products," *IBM Research Journal*, Vol. 28, No.1, 1984, pp. 2–14.

[45] Fenton, N. E., P. Krause, and M. Neil, "Software Measurement: Uncertainty and Causal Modeling," *IEEE Software*, July/August 2002, pp. 116–122.

[46] Weyuker, E. J., "Evaluating Software Complexity Measures," *IEEE Trans. on Software Engineering*, Vol. 14, No. 9, 1988, pp. 1357–1365.

[47] Oviedo, E. I., "Control Flow, Data Flow and Program Complexity," *Proc. of IEEE COMPSAC*, Chicago, IL, November 1980, pp. 146–152.

[48] Compton, T., and C. Withrow, "Prediction and Control of Ada Software Defects," *Journal of Systems and Software*, Vol. 12, 1990, pp. 199–207.

[49] Lind, R., and K. Vairavan, "An Experimental Investigation of Software Metrics and Their Relationship to Software Development Effort," *IEEE Trans. on Software Engineering*, Vol. 15, 1989, pp. 649–653.

[50] El Emam, K., et al., "The Optimal Class Size for Object-Oriented Software," *IEEE Transactions on Software Engineering*, Vol. 28, No. 5, 2002, pp. 494–509.

[51] Lorenz, M., *Object-Oriented Software Development: A Practical Guide*, Upper Saddle River, NJ: Prentice Hall, 1993.

[52] Weller, E. F., "Practical Applications of Statistical Process Control," *IEEE Software*, May/June 2000, pp. 48–55.

[53] Florac, W. A., A. D. Carleton, and J. R. Barnard, "Statistical Process Control: Analyzing a Space Shuttle Onboard Software Process," *IEEE Software*, July/August 2000, pp. 97–106.

CHAPTER

17

Verification of quality for component-based software

Contents

Systematic reuse of software components across multiple application software systems is still in its infancy. Some researchers are skeptical about the quality and productivity potentials associated with widespread development and use of software components. Some of this skepticism has been based on empirical observations and some on theoretical reasoning. This chapter begins with a brief discussion of such skepticisms and the companion constructive suggestions. Because of the focus on testing in this book, the focus here will be on those skepticisms based on a testing perspective. We then discuss ways in which the related critical issues can be alleviated. The chapter concludes with a few remarks on some important research tasks still ahead of us.

From the perspective of conventional software testing, critical issues regarding reuse of software components include the unavailability of components' source code to the user and the development of the components by their producers without the ability to envision all possible assembly, porting, and usage scenarios. Such issues may result in an excessive amount of testing effort on the part of the user, which in turn may significantly offset or even exceed the possible savings due to component reuse. In Parts I through III, we discussed, among other things, technical approaches to tackling these difficult testing issues. In addition to the technical aspects of testing software components, we also addressed some organizational aspects of component testing. For example, as pointed out in our discussion of standards and certification in Chapter 15, in addition to the obvious fact that development and use of a software component may economize the effort required for developing software systems, third-party testing or certification of a software component may economize the total effort otherwise required for testing the component by all the users.

This chapter turns attention to those aspects of quality verification for component-based software that go beyond testing. We first address reuse life-cycle processes and then modes and causes of reuse failures.

IEEE 1517 [1] addresses reuse of software components, although it is put in the context and language of software use in general. It builds on and supplements IEEE 12207 [2], which is the IEEE standard for software life cycle processes. We will discuss IEEE 1517, with IEEE 12207 as the backdrop. We also draw material from C. McClure [3]; McClure is the technical contact for IEEE 1517.

IEEE 1517 provides much guidance for development and quality assurance of component-based software but only briefly mentions the development of software components. This is why we discussed in much more detail the development and QA for software components (in Chapter 13) than the development and QA for component-based software systems (in Chapter 14). We proposed a detailed integrated process for the development and QA of software components in Chapter 13. The reader can use IEEE 1517 to develop an integrated process for the development and QA of component-based software systems by tailoring the requirements specified in IEEE 1517 to the life-cycle model already used by his or her organization.

Since good reuse processes (including predevelopment or cross-development domain engineering, development, and quality assurance processes) do not guarantee product quality, high-level process control should be accompanied by concrete and specific design, coding, and testing discipline. Detailed failure mode and effects analysis has been a common practice in development and quality assurance of hardware. There exists a very small number of published reports on software reuse failure modes and causes; the failure modes and causes having been reported in the literature tend to be highly aggregated.

Component reuse across multiple application systems implemented on different platforms, including operating system and hardware, is a relatively recent phenomenon, and the vast majority of the "infrastructure" of computing has been developed with no anticipation of such reuse. Therefore, many ad hoc "dos" and particularly "don'ts" may exist, and quality assurance for software reuse must go beyond controlling the reuse processes according to high-level guidelines and move toward also specifying design, coding, and testing discipline to address the many possible assembly, usage, or porting failures. [See Parts I, II, and III (e.g., Chapters 3 and 4) for some examples of such failures.]

We have made similar points in previous chapters of Part IV. In the context of testing, the exact causes of software failures resulting from improper reuse or improper assembly of software components should be identified, collected, and archived in a repository, perhaps like the Aviation Accident Database [4] developed and maintained by the National Transportation Safety Board to track air disasters and to help identify and prevent their causes. Some efforts are being made to avoid or to reduce possible assembly and usage failures through the development of an integrated operating system and component model, but such ad hoc discipline is necessary before the situation is significantly improved.

This chapter is partitioned into six sections. Section 17.1 addresses some skepticisms about the quality and productivity potentials associated with reuse of software components and offers some companion constructive suggestions. Section 17.2 discusses the essence of IEEE 1517, with IEEE 12207 as the background. Although many software professionals have complained about the proliferation of software standards, more standards may be needed for software components. Section 17.3 discusses possible areas of standardization. Section 17.4 addresses high-level success and failure factors for software reuse. Section 17.5 discusses failure modes and failure causes for component-based software. Closing remarks are given in Section 17.6.

17.1 Some skepticisms and the companion constructive suggestions

Certainly, there have been published reports of success in software component reuse (e.g., [5]), as well as anecdotes. However, there is no empirical evidence demonstrating the fulfillment of the productivity and quality potentials yet. In fact, it has been determined that some disasters have been caused by improper software reuse. For example, during the maiden voyage of the European Ariane 5 launch vehicle (Flight 501) in June 1996, the vehicle veered off course and exploded shortly after takeoff. It was later determined that the explosion was caused by the failure of a piece of software that had been reused from the Ariane 4 Project without adequate testing. The test inadequacy resulted from the decision that the Ariane 5 and Ariane 4 projects were so similar that substantial retesting of some reused Ariane 4 software components was not necessary. Weyuker [6] discussed this particular failure from the perspective of testing software components for reuse and used it to highlight the importance and difficulty of such user testing.

Details about the Ariane 5 disaster are provided in [7]. We discuss briefly the crux of the failure. The disaster resulted from a failure of the software controlling the horizontal acceleration of the Ariane 5 rocket. That piece of software contained a small computer program that converts a 64-bit floating-point real number related to the horizontal velocity of the vehicle to a 16-bit signed integer. The software was tested and used for the Ariane 4 Project without any problems. However, Ariane 5 is a much faster vehicle than Ariane 4, and the 16 bits allocated for the converted signed integer was no longer sufficient. This overflow error confused the control system and caused it to determine that a wrong turn had taken place. As a result, an abrupt course correction that was not needed was triggered.

Weyuker [6] examined critically the quality and productivity potentials of reuse of software components from the testing perspective and succinctly pointed out several critical issues associated with testing and using software components. She partitioned reused software into three broad categories:

- COTS software components;
- Software components specifically designed for reuse and residing in in-house software repositories;
- Custom-designed components originally written explicitly for one system with no intent for reuse.

We first address the third category, and then focus on the first two. In the language of the software reuse community, software products that can be reused are referred to as reusable assets or simply assets. Legacy systems are often the sources of assets of the third category. Recently, the concept of product line receives much attention [8]. L. M. Northrop [8] stated, "A software product line is a set of software-intensive systems that share a common, managed feature set satisfying a particular market segment's specific needs or mission and that are developed from a common set of *core assets* in a prescribed way. Core assets form the basis for the software product line. Core assets often include, but are not limited to, the architecture, reusable software components, domain models, requirements statements, documentation and specifications, performance models, schedules, budgets, test plans, test cases, work plans, and process descriptions. The architecture is key among the collection of core assets." Core assets include in-house developed and COTS software components. The concept of product line is actually a special type of systematic reuse; the emphasis of a product line is on a company's exploiting their software products' commonality to achieve economies of production.

Product line is sometimes also referred to as product family [8]. (However, F. van der Linden [9] pointed out that the term "product family" was used in Europe to mean something different. It had been used in Europe to describe a collection of products that are based on the same technology.) The concept of product line has a high potential because companies mostly compete in particular domains and many different products may share common components, including not only software components but also system components or subsystems. As mentioned earlier, in-house repositories of reusable components and in-house legacy systems or subsystems are major sources from which assets can be developed. In fact, a method advocated by some for developing reusable assets and building product lines is to mine the legacy systems. Although product lines may include COTS software components, they seem to involve mostly in-house developed software components.

In Part IV of this book, we have focused mostly on the first two categories of software components, and a commonality between the two is that the assets have been developed explicitly for reuse. The discipline that we have advocated so far is meant for all three of these categories, and the major difference between the third and the first two categories is that, for the third category, a software product already exists and must be transformed into one that can be reused to increase software quality and development productivity. Reuse of "components" of the third category in an ad hoc fashion can be very dangerous. For the rest of this chapter, we focus on the first two categories. For ease of discussion, we refer to these two categories of software reuse as COTS reuse and in-house reuse, and the software components as COTS (software) components and in-house developed (software) components, respectively. There is a big difference between the two kinds of reuse because reuse of COTS components incurs much more development complexity than reuse of in-house developed components, primarily due to the inaccessibility of the user to the internals of the COTS components and the companion technical expertise and support.

Weyuker [6] explored the extent to which traditional software testing approaches can be used in testing software containing COTS components. She doubted that the traditional software testing approach sufficed for the success of reuse of COTS components and called for new research into user testing of COTS components as well as user testing of systems containing COTS components. One fundamental difference between testing conventional software and testing software containing COTS components (without access to the source code) is that unit testing, as it is traditionally performed, is infeasible for testing COTS components. Since the producer of a software component cannot be expected to envision every possible use scenario, Weyuker argued, "Prudent users of software components will definitely want to spend a significant amount of resources testing the component, both in isolation and once it is integrated into their systems. ... This might well offset most or all of the savings made from using the component, because testing and debugging without the source code or a detailed design specification might be necessary, thereby making testing significantly more expensive than if a traditional custom-designed development paradigm was used." She continued, "The lesson to draw [from the European Ariane 5 Project] is that if component-based software construction becomes standard, with associated routine reuse of components, new ways to test the resulting software must be developed so that risk of similar disasters will be significantly reduced." We certainly agree with her, and her concerns actually have in part motivated our interest in writing this book and conducting research in related areas. For example, this difficulty in unit testing motivated our research into testability as well as our dedicating a chapter to the testability of software components (Chapter 5).

Weyuker made the following suggestion to make testing software components easier. "To improve the quality of the resulting software, software specification and test suites should be stored with each software component and care should be taken to update them appropriately whenever modifications are made to the component. In addition, three sets of links between specific entries in the documents and code should be maintained." The three sets of links are as follows:

- Each individual specified requirement and its implementation in both directions;
- Each individual test case and the portion or portions of the code that the test case was intended to validate;
- From a portion of the specifications to individual test cases that were included to validate the functionality described in that portion of the specifications, as well as pointers from the test cases into the specifications.

We note that since the code of a software component may not be available for the user of that component, the user may be able to see only the third set of links. The three sets of links combined together would definitely help the producer in testing the components, but the first two may not help the user's unit testing, which is where the fundamental problem of applying the traditional testing techniques to component-based testing lies.

Weyuker called for research into testing software components. We note, however, that computer code would be accessible to third-party certifiers, and the certifiers can certainly benefit from these links.

Testing is the primary subject of this book and has been discussed in detail in Parts I through III. Test coverage for the purpose of verifying the quality of a component-based software system by its developer during integration testing was discussed in Chapter 9. (Test coverage for the purpose of verifying the quality of a software component was discussed in Chapter 7, in the context of unit testing performed by the component developer.)

17.2 Minimum requirements for the life cycle of component-based software

Both ISO/IEC 12207 and IEEE/EIA 12207 [2] do not explicitly define reuse process requirements for the software life cycle, and therefore, they are not sufficient by themselves to enable and support the practice of software reuse.

17.2.1 Overview of IEEE 1517

IEEE 1517 [1] was published in 1999 as a supplement to IEEE/EIA 12207 [2]. Its purposes are to:

- Provide a framework for preparing reuse within the software life cycle;
- Specify the minimum set of processes, activities, and tasks to enable the practice of reuse when developing or maintaining software applications and systems;
- Explain how to integrate these reuse processes, activities, and tasks into the ISO/IEC and IEEE/EIA 12207 software life-cycle framework;
- Clarify terminology regarding software reuse in general and regarding reuse processes, activities and tasks in particular;
- Help promote and control the practice of software reuse in developing and maintaining software products.

IEEE 1517 does not define a specific software life cycle for software reuse. It is intended as a specification of the minimum requirements that a software life cycle model must meet to enable and support proper reuse. In other words, this standard is a requirements specification rather than a process specification. As such, it is intended to be used with virtually any software life cycle model, including waterfall, rapid prototyping, and iterative model [1, 3].

The purposes of ISO 12207 include:

- Providing a common framework for the life cycle of a software system or product;
- Providing a common terminology to clarify and improve communication among all parties involved in the life cycle of a software system or product;

- Defining the processes that are applicable during the complete life cycle of a software system or product starting from acquisition through retirement.

Like IEEE 1517, ISO 12207 does not define a specific software life cycle and is intended to be used with virtually any software life cycle model. As discussed in Chapter 15, ISO 12207 consists of three categories of processes; each of the processes consists of a number of activities; each of the activities consists of a set of tasks. Table 15.11 summarizes all ISO 12207 processes and their relationships.

IEEE 12207 states, "The standard contains a set of well-defined building blocks (processes); the user of this standard should select, tailor, and assemble those processes and their activities and tasks that are cost-effective for the organization and the project." As mentioned earlier, however, IEEE 1517 is intended as a specification of the minimum requirements that a software life-cycle model must meet to enable and support proper reuse.

17.2.2 IEEE 1517 processes, activities, and tasks

IEEE 1517 organizes the required life-cycle processes in four categories:

- Cross-project;
- Primary;
- Supporting;
- Organizational.

Note that the last three of the four categories are adapted from ISO 12207, and the cross-project category is added to accommodate software reuse. We discuss these four categories of process in this section.

Domain engineering is the only process categorized as a cross-project process, where domain is defined by IEEE 1517 as "a problem space" (e.g., the business process of procurement or that of sales). Domain engineering covers the life cycle of an asset and provides the assets that can be used by the primary processes. This process of domain engineering consists of the following five activities [1, 3]:

- *Process implementation:* to develop the domain engineering plan;
- *Domain analysis:* to produce the definition of the domain, domain vocabulary and domain models;

- *Domain design:* to produce the domain architecture and asset design specifications;
- *Asset provision:* to develop or acquire domain assets;
- *Asset maintenance:* to maintain domain assets.

Note that a software component is only one of the many possible types of assets provided by domain engineering; others include domain architectures and domain models. As mentioned earlier, IEEE 1517 does not address in detail the process requirements for the development and quality assurance of a software component or any other type of reusable asset; Chapter 13 provided an integrated process for the development and QA of a software component.

IEEE 1517 requires the same five primary processes as required in ISO 12207: acquisition, supply, development, operation, and maintenance. However, in IEEE 1517, every primary process includes additional tasks related to the use of reusable "assets" in producing the process deliverables. Such assets include domain architectures, domain models, domain standards, requirements, design specifications, software components, and test suites.

IEEE 1517 requires that reuse be made explicit in every primary process by specifying tasks to [1, 3]:

- Search for available assets from sources including domain architectures, domain models, domain standards, internal and external reuse libraries, COTS software products, other ongoing projects in the organization, existing legacy systems and subsystems, and object libraries;
- Select appropriate assets according to cost, expected benefit, quality, domain, availability, constraints, usage requirements, and other criteria;
- Incorporate assets into the deliverables;
- Evaluate the value and impact of assets according to their compliance with the organization's reuse standards, usability, and reusability in multiple contexts.

The category of supporting processes specified in IEEE 1517 consists of the eight supporting processes specified in ISO 12207 plus an additional process of *asset management*. The eight ISO 12207 processes are: documentation, configuration management, quality assurance, verification, validation, joint review, audit, and problem resolution. The purpose of the new process of asset management is to ensure that potential users of an asset can know its existence, can easily locate the current version of the asset, and can easily understand its purpose, limitations, status, and quality. The asset management process includes the following activities:

- Process implementation activity;
- Asset storage and retrieval definition activity;
- Asset management and control activity.

They are conducted to create and implement an asset management plan, to create and maintain an asset storage and retrieval system (e.g., reuse library), to create and maintain an asset classification scheme and certification procedures, to evaluate and accept new candidate assets and updated or new versions of current assets into the asset storage and retrieval system, and to manage asset storage, usage tracking, and problem tracking.

IEEE 1517 includes five organizational processes: the four specified in ISO 12207 plus the process of *reuse program administration*, which is used to plan, establish, and manage a reuse program. The four ISO 12207 organizational processes are management, infrastructure, improvement, and training. This new process of reuse program administration consists of six activities at the minimum:

- Initiation activity;
- Domain identification activity;
- Reuse assessment activity;
- Planning activity;
- Execution and control activity;
- Review and evaluation activity.

They are conducted to define the organization's reuse strategy, to assess the organization's capability to practice systematic reuse, to define and implement a reuse program plan, to establish a management and organizational structure to support the reuse program, and to monitor, evaluate, and improve the reuse program.

17.3 Areas for component standardization

Since verification or testing of software against existing standards is important, quality assurance and testing personnel need to understand existing asset standards. Moreover, standards are often developed because errors will more easily occur otherwise, and areas not yet standardized but important for standardization may point to where and the degree to which tests should be performed. There does not yet seem to be a consensus regarding asset areas for standardization. R. Weinreich and J. Sametinger [10] and McClure [3] combined seem to cover most of the areas for standardization discussed in the literature. Weinreich and Sametinger [10] proposed the following areas for standardization as fundamental elements of a component model:

- *Interface:* Definition of an *interface definition language* (IDL) for the purpose of specifying a "contract" between a component and its clients;
- *Naming:* How to generate a globally unique identifier for every component;
- *Metadata:* How metadata (i.e., information about components, interfaces, and their relationships) is described and how it can be obtained;
- *Interoperability ("wiring" or "connection"):* How to conduct communication and data exchange for components implemented in different programming languages, operating systems, and hardware devices; how to conduct communication and data exchange among components across different processes on the same computer (e.g., through interprocess calls) or over the network (e.g., through secure RMC); how to bridge communication and data exchange among implementations of different component models;
- *Customization:* How to customize a component through the interface of the component (a component may come with more functionality than what any one user might need and may be configured to satisfy the needs of a particular user. Such a component may be configured using clearly defined customization interfaces. Customization is different from wrapping.);
- *Composition:* How to create a larger structure by connecting components and how to insert or substitute components within an existing structure; what level of abstraction (e.g., all-purpose programming languages, scripting, component infrastructure designed for a specific domain);
- *Evolution support:* How should components or interfaces be replaced with newer versions and what should the requirements be for such replacement?
- *Packaging and deployment:* How to package a software component for independent deployment; the process for installation and configuration; provision of additional software and other supporting material that is

required for installation and configuration but is likely not part of the user's component infrastructure.

McClure [3] suggested some of the above plus the following areas:

- Analysis and design modeling language (e.g., UML);
- Data representation;
- Invoking, controlling and terminating a function;
- Error handling;
- Query processing;
- User interface (e.g., common user dialogue);
- Documentation of information and format:
  - General component description: name, author, version number, date created, parameter descriptions, data definitions, documentation template for different types of assets;
  - Reuse information: for example, software products in which the software component has been used, reuse successes and failures, reuse guidelines;
  - Quality/certification information: for example, prerelease defect density, process certification, and product certification;
  - Detailed component information: for example, interface requirements, platform requirements, domain, unresolved problems and defects, legal aspects (warranty, disclaimer, licensing), test plan, test cases, test data, and test results.
- Help.

17.4 Success and failure factors for reuse of in-house developed components

Before beginning to discuss success and failure factors for software reuse, we would like to remind the reader of the difference between reuse of COTS components and reuse of in-house developed components. Note again that the technical issues associated with COTS reuse are much more complex than their in-house counterparts. For ease of discussion, we have referred to them as COTS reuse and in-house reuse. Some research findings have been cast in terms of software reuse, but are really about in-house reuse and not about COTS reuse.

In this section, we focus on in-house reuse and discuss its success and failure factors. Product line development in practice seems to mostly involve in-house reuse, although in theory it may also involve acquisition and reuse of COTS software components.

To successfully implement in-house reuse, many technical and nontechnical obstacles must be overcome. Although the technical complexity of in-house reuse is much lower than its COTS counterpart, the technical challenges for implementing in-house reuse are substantial. However, most published reports about its success and failure factors come to a common conclusion that nontechnical obstacles have been demonstrated to be more difficult to overcome than the technical ones. As a result, the discussion of this section will be primarily about nontechnical factors. Before we continue, the reader is reminded that the Ariane 5 disaster was caused by improper reuse of in-house developed software components, having nothing to do with COTS software components.

Note that the lessons learned in this section about in-house reuse are applicable to COTS reuse because the corresponding issues are also present in COTS reuse. We will discuss success and failure factors for COTS reuse in the next section. Due to the much more complex technological challenges associated with COTS reuse, the discussion there will be focused on technical factors, particularly on failure modes and causes.

M. Morisio et al. [11] reported very recently nontechnical success and failure factors in adopting or running a company-wide reuse program and summarized a number of previous studies of similar nature (e.g., [5, 12–15]). Although not explicitly stated, they seem to have primarily dealt with in-house reuse, judging from their approach, data, and conclusion. All these studies seem to have come to very similar conclusions. The uniqueness of Morisio et al.'s study, however, is that their empirical study was based on face-to-face interviews with program managers responsible for software reuse, rather than mail surveys.

The European Commission has been funding *process improvement experiments* (PIEs), each of which is a technology transfer project in which a specific technology is applied in a company. Morisio et al. [11] developed a questionnaire and conducted interviews with the managers of 24 PIEs pertaining to software reuse. Based on the interviews, they organized the results in three sets of categorical data, some of which have ordinal scales. The three sets of data correspond to three different types of attributes or variables: state variables (attributes or variables that describe the current status of a company and cannot be changed by the company easily and quickly—for example, size, application domain, and staff experience), high-level control variables (those high-level variables that a company can control), and low-level control variables (those low-level variables that a company can control). The state variables consist of size of software staff, size of overall staff, type of software production (e.g., product line/family or isolated), software and product (e.g., software embedded in a hardware product, software alone), software process maturity, application domain, type of software (e.g., embedded real-time, business DBMS), size of baseline reuse project, development approach (e.g., object-oriented or procedural), and staff experience. The high-level control variables consist of top management commitment, key reuse roles introduced, reuse processes introduced, nonreuse processes modified, repository established, and human factors addressed. The low-level control variables consist of reuse approach (e.g., reusable products loosely coupled and to be reused in isolation versus tightly coupled and to be reused with others as a group), type of reusable asset (code, design and/or requirements), domain analysis (yes versus no), origin (developed from scratch, reengineered from legacy products, existing products without modification), independence of asset development from system development (versus development and reuse of asset by the same project team), timing of asset development (before the need versus just-in-time), qualification of an asset as reusable via a defined process, configuration management, reward policy to promote reuse, and the number of reusable assets in the company repository. We note that the data reflects the fact that the aim of their study was to identify managerial issues, rather than technological issues.

With a caution to the reader about small sample size, they derived key management tasks required for successful (in-house) reuse based on the data and suggested a reuse introduction decision sequence:

- *Evaluate reuse potential.*
- *Build reuse capability:* Get commitment of top management to obtain resources and power to (1) change nonreuse-specific processes, (2) add reuse-specific processes, (3) address human factors, and (4) set up a repository.
- *Reuse implementation:* Perform (1) through (4).

In theory, absence of any of these tasks would cause a failure. Morisio et al. [11] were able to attribute the failures among the 24 projects to the absence of one or more of these tasks. They found that, given a productivity potential in terms of significant commonality among different applications within the same company, the success of initiating a reuse program depends on a mix of factors:

- Succeeding in a reuse initiative is a technology transfer endeavor, which requires commitment of management.
- The successful approaches seem to be similar in the sense of requiring the consideration of a common set of elements: (1) through (3) defined above.
- Methods of implementation need not be the same, however. The way in which each element defined above should be deployed depends on company context.

Other researchers had performed surveys prior to their study, but those were mail surveys [13, 14], rather than interviews. The reader is referred to M. Paci and S. Hallsteinsen [16] for experiences accrued from 15 reuse projects in European companies, particular information on economic results from long-term application of reuse and planned evolution and maintenance. There have also been published reports on company experiences [5, 12]. A common theme seems to have emerged (for in-house reuse): The most important reuse success or failure factors are economic and cultural, not technological [17]. For example, M. L. Griss [18] reported, "Reuse is a business issue that involves technology transition and organizational change. Instituting a reuse culture, providing training, adhering to standard, and securing management commitment are the key success factors." D. Card [17] stated, "The most important obstacles to reuse are economic and cultural, not technological." Having said this, it is important to remind the reader that some reuse disasters (e.g., the Ariane 5 Project) were caused by technical problems and have little to do with the maturity of the software development and quality assurance processes. We will discuss some studies conducted to uncover both managerial and technical issues in the next section.

These findings complement the earlier discussion of IEEE 1517 in at least two important ways. First, these address empirical reuse work. Perhaps more importantly, IEEE 1517 seems to be focused more on the required processes to enable and support software reuse. These findings complement the standard in several ways. First, they address the study of economic potentials that may warrant reuse. Second, they address the organizational requirement of management commitment to the cultural change from a conventional to a reuse development organization. Third, they address the human factors issues involved in making the changes.

17.5 Failure modes and failure causes

Nontechnical factors that may influence the success of reusing in-house developed software components and that can be controlled by a company were the primary focus of the previous section. In this section, we look at not just in-house reuse but also COTS reuse, which, as mentioned earlier, involves much more complex technical issues than in-house reuse. Obviously, those nontechnical success factors combined with standard reuse development and QA processes like those specified in IEEE 1517 do not guarantee the quality of a component-based software system. This section focuses on technical failures, with a particular focus on what testing personnel can do to help verify or improve the quality of such a system. We will discuss some recent findings reported in the literature. However, from the testing perspective, these modes and causes are still at too high a level. We will discuss some research needs for detailed studies of failure modes and causes later in this section.

Failure mode and effects analysis (FMEA) has been an integral part of development and QA of hardware systems and hardware components [19]. Very detailed reports of required operating conditions, operating characteristics, and failure modes have been widely reported by researchers and producers, and these have been routinely used by the users. This large body of work has accumulated for decades through scientific and engineering studies. The software reuse community may need to be prepared to accept the need for a large accumulated body of very detailed information regarding software reuse going much beyond the nontechnical success factors and standard reuse processes. Such a need is already alluded to in the existing literature. We discuss reuse failure modes and causes already reported in the literature first and then some such allusions.

W. B. Frakes and C. J. Fox [15] developed a reuse success chain and, based on it, developed a sequential model for reuse failure modes. They focused on code reuse (i.e., reuse of software components) and did not address reuse of any other type of reusable assets (e.g., requirements). The success of software component reuse depends on the success of each of the seven steps in the chain. The seven steps in their original terminology are: try to reuse, part exists, part available, part found, part understood, part valid, and part integratable. Note that testing plays a critical role in the last three steps.

A failure of any step in the chain will lead to a reuse failure. The sequential nature of the chain naturally gives rise to seven reuse failure modes, again in their original terminology: no attempt to reuse, part does not exist, part not available, part not found, part not understood, part not valid, and part not integratable. Note that the last three failure modes capture technical failures. (Morisio et al. [11] also identified several reuse failure causes but they are all of management nature—for example, failure to create new reuse processes.)

As a part of a bigger survey [15] conducted in 1991 and 1992, 113 software professionals responded to the question "What problems have you had when trying to reuse software?" All the responses were analyzed to identify possible failure causes. A response might have cited more than one failure cause; different responses may have cited a common failure cause. All failure citations were counted, with a total of 131 counts. Each of the failure causes was analyzed to determine which of the seven failure modes fits the cause best. The top four failure modes are: no attempt to reuse, part not integratable, part not understood, and part not valid. Of the 131 counts of failure causes, 32, 22, 21, and 19 fell in these four modes, respectively. (Thirty-seven counts fell in the other three modes, which relate to the existence and the availability of reusable software components.)

From the perspective of verification of the quality of a component-based software system and particularly from the perspective of testing, knowing what to test for is critical. Therefore, the three technical failure modes and related failure causes are particularly relevant to testing. We now provide some details for these three technical failure modes.

Failure causes under the failure mode of part not integratable include (with the counts shown in the parentheses following the causes): environment incompatibilities (11), improper form (7), hardware incompatibilities (4), and too much modification required (3). Failure causes under the failure mode of part not understood include: inadequate documentation (18) and part too complex (2). Failure causes under the failure mode of part not valid include: poor inspection, testing, or verification (6); poor support from producer (5); inadequate performance (5); and lack of standards (2).

Of these failure causes, the following are of a technical nature and deserve special attention from testing personnel (with their correspondences to the categories of software requirements listed in Section 13.4.2 included in the parentheses): environment incompatibilities (assembly into different applications and operating environments—operating systems), improper form (interface), hardware incompatibilities (operating environments—hardware), too much modification required (functionality, interface, and so forth), inadequate documentation (documentation), poor inspection, testing, verification (quality requirements), inadequate performance (performance), and lack of standards (standards).

There seems to be much less discussion in the literature about concrete examples of such failures, not to mention a repository of such failures. Such failures may involve complex technical details.

Frakes and Fox [15] alluded to the level of detail at which software components may not be assembled properly into a software system and stated, "A component might have the right concept, but not the right content or context.

For example, it might be in the wrong language, use the wrong version of the compiler, have the wrong parameter list, etc." Note the reference to the detail level of compiler version. Moreover, Weinreich and Sametinger [10] stated, "If the component model allows the implementation of components in different programming languages, calling conventions must be standardized at the binary level to ensure interoperability of these components. Even if component implementations share the same language, the binary layout of the interfaces and parameter types may still be different. Interoperability of components within a process address space is possible if the component model defines the binary interface structure and calling conventions." Note the reference to the binary interface structure. These provide hints to the testing complexity. We now look at this issue in a more systematic way.

Griss [18] proposed a four-layer component architecture containing three software-component layers:

- Applications (built for, for example, banking or other domains);
- Business-specific components (e.g., components built for banking or other domains);
- Middleware components [e.g., cross-domain components, for example, *object request brokers* (ORBs), databases, GUIs];
- System software components (e.g., platform-specific hardware or software such as operating systems, networking protocols, and hardware devices).

Note that in this layered view, an application software system is viewed as an assembly of components at all three bottom layers plus additional software products built specifically for the application software (at the top level). (Even an operating system is considered as a component.) In the literature, a component infrastructure or reuse infrastructure usually refers to such an architecture, the particular components included in the layers, and frameworks organizing them into organizational assets.

Since component reusability across different application systems implemented on different platforms, including operating system and hardware, is a relatively recent phenomenon and the vast majority of the components of the general computing "infrastructure" have been developed with no anticipation of such reuse, many ad hoc dos and particularly don'ts may exist. More importantly, not all don'ts are publicized or even known. In fact, many unknown don'ts may exist. Therefore, reuse success may not be predictable, and, to be safe, a software component should be considered unfit for reuse in a software system until proven otherwise through thorough testing, by the developer, the certifier, and/or the user.

Weyuker [6] also discussed the possible requirement for an excessive amount of possible test cases and stated, " Based on my personal experience, I conclude that developers and potential users should be aware that components likely will not have been adequately tested for their infrastructure. It may also be difficult to anticipate what is known about the expected behavior and performance once it is installed in a project's infrastructure. This implies that the component might have to be tested as if it has not yet undergone any testing at all. In some cases, such as when the source code is unavailable to the potential new users and there is no low-level understanding of the implementation, there may be no way to adequately test and debug the software."

Weyuker's concern is real. It should be clear from Griss' layered infrastructure that many available choices of components at all three layers exist, and the complexity may be increasing exponentially in the number of available choices at these layers. Her concern can be alleviated if the producer specifies exactly in what infrastructures the component has been developed and thoroughly tested. However, this may limit the market. The producers may have to make a strategic decision. Without such exact specifications, their products may not be trusted and hence not be purchased. Some have suggested not purchasing any software component if the producer does not specify that the component has been developed and tested for the potential purchaser's specific platform. Recall the detail level of compiler version number alluded to in the context of software reuse failure modes by Frakes and Fox [15].

17.6 Summary

Component models have been developed to provide a more coherent infrastructure, but at different levels. Weinreich and Sametinger [10] stated, "Typically, component model implementations exist on top of an operating system. However, some operating systems, such as Microsoft Windows, have already begun to incorporate component model implementations. Eventually, operating systems may directly serve as component model implementations for CBSE." Before such integration is completed and has matured, users of software components must be cautious, and testing could be a challenge. Collecting and compiling failure cause information in a repository would facilitate training of both developers and testers and may help prevent failures. Failures and their causes may very well be ad hoc in nature, and there may be a large number of them. If such a failure-cause repository is too large to handle, widespread use of software components may not be feasible. Such a repository would archive not only interoperability or portability failures (and their causes) but also failures (and their causes) like the software error that caused the Ariane 5 disaster. Without such a repository, testing is even more challenging and testers must become knowledgeable through other means about a possibly large number of problems that could occur. In any case, testing software containing software components, like developing it, requires much discipline, and personnel certification may be a good idea.

Constructing such repositories may take time, and this leads to an issue of deployment strategy. Perhaps, reuse should be first practiced on less reliability or safety-critical projects, and then on more critical projects once reuse technology and culture have matured.

As we argued in Chapter 14, when compared to hardware QA, software QA must penetrate deeply into design, coding, and testing if it is to make a contribution to the quality of software that is as significant as the contribution of QA to hardware quality. If no public-domain failure-cause repositories exist, software QA and test personnel should compile for their own companies failure modes and detailed failure causes for software reuse, based on their own experience in unit, integration, and system testing, particularly regarding the additional complexities incurred by using software components: interface for integration, assembly into application systems (interoperability), and platform portability.

Given the current state of reuse technology and infrastructure, ad hoc testing continues to be very valuable. Books on ad hoc testing for conventional software include [20, 21].

Dromey's product quality model offers a systematic approach to building quality into software and may be further developed to help improve reusability of software components and quality of component-based software. See Section 14.6 for his definition of reusability and a brief introduction to his approach; see [22, 23] for details.

This book seems to be the first book on the market that addresses testing and quality assurance for component-based software. We addressed systematic reuse and testing. Much more research is required. Systematic testing for functionality and performance has received much attention. Reuse of software components drastically increases the acuteness and seriousness of the interoperability and portability issues, and hence requires urgent research attention regarding assembly of software components into various application software systems, integration of software components with other software components (including middle-ware components), and portability to various platforms (including operating systems and computer hardware). These issues tend to be ad hoc in nature, and efforts aimed at their resolution may not be considered glamorous. However, such efforts may be required at least before a fully integrated component infrastructure has been developed and becomes widely available, in the sense that a component model implementation is incorporated in an operating system and the operating system can directly serve as component model implementations.

References

[1] IEEE, "IEEE Standard for Information Technology—Software Life Cycle Processes—Reuse Processes (IEEE 1517-1999)," Piscataway, NJ, 1999.

[2] IEEE/EIA, "IEEE Standard for Information Technology—Software Life Cycle Processes (IEEE 12207.0 - 1996)," Piscataway, NJ, 1999.

[3] McClure, C., *Software Reuse: A Standards-Based Guide*, Los Alamotos, CA: IEEE Computer Society, 2001.

[4] National Transportation Safety Board, Aviation Accident Database, http://www.ntsb.gov/ntsb/query.asp, accessed February 2003.

[5] Lim, W. C., "Effects of Reuse on Quality, Productivity, and Economic," *IEEE Software*, September 1994, pp. 23–30.

[6] Weyuker, E. J., "The Trouble with Testing Components," in *Component-Based Software Engineering: Putting the Pieces Together*, G. T. Heineman and W. T. Councill, (eds.), Reading, MA: Addison-Wesley, 2001.

[7] Lions, J.-L., et al., "Ariane 5 Flight 501 Failure: Report by the Inquiry Board," European Space Agency (ESA), Paris, France, July 19, 1996, http://java.sun.com/people/jaf/Ariane5.html, accessed February 2003.

[8] Northrop, L. M., "SEI's Software Product Line Tenets," *IEEE Software*, July/August 2002, pp. 32–40.

[9] van der Linden, F., "Software Product Families in Europe: The Esaps & Café Projects," *IEEE Software*, July/August 2002, pp. 41–49.

[10] Weinreich, R., and J. Sametinger, "Component Models and Component Services: Concepts and Principles," in *Component-Based Software Engineering: Putting the Pieces Together*, G. T. Heineman and W. T. Councill, (eds.), Reading, MA: Addison-Wesley, 2001.

[11] Morisio, M., M. Ezran, and C. Tully, "Success and Failure Factors in Software Reuse," *IEEE Trans. on Software Engineering*, Vol. 28, No. 4, April 2002.

[12] Griss, M. L., and M. Wosser, "Making Reuse Work at Hewlett-Packard," *IEEE Software*, January 1995, pp. 105–107.

[13] Rine, D. C., and R. M. Sonneman, "Investments in Reusable Software: A Study of Software Reuse Investment Success Factors," *Journal of Systems and Software*, Vol. 41, 1998, pp. 17–32.

[14] Lee, N. Y., and C. R. Litecky, "An Empirical Study of Software Reuse with Special Attention to Ada," *IEEE Trans. on Software Engineering*, Vol. 23, No. 9, September 1997, pp. 537–549.

[15] Frakes, W. B. and C. J. Fox, "Quality Improvement Using a Software Reuse Failure Modes Model," *IEEE Trans. on Software Engineering*, Vol. 22, No. 4, April 1996.

[16] Paci, M., and S. Hallsteinsen, (eds.), *Experiences in Software Evolution and Reuse: Twelve Real World Projects*, Berlin, Germany: Springer, 1997.

[17] Card, D., "Why Do So Many Reuse Programs Fail?" *IEEE Software*, September 1994, pp. 114–115.

[18] Griss, M. L., "Software Reuse: Objects and Frameworks Are Not Enough," *Object Magazine*, February 1995, pp. 77–87.

[19] Kolarik, W., *Creating Quality: Concepts, Systems, Strategies, and Tools*, New York: McGraw-Hill, 1995.

[20] Kaner, C., J. Bach, and B. Pettichord, *Lessons Learned in Software Testing: A Context-Driven Approach*, New York: John Wiley, 2002.

[21] Whittaker, J. A., *How to Break Software: A Practical Guide to Testing*, Reading, MA: Addison-Wesley, 2003.

[22] Dromey, R. G., "A Model for Software Product Quality," *IEEE Trans. on Software Engineering*, Vol. 21, No. 2, February 1995, pp. 146–162.

[23] Dromey, R. G., "Cornering the Chimera," *IEEE Software*, January 1996, pp. 33–43.

About the authors

Jerry Zeyu Gao is an associate professor in the Department of Computer Engineering at San José State University. He received an M.S. and a Ph.D. from the University of Texas at Arlington. Dr. Gao has more than 10 years of industry development experience and management experience with software development, testing, and Internet applications. Dr. Gao joined the Department of Computer Engineering at San José State University in 1998. His research areas include software engineering, component-based software engineering, and Internet computing, e-commerce, and wireless Web applications. He has published widely in IEEE/ACM journals, magazines, and international conferences. He has published more than 40 technical papers and has written a book, *Object-Oriented Software Testing* (IEEE Computer Society Press, 1998).

His teaching subjects include programming in C/C++/Java, software engineering I and II, data structure and algorithms, object-oriented technology, and software testing and quality assurance. In addition, he has developed and taught a number of new courses on emerging technology, such as Design of Web-Based Application Systems, Design of E-Commerce Systems, and Design of Mobile-Based Software Systems.

H.-S. Jacob Tsao has been an associate professor of industrial and systems engineering at San José State University since 1999 and has taught courses in quality assurance, reliability engineering, information engineering, operations research, and statistics. He has numerous technical publications, including more than 30 refereed journal papers and a book, *Entropy Optimization and Mathematical Programming* (Kluwer, 1997). He received a B.S. in applied mathematics from the National Chiao-Tung University in Taiwan in 1976, an M.S. in mathematical statistics from the University of Texas at Dallas in 1980, and a Ph.D. in operations research from the University of California at Berkeley in 1984. Dr. Tsao has worked for Consilium Inc. (currently an Applied

Materials company) as a software development engineer on computer-aided manufacturing, and for AT&T Bell Laboratories and Bell Communication Research as a systems engineer on large-scale software systems designed to automate circuit provisioning and capacity expansion planning. He joined the Institute of Transportation Studies of University of California at Berkeley in 1992 and researched concept development and evaluation for intelligent transportation systems.

Ye Wu received a B.S. and an M.S. in computer science from Shandong University, China, in 1993 and 1996, respectively. He received a Ph.D. in computer science from the State University of New York at Albany in 2000. Dr. Wu joined the faculty of George Mason University in 2000 as an assistant professor in the Department of Information and Software Engineering. His research interests include software testing, software maintenance, and software architecture analysis. Currently, his research focuses on testing and maintaining component-based and Web-based software.

Index

D

Q

R

U

V

W

X

Recent Titles in the Artech House Computing Library

Advanced ANSI SQL Data Modeling and Structure Processing, Michael M. David

Advanced Database Technology and Design, Mario Piattini and Oscar Díaz, editors

Action Focused Assessment for Software Process Improvement, Tim Kasse

Building Reliable Component-Based Software Systems, Ivica Crnkovic and Magnus Larsson, editors

Business Process Implementation for IT Professionals and Managers, Robert B. Walford

Data Modeling and Design for Today's Architectures, Angelo Bobak

Developing Secure Distributed Systems with CORBA, Ulrich Lang and Rudolf Schreiner

Future Codes: Essays in Advanced Computer Technology and the Law, Curtis E. A. Karnow

Global Distributed Applications with Windows® DNA, Enrique Madrona

A Guide to Software Configuration Management, Alexis Leon

Guide to Standards and Specifications for Designing Web Software, Stan Magee and Leonard L. Tripp

Implementing and Integrating Product Data Management and Software Configuration, Ivica Crnkovic, Ulf Asklund, and Annita Persson Dahlqvist

Internet Commerce Development, Craig Standing

Knowledge Management Strategy and Technology, Richard F. Bellaver and John M. Lusa, editors

Managing Computer Networks: A Case-Based Reasoning Approach, Lundy Lewis

Metadata Management for Information Control and Business Success, Guy Tozer

Multimedia Database Management Systems, Guojun Lu

Practical Guide to Software Quality Management, Second Edition, John W. Horch

Practical Process Simulation Using Object-Oriented Techniques and C++, José Garrido

Risk-Based E-Business Testing, Paul Gerrard and Neil Thompson

Secure Messaging with PGP and S/MIME, Rolf Oppliger

Software Fault Tolerance Techniques and Implementation, Laura L. Pullum

Software Verification and Validation for Practitioners and Managers, Second Edition, Steven R. Rakitin

Strategic Software Production with Domain-Oriented Reuse, Paolo Predonzani, Giancarlo Succi, and Tullio Vernazza

Successful Evolution of Software Systems, Hongji Yang and Martin Ward

Systematic Process Improvement Using ISO 9001:2000 and CMMI®, Boris Mutafelija and Harvey Stromberg

Systematic Software Testing, Rick D. Craig and Stefan P. Jaskiel

Systems Modeling for Business Process Improvement, David Bustard, Peter Kawalek, and Mark Norris, editors

Testing and Quality Assurance for Component-Based Software, Jerry Zeyu Gao, H. -S. Jacob Tsao, and Ye Wu

User-Centered Information Design for Improved Software Usability, Pradeep Henry

Workflow Modeling: Tools for Process Improvement and Application Development, Alec Sharp and Patrick McDermott

For further information on these and other Artech House titles, including previously considered out-of-print books now available through our In-Print-Forever® (IPF®) program, contact:

Artech House
685 Canton Street
Norwood, MA 02062
Phone: 781-769-9750
Fax: 781-769-6334
e-mail: artech@artechhouse.com

Artech House
46 Gillingham Street
London SW1V 1AH UK
Phone: +44 (0)20 7596-8750
Fax: +44 (0)20 7630-0166
e-mail: artech-uk@artechhouse.com

Find us on the World Wide Web at:
www.artechhouse.com